THE SILENT DEEP

THE SILENT DEEP

The Discovery, Ecology and Conservation of the Deep Sea

TONY KOSLOW

THE UNIVERSITY OF CHICAGO PRESS

The University of Chicago Press, Chicago 60637
The University of Chicago Press, Ltd., London
The University of New South Wales Press, Sydney NSW 2052
 Published 2007

16 15 14 13 12 11 10 09 08 07 1 2 3 4 5

ISBN-13: 978-0-226-45125-1 (cloth)

ISBN-10: 0-226-45125-9 (cloth)

Library of Congress Cataloging-in-Publication Data

Koslow, Julian Anthony.
The silent deep : the discovery, ecology, and conservation of the deep sea / Julian Anthony Koslow.
p. cm.
Includes bibliographical references and index.
ISBN-13: 978-0-226-45125-1 (cloth : alk. paper)
ISBN-10: 0-226-45125-9 (cloth : alk. paper)
1. Deep-sea ecology. 2. Marine resources conservation. I. Title.
QH541.5.D35K67 2007
578.77—dc22

2006022282

♾ The paper used in this publication meets the minimum requirements of the American National Standard for Information Sciences—Permanence of Paper for Printed Library Materials, ANSI Z39.48-1992
Printed in China

Contents

"I don't know why I don't care about the bottom of the ocean, but I don't."

Preface

Only 150 years ago the prevailing scientific paradigm held that the deep sea was lifeless. It was assumed that conditions at the bottom of the ocean – the cold, the perpetual darkness, and immense pressures – were as inimical to life as today we might view conditions on our neighboring planets.

But as soon as men (and they were mostly men in those days) put nets and dredges to the bottom of the sea, they encountered life. And ever since, each new means to sample life in the oceans, whether bigger nets to capture bigger creatures or smaller meshes to capture smaller ones – or cameras to capture what can't be caught! – has uncovered new life forms. Entirely new biological communities have been discovered in the deep sea each decade now for the past 40 years. The vent fauna, including giants clams and two-meter high tubeworms, sustained through microbial symbiosis by otherwise highly toxic hydrogen sulfide, is only the best known and most spectacular. Each month the pages of the leading scientific journals, *Science* and *Nature*, cover major new deep-sea discoveries: new life forms, new understandings of their ecological workings.

But the deep sea is more than a source of scientific wonder; it is also the last great human frontier on this planet. The ecologist's pace of discovery has been matched by fishers' and geologists' discoveries of the oceans' last virgin fisheries and of vast deep-sea hydrocarbon and mineral resources.

For decades the fishers and ecologists pursued their business in the deep sea along apparently parallel tracks that never actually met. Deep-sea ecologists spent the 1960s grubbing about in the mud. They had discovered a new fauna: untold thousands, if not millions, of new species, tiny worms and crustaceans living within the sediments of the deep seafloor. This absorbed their attention until the end of the 1970s, when they discovered the vent fauna. They were not much interested in deep-sea fisheries, in part because the prevailing scientific paradigm held that they were an oxymoron:

the deep sea was too depauperate to support significant fisheries.

Fishers tend to be a secretive lot, particularly when the fishing is good. If, as some believe, the Basques were hauling in Grand Banks cod before Columbus stumbled upon the New World, they never proclaimed it. And so, when the massive Soviet and Japanese post-war fishing fleets outstripped their traditional coastal fishing grounds and discovered huge unfished aggregations over North Pacific seamounts – extinct underwater volcanoes – in the 1960s, they quietly set to work. The seamounts were near Hawaii but in international waters, so there was no one watching and no one to regulate them. By the time American fishery scientists took notice and decided to hold a symposium, the fish had all been caught. The Soviets didn't even bother to attend.[1] Within the decade they had landed almost a million tonnes of pelagic armorhead and were moving on to the seamounts of the South Pacific, where they and the New Zealanders soon discovered orange roughy.

It was the mid-1990s before fishery scientists, such as myself, asked what else might be on the seamounts. In collaboration with a score of taxonomists, I discovered coral reefs on the seamounts south of Tasmania at a depth of 1000 meters, whose diversity rivaled that of shallow-water tropical reefs. Of the hundreds of species that we encountered, about a third were new to science and apparently restricted to those local seamounts. But where the trawlers had been working, there was nothing: only coral rubble and bare rock.

At about the same time, extensive deepwater coral grounds were discovered along the continental margins of the North Atlantic and North Pacific – also under immediate threat. Suddenly Norway found itself home to about 2000 square kilometers of deepwater coral reef, with up to half already destroyed by deepwater trawling. Extensive soft coral and sponge 'gardens' were discovered adjacent to rich Alaskan fishing grounds around the Aleutian Islands.

Within a few years, conservation groups around the world – the International Union for the Conservation of Nature and Natural Resources (IUCN), the World Wildlife Fund (WWF), Greenpeace, the Natural Resources Defense Fund, Oceana, and others – united to call for the protection of deepwater corals throughout the world's oceans. Australia, Norway, and New Zealand moved to protect some of the deepwater corals within their Exclusive Economic Zones (EEZ), and in 2002 the United Nations General Assembly passed a resolution to protect the biodiversity of the high seas. At this time the eventual outcome remains uncertain: while the international community deliberates on how to best manage and conserve the global deep ocean, fishers continue to sweep across it in search of the last remaining new grounds.

I set out initially to write a book simpler than the present one, concerned mostly with the story of the discovery of deepwater coral habitats, the struggle to conserve them in the face of opposition from a few fishing groups, and the indifference of much of the rest of the world. I realized, however, that the story of deep-sea discovery is far richer, its diversity far greater and the threats facing it far broader than I had originally conceived. We may think of the deep sea as pristine, but in fact no portion of the deep sea is today unaffected by human activities.

I realized also that if the New York matron in Charles Saxon's cartoon (opposite p. 1) is ever to care about the bottom of the ocean, she needs to see more of what is there and know more about what is at stake. After all, for most of the world the deep sea is out of sight; isn't it safely out of mind as well? What really could be happening in the deep sea to compare with all the other global issues that press upon our attention?

The deep sea represents one of the last great frontiers for humankind. However, we humans tend to devalue what we do not know, and the deep sea is no exception. From time immemorial, scientists and lay people alike have underestimated the diversity and richness of the deep sea. Consider Pliny, the classical authority who believed that everything of significance was already known about the deep sea, or Edward Forbes, the eminent 19th century marine naturalist who predicted that the deep sea would prove lifeless. If today we do not think of the deep sea as being entirely without life, our image of it is still largely of a vast and depauperate realm: unrelieved sediment-covered plains, the utter stillness punctuated only occasionally by the passage of some very slow and strange fish.

Were Adam and Eve cast out of Eden, or did they (and their descendents) destroy it themselves? The dark underside of the frontier saga is that we so often despoil the new worlds that we stumble upon, a pattern that far predates Western societies of the past 500 years. The largest of the North American mammalian megafauna – mammoths, mastodons, giant buffalo and a host of others – all vanished in a wave of extinctions that coincided with the arrival of the first humans, the Clovis hunters, at the end of the last Ice Age about 12 000 years ago. More recently, New Zealand's eleven species of giant flightless moas were hunted to extinction within a century of the Polynesian Maoris settling there in the late 13th century.[2]

The rapid demise of these faunas at the hands of Stone Age hunters was facilitated by their long isolation, which rendered them tame and unprepared for the human onslaught. Seamount faunas may be similarly vulnerable to extinction, given their restricted distributions and peculiar life histories, which evolved in an environment with few predators and little natural disturbance.

Until very recently, as little was known of the deep ocean as of the backside of the moon. Human impacts were negligible. But a host of new technologies – acoustics, the satellite-based Global Positioning System, broad-swath mapping of the seafloor, computerized track plotters and so on – now render the deep sea accessible even to small operators. Fishers in 25-meter boats can level a pristine coral reef at 1000 meters depth as readily as loggers with chain saws and bulldozers can clear an old-growth forest.

In the past 35 years, dramatic new vistas have opened on life in the deep sea. This is one of the great scientific voyages of discovery, one that humankind has only just embarked upon. Scientists have discovered that the Lilliputian fauna of the deep-sea abyssal plains is as diverse as the life of tropical rainforests; that there are as many species of stony corals in the deep sea as in the reefs of the Indo-Pacific, Red and Caribbean Seas. Major fisheries have developed at depths of 1000 meters and more. To the fisheries resources and mineral potential of the deep sea – manganese nodules, cobalt-rich crusts, black smokers containing rich gold and other ore deposits – has now been added the prospect of vast hydrocarbon resources in the form of methane hydrates. There is the potential for new biotechnologies based on hitherto unknown life-forms, the archaea, which are spewed in great profusion from deep-sea hydrothermal vents. Thriving in the vent waters at temperatures around 100°C, the archaea and hyperthermophilic bacteria have revolutionized our understanding of the range of conditions in which life may thrive; indeed, they may hold the key to the origin of life.

Conservation of the deep is inextricably linked to understanding and appreciating what is there. As Bishop Berkeley famously asked, how do we know that a tree has fallen in the forest if we do not see or hear it? This book is written to tell the story of our unfolding understanding of the deep sea, as well as to point the way toward the new stewardship arrangements that are urgently required, if the diversity and richness of the deep sea are to be conserved for future generations.

Acknowledgements

This book was conceived some six years ago, based on a suggestion from Nicola Young, then an editor at UNSW Press, following publication of my paper with Bertrand Richer de Forges and Gary Poore in *Nature* on the diversity and endemism of the seamount fauna in the Tasman and Coral seas. It has been a long journey since then, assisted by many along the way.

I would like to thank, first, John Elliot, publisher at UNSW Press, for his long-standing support and many helpful suggestions, as well as his patience through many stops and starts and missed deadlines. Catherine Page ably served as editor, as well as indexer, helping to polish the final draft of the manuscript.

Several librarians patiently responded to innumerable requests for reference material with unfailing patience and good humor: Denis Abbott at CSIRO's Marine Library in Hobart, and Jane Pollock and Bernadette Waugh at CSIRO's library in Floreat, Western Australia. Louise Bell and Lea Crosswell assisted with several of the figures.

A number of people read drafts of various chapters, suggesting countless improvements and corrections: Alan Butler, Malcolm Clark, Clive Crossley, Kristina Gjerde, John Gordon, Eric Mills, Alan Pearce, Joanna Strzelecki, and Karen Vincent. Hjalmar Thiel and Craig Smith, long-time friends and collaborators, read through and commented on the entire manuscript, and Kristina Gjerde collaborated with me in writing the final chapter.

This book would have taken even longer without the support of a Pew Fellowship, shared with Craig Smith, which allowed me to take time from my day-job at the CSIRO to work on the book. John Keesing, Director of the Strategic Research Fund for the Marine Environment, was unfailingly patient and understanding of the time and effort required, so much greater than I initially conceived.

Finally I would like to thank Karen Vincent, my wife and constant companion. I owe her a great debt for all the evenings and weekends that I spent secluded in my study. Her love, support and good humor added a brimming measure of pleasure and delight to a long and arduous journey.

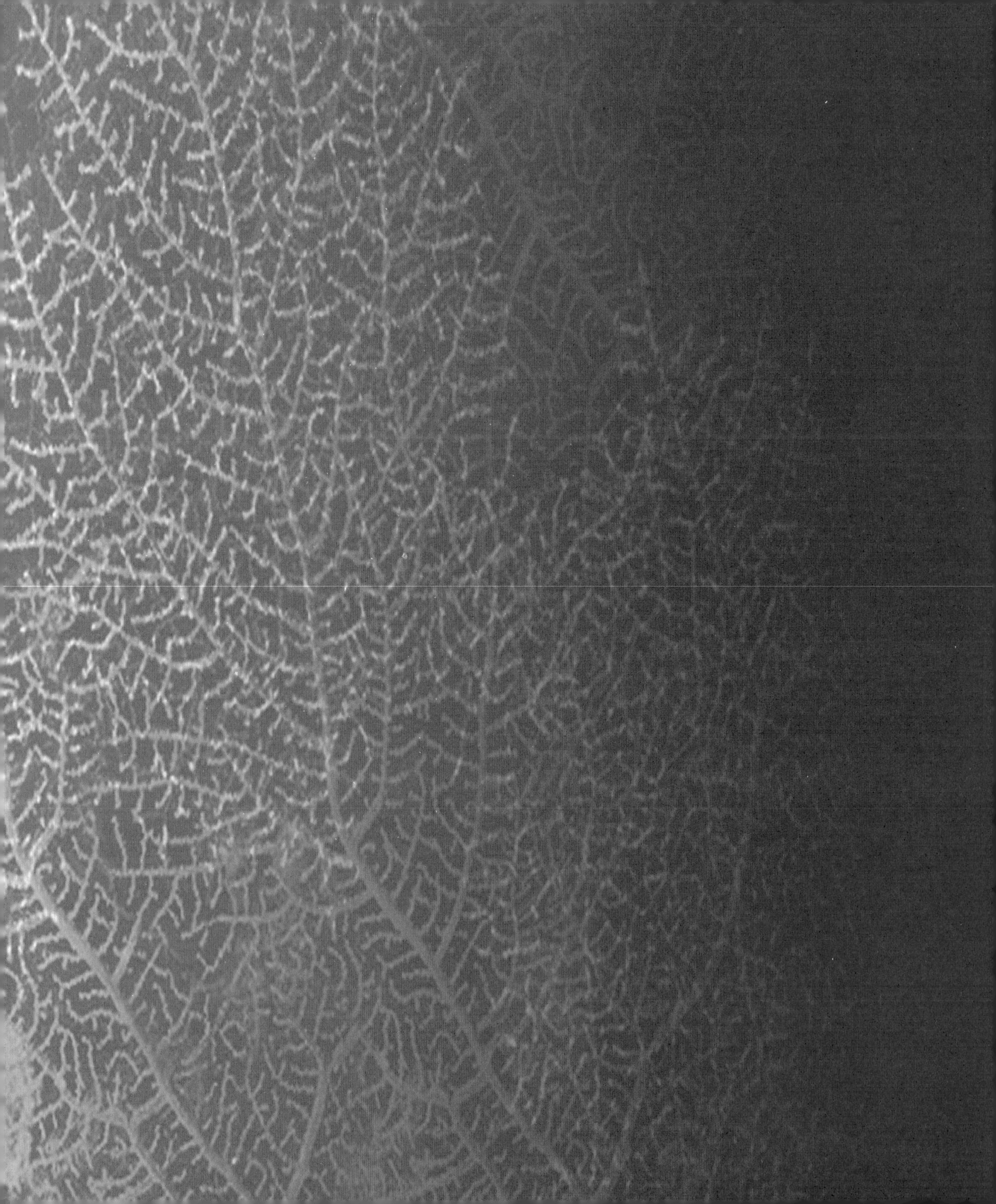

Part I

The Early History of Deep-Sea Exploration

CHAPTER 1

The Rise of Deep-Sea Exploration: Early Paradigms

The people along the sand
All turn and look one way.
They turn their back on the land.
They look at the sea all day.

They cannot look out far.
They cannot look in deep.
But when was that ever a bar
To any watch they keep?

Robert Frost,
Neither Out Far Nor In Deep

The ocean in the ancient world

Some of the earliest and most remarkable testaments to man's appreciation of the sea come down to us from the Minoans at the Bronze Age dawn of Western civilization. The ruins of the palace of Knossos, built around 1500BC, contain wall paintings of dolphins and fish that still strike the observer as light, fresh, sensuous and utterly marvelous (p. 10). Pottery from the period is adorned with octopus rendered with all the fluidity of their watery world (p. 11). Little today is known of these people – not even whether their language was within the Indo-European family (their original script, known as Linear A, remains undeciphered) – other than that they were among the first great maritime civilizations and established trade around the Mediterranean. But their art speaks of a people whose daily life and aesthetic imagination were deeply imbued with the sea, a people who were keen observers of its nature, frankly appreciative of its exuberance and the beauty of its forms. The history of Western art is replete with seascapes, harbor scenes, sailing ships, still lifes of fish draped across tables – but where again do we find artist and audience face to face with the life of the sea, in celebration of the natural world in and of itself?

The Minoan civilization fell victim to natural and human disasters – earthquakes, fire, and invasion by the Mycenaean people from the mainland – disappearing about 500 years before Homer and a millennium before the full flowering of classic Greek civilization in Periclean Athens. The Mediterranean, enclosed and tideless, so conducive to navigation by early seafaring peoples, was next dominated by the Phoenicians. These people, the Canaanites and Philistines of the Bible, had settled along the coast of what is today Lebanon, Syria, and Israel, following an ancient migration, according to Herodotus, out of the Persian Gulf region.[1] They went on to colonize and trade around the rim of the Mediterranean, eventually (along with their Carthaginian descendents) venturing beyond the Pillars of Hercules into the Atlantic. Little trace remains of these voyages, although they apparently explored the fog-bound waters off northwestern Europe, sufficiently far west to encounter seas covered with *Sargassum*, and southward along the coast of Africa; farther probably than Europeans, with the exception

Fresco of dolphins and fish from the palace of Minos at Knossos, Crete, circa 1500BC. (© 2002, Kwan Choi)

of the Vikings, were to venture until the Age of Exploration, some 2000 years later. However there is no evidence that the Phoenicians conducted their voyages other than to seek trade and resources; the development of marine science was, so far as we know, first associated with the Greeks of the classic period.

When Greek philosophy and science underwent their great flowering in the Age of Pericles (circa 400BC), the Minoan culture was part of a long-vanished antiquity. Although the Greeks, through trade and colonization, came to dominate the eastern Mediterranean, the sea and its natural history never provided dominant motifs of the popular and creative imagination as once had been the case among the Minoans. The focus of Greek art and the newly developed dramatic and philosophical genres shifted to man and the heroes and gods of their anthropocentric mythology.

Plato, source of so much in Western philosophy, had little to say about the marine world – frankly, it did not interest him. But his very indifference was archetypal, representing an attitude toward the hidden world beneath the waves that resonates down to the present day. One of his few references to the oceans is set in the *Phaedo*. As Socrates prepares for death, he argues that the heavenly realm before him will prove as superior to our own as the terrestrial world is to the world beneath the sea, where:

> everything is corroded by the brine, and there is no vegetation worth mentioning, and scarcely any degree of perfect formation, but only caverns and sand and measureless mud, and tracts of slime wherever there is earth as well, and nothing is in the least worthy to be judged beautiful by our standards.[2]

This view of the oceans appears extreme to our post-Cousteau sensibility, informed by underwater nature videos and readily available holidays to Australia's Great Barrier Reef and the Caribbean. However in one form or another, this bleak perspective served as the backdrop to deep-sea exploration until at least the 19th century; and strains of this worldview persist to the present day, even though marine mammals have become the sacred cows of the Western world.

Mankind is firmly at the center of the Western cultural heritage. Humans were the measure of all things in the Greek worldview, and the key compact set out in Genesis is that a God conceived in man's image created the universe and conveyed to man dominion over all living things. Of what interest is the marine world – and, in particular, the hidden world

Minoan 'octopus vase' and marine-style vessel from circa 1500 BC, held at the museum in Heraklion, Crete. (Courtesy of Misha Nedeljkovich, University of Oklahoma.)

of its depths – in such a worldview? The roots of Western civilizations lie in the development of agriculture and pastoralism: the cultivation of the olive, the vine, and grains; the shepherding of sheep, goats, and cattle. Although there has been no shortage of great maritime powers in Western history – from the Phoenicians in the ancient world to the Spanish, Portuguese, Dutch, and British in the modern – the sea served them primarily as the means to establish their trade routes and dominion over distant lands. The life within the sea was of little interest. It was left to fishermen to ply their lowly trade in the belief, held until the close of the 19th century (and beyond), that man could have no significant impact upon the world's great sea fisheries.[3]

The exception to this attitude in the classic and pre-modern world was Plato's great successor and architect of the Western scientific enterprise, Aristotle, who turned in mid-career from metaphysics to science and laid the foundations for Western physics and biology. Aristotle took a keen interest in the sea and wrote extensively about it. Indeed, Galileo himself, arguably Aristotle's successor in the rebirth of science in the 17th century, perpetuated the legend that, frustrated by his inability to explain the tides after long observation of the tide races in the Strait of Euperbus, Aristotle ended his life by hurling himself into the sea from the cliffs of Euboea.[4] There seems little factual basis for this – Aristotle apparently died from a stomach ailment – but the legend conveys his commitment to empirical science generally and his fascination with the oceans in particular. In this he proved an outstanding exception to all who were to follow until the dawn of the modern era when men such as Galileo began again to systematically observe nature and to fit their observations into models of reality.

Through contacts with fishermen and his own observations and experiments, Aristotle recorded some 180 marine species from the Aegean Sea, which in *Historia Animalia* he worked into a systematic classification that included all known forms of life, both marine and terrestrial. Many aspects of Aristotle's classification persisted with little change for almost 2000 years, including his division of living things into plants and animals, and the animals into vertebrates and invertebrates, which he distinguished on the basis of those with blood (mammals, birds, reptiles, amphibians, and fishes) and those that are 'bloodless.' Aristotle then proceeded to divide the invertebrates into the cephalopods, higher crustaceans, insects, and the 'testaceans,' a grab-bag of creatures with shells, such as the majority of mollusks, but also including the sea urchins and ascidians. Today this is of mostly historical interest, but Aristotle's outlook – his emphasis on observation and experiment, his development of a sophisticated theoretical framework for his observations – was essentially that of a modern scientist. Today he is recognized as the first great zoologist, and one of the greatest of all time: he was the first to understand the close links between form, function, ecology, and behavior; and the first to develop a systematics based on multiple rather than simple characteristics (such as the presence of wings).

The concept of progress is firmly embedded in contemporary consciousness. Given the rapidity, indeed the accelerating rate of scientific and technological change in modern times, it is difficult today to appreciate the apparent stasis of the pre-modern world. Pliny, writing 400 years after Aristotle at the height of Roman power and influence in the first century AD, produced an encyclopedic 37-volume natural history (the *Historia Naturalis*), which served as the principle scientific source-book in western Europe throughout the Middle Ages. But Pliny was a compiler, not a scientist, and most of what he wrote about marine animals was derived from Aristotle; indeed, he recorded four fewer species from the entire world of the Roman Empire than did his great predecessor working in the Aegean alone. Still, Pliny wrote with a remarkable blend of complacency, hubris, and naïveté:

> By Hercules, in the sea and in the ocean, vast as it is, there exists nothing that is unknown to us, and a truly marvelous fact, it is with those things that nature has concealed in the deep that we are best acquainted![5]

Aristotle was not available to European scholars until the 12th century, when Greek works were translated into Latin; but more was lost during the European Middle Ages than just the body of factual material – the Greek spirit of inquiry was missing too. The scientific method was eclipsed within the Roman and Christian worlds for almost two millennia. Until the Renaissance, western European scholars followed Pliny's model, synthesizing past studies in encyclopedic works.

The history of marine science – and the science of the deep ocean, in particular – appears to have developed as a dialectic between the inquiring spirit of Aristotle and the more common Platonic/Plinian view that the deep is not simply unknown – a frontier to be explored – but rather an inferior world, such that what little we know of it is all there is to be known, and even that is hardly worth knowing. How different from the Socratic perspective with respect to the quest for self-knowledge: that the beginning of wisdom was to know that one knows nothing!

Following Aristotle, interest in the ocean depths appears not to have been rekindled until the Age of Exploration, when Magellan (unsuccessfully) attempted to sound the ocean depths between two coral islands in the Pacific. Even to ascertain the depth of the open ocean proved intractable until the 19th century, when interest in laying a trans-Atlantic telegraph cable led to the development of massive sounding 'machines': drums with thousands of meters of sounding line that overcame the twin hurdles of sufficient strength not to break under the strain of their own weight, yet sufficient sensitivity to enable the moment of impact on the bottom to be noted. Mariners' use of sounding lines is recorded as far back as Herodotus, who begins 'a general description of the physical features of Egypt [noting that] if you take a cast of the lead a day's sail off-shore, you will get eleven fathoms, muddy bottom – which shows how far out the silt from the river extends.'[6] Soundings at the time of Magellan and well after were generally limited to the depths of the continental shelf – about 100 fathoms or 200 meters – but once again the ancient Greeks astonish us. Posidonius in the first century BC pointed to the Mediterranean Sea off Sardinia as the deepest of known seas, with a depth measured to about 2000 meters.[7] Unfortunately the method of measurement was not recorded.

Deep-ocean science: tentative beginnings

At the dawn of the modern scientific era, the 17th century physicist, Robert Hooke, argued (mostly unsuccessfully) to revive support for deep ocean studies. He speculated that the depths may be as fully populated as the land, 'only we are less knowing of them, because they are out of our Element, and we want *Nuntii* or Messengers, to send thither to bring us back Information, and also the Productions and Commodities that this *Terra incognita*, or unknown World, does afford.' He went on to note the objections that

> most will be apt to make, that Animals and Vegetables cannot be rationally supposed to live and grow under so great a Pressure, so great a Cold, and at so great a Distance from the Air, as many Parts at the Bottom of very deep Seas are liable and subject to; I say, I think that these Objections may be easily answer'd, by shewing, that they all proceed from wrong Notions that Men have entertain'd, from the small Experience they had had of the Effects, and Powers, and Methods of Nature, and a few Trials will easily convince them of the Erroneousness of them. We have had Instances enough of the Fallaciousness of such immature and hasty Conclusions. The Torrid and Frigid Zones were once concluded uninhabitable; and to assert *Antipodes* was thought atheistical, heretical, and damnable; but Time has discover'd the Falsity and Narrowness of those hasty Conclusions.

In what might serve as the archetypal scientist's lament, Hooke predicted that 'the Harvest [from the projected marine science] is great, but the Labourers are few; and without Hands and Heads too, little can be expected; and to rely only upon Time and Chance, is, probably the most likely Way to have all our Hopes frustrated.'[8] Indeed little was to come of Hooke's vision, his exhortations and efforts, until the 19th century, which marked the first century of deep-sea exploration.

Deep-sea science in the Age of Naturalists

How deep is the ocean? Is there life at the bottom of the sea? At the beginning of the 19th century no more was known about such matters than was known to the ancients, an anomaly in that era of scientific and geographic exploration noted even by contemporaries. As Wyville Thomson, the pioneering scientist who did more than any to open the deep sea to view in the 19th century, stated in the introduction to his classic, *The Depths of the Sea*, in 1874:

> Every gap in the noble little army of martyrs striving to extend the boundaries of knowledge in the wilds of Australia, on the Zambesi, or towards the North or South Pole, was struggled for by earnest volunteers, and still the great ocean slumbering beneath the moon covered a region apparently as inaccessible to man as the 'mare serenitatis' [one of the so-called 'seas' of the moon].

The early 19th century was the era of the amateur naturalist. Gentlemen collected shells and fossils. Yachtsmen and other amateurs enthusiastically took to using the 'naturalist's dredge,' introduced by a Dane, O.F. Muller, in 1799. As Edward Forbes later related, 'the chief purpose for which the dredge [was initially] employed on our coast was for the procuring of specimens of shells of the Mollusca.'[9] Collectors working for profit or imbued with misguided patriotism soon swelled the lists of the so-called British fauna. However by the 1830s, a small group of more dedicated naturalists – including Henri Milne-Edwards in France and Michael Sars in Norway – had begun to put the dredge to more serious scientific use. In Britain, at the young age of 24, Forbes led a group of similar-minded enthusiasts in forming a 'dredging committee' in 1839; they obtained a grant of £60 from the British Association 'for researches with the dredge.'[10]

The apotheosis of this pre-professional scientific tradition was Charles Darwin, who set out as naturalist at the age of 21 on the circum-global voyage of the *Beagle* (1831–36) and returned with the material to develop the theory of evolution over the next 20 years, without academic or other appointment.

But deep-sea science had more tentative origins. Edward Forbes, the most influential marine naturalist in this early period, today cuts a rather unfortunate transitional figure, joining the ranks of those in the history of science who, like Lamarck, are best remembered for the influential but completely wrong paradigm that they championed.

In his relatively brief career – he died in 1854 at only 39 years of age – Forbes completed a substantial body of work describing the marine faunas of European seas. His pioneering contribution, based on his dredging studies, described the major biogeographic provinces from the Arctic to the Mediterranean and Caspian Seas, and, more importantly, their vertical zonation. In this work he raised dredging researches from the realm of shell-collecting – or at best, taxonomy, the discovery and description of new species – and firmly established a new science of marine benthic ecology: the description and understanding of pattern in species' distributions. Naturalists working on land had previously described how plant communities change with altitude, for example from coastal to alpine regions. Forbes observed corresponding changes with depth on the seafloor, presumably due to the effects of tidal exposure and diminishing light: first an intertidal or littoral zone, inhabited by seaweeds and animals adapted to withstand periodic exposure to the air; below that, a subtidal Laminarian zone dominated by kelps, extending to about 25 meters; a Coralline zone down to a depth of about 100 meters, where the vegetation gave way to plant-like hydroids and bryozoans; and below that a zone dominated by deep-sea corals.

Forbes' deepest and most extensive fieldwork was carried out in the Aegean Sea, where he served as naturalist for 18 months in 1841–42 aboard the British survey vessel H.M.S. *Beacon*. Like Darwin ten years earlier, he seized upon the opportunity to serve as shipboard naturalist. Classically educated, he was well aware that he would be revisiting the very region and fauna that Aristotle had first described more than 2000 years before. With his naturalist's dredge, Forbes was able to extend his investigations far deeper than his great predecessor, dredging about 100 times from near-surface depths down to about 240 meters (130 fathoms) and several times as deep as 420 meters (230 fathoms).

But here Forbes' fortune and perspicacity ran out. The Aegean is a particularly unproductive sea, and although Forbes never encountered a depth where his dredge did not yield living creatures, the fauna grew ever sparser the deeper he went. Forbes held to the popular and apparently obvious notion, critiqued by Robert Hooke 150 years previously, that conditions in the deep sea were inimical to life. 'As we descend deeper and deeper in this region, its inhabitants become more and more modified, and fewer and fewer, indicating our approach towards an abyss where life is either extinguished, or exhibits but a few sparks to mark its lingering presence,' he wrote in his popular and influential work (published posthumously), *The Natural History of the European Seas*. Extrapolating from his sampling to 230 fathoms, he went on to conclude that life in the sea likely ceased altogether below about 550 meters (300 fathoms).[11]

Forbes' so-called *Azoic hypothesis* rested on such a strong foundation of common sense that it was accepted for decades, despite considerable evidence to the contrary. Conditions in the deep sea were cold – just a few degrees above freezing – eternally dark and sustained huge pressures. (Pressure increases by an atmosphere – 1.03 kilograms per square centimeter [14.7 pounds per square inch] – with every ten meters of depth, such that the pressure at 1000 meters is 100 kilograms per square centimeter, an area about a fifth the size of a small postage stamp.) As light diminished with depth, Forbes observed a dramatic decline in plant life; surely animal life would prove impossible not far below. In addition, the depths of the

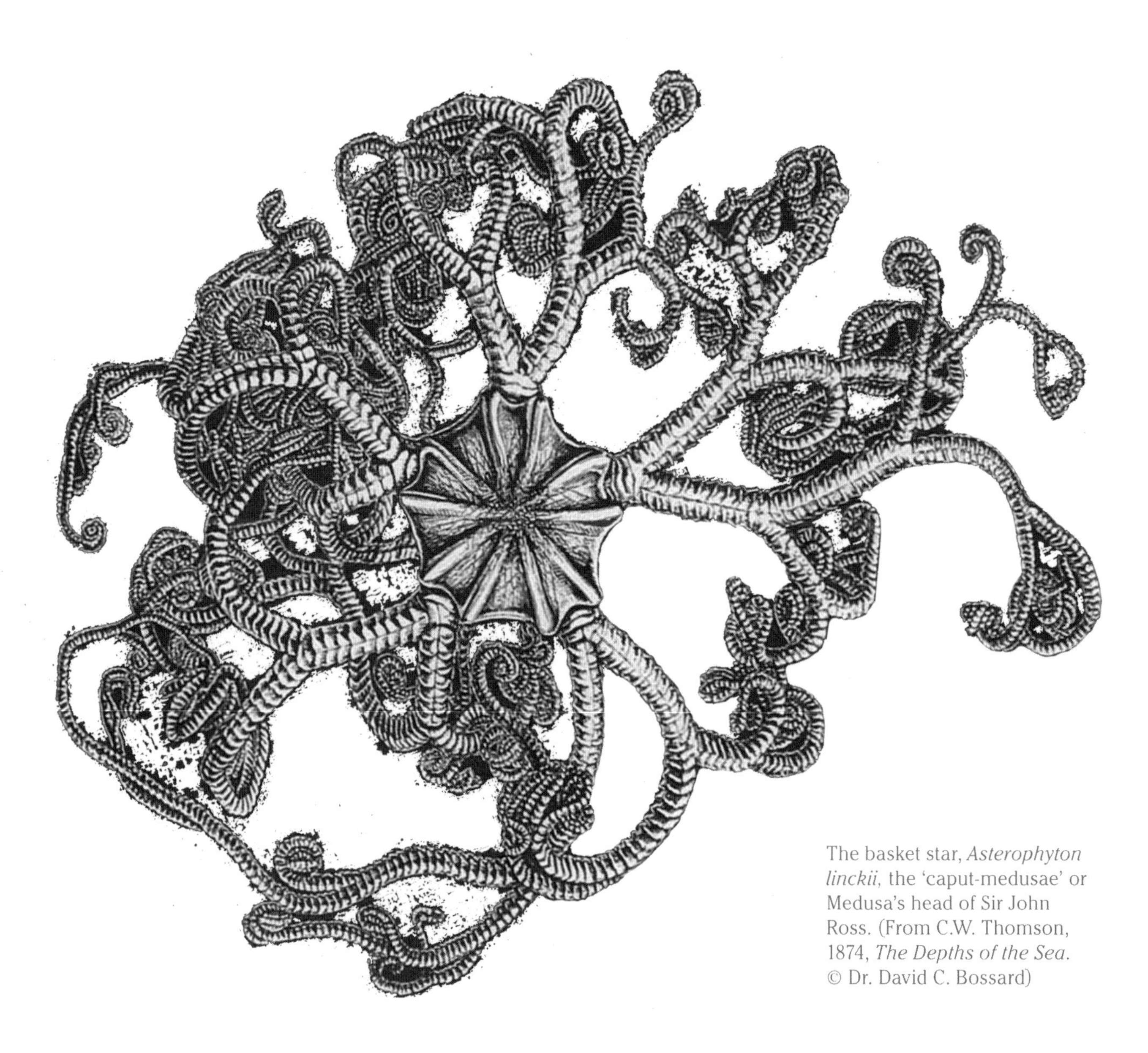

The basket star, *Asterophyton linckii*, the 'caput-medusae' or Medusa's head of Sir John Ross. (From C.W. Thomson, 1874, *The Depths of the Sea*. © Dr. David C. Bossard)

sea were assumed to be stagnant and hence anoxic: yet another factor inimical to all forms of higher life.

As Wyville Thompson, who began his career dredging with Forbes in the Firth of Forth, wrote in 1874, Forbes' 'experience was much wider than that of any other naturalist of his time; the practical difficulties in the way of testing his conclusions were great, and they were accepted by naturalists generally without question.'[12]

In fact there was evidence of life at depths of over 2000 meters well before Forbes dredged in the Aegean, as Thomson and his successor, John Murray, recount in their histories of early deep-sea exploration.[13] In

1818, Captain John Ross led an expedition to seek out a Northwest Passage north of Canada. In keeping with the spirit and capabilities of the time, he was provided gear to sample the bottom, but it failed to work. Ross persisted, engaging the blacksmith on board to construct a 'deep-sea clamm,' with which he brought up four samples from 850 to 2000 meters. His samples contained, variously, crustaceans, corals, shellfish and worms. At one of the deepest soundings, 1000 fathoms or almost 2000 meters, about six times the depth to which Forbes later dredged, the sounding weight sank several feet into the soft ooze on the bottom and the line came up with 'a magnificent *Asterias caput-medusae*' entangled in it – the Medusa's head or basket star. But the paradigm of the lifeless deep had such a hold on the naturalists of the day that these findings were ignored as anomalous; some conjectured that the basket star may have swum onto the sounding line![14]

Twenty years later, Sir James Clark Ross, for whom the Ross Sea is named, a nephew of Sir John Ross and a contemporary of Forbes, led the British Antarctic Expedition of 1839–43 in *Erebus* and *Terror* and carried out some of the first deepwater soundings and dredging near the Antarctic continent. He noted 'some beautiful specimens of Coral, Corallines, Flustrae [mat-like bryozoans], and a few Crustaceous animals' from a dredge haul to 400 fathoms (750 meters) and concluded:

> Although contrary to the general belief of naturalists, I have no doubt that from however great a depth we may be enabled to bring up the mud and stones of the bed of the ocean, we shall find them teeming with animal life; the extreme pressure at the greatest depth does not appear to affect these creatures; hitherto we have not been able to determine this point beyond a thousand fathoms, but from that depth several shellfish have been brought up with the mud.[15]

Sir James Ross recognized some of his Antarctic species as being similar to those dredged from high northern latitudes, which led him to conclude, on the basis of the likely limited temperature tolerance of these benthic creatures, that the temperature at the bottom of the oceans must be uniform for them to be able to pass from one pole to the other.

But neither John nor James Ross were zoologists, and they did not publish in the zoological journals. By the time of his death, Sir James Ross' specimens had deteriorated due to improper preservation, and his findings and keen observations were largely forgotten. Forbes was familiar with the Ross' explorations and findings – he had examined James Ross' deepest dredge sample[16] – yet he clung tenaciously to the view that his own sampling to only a few hundred fathoms approached the depth limits to life in the ocean. Forbes stands in the history of science as one of those enigmatic figures who, through an excess of antiquated intellectual baggage, comes close yet fails to grasp greatness.

In the decades following publication of Forbes' synthesis, evidence of life in the deep sea continued to accumulate, largely arising from very practical efforts to develop one of the world's first communications networks, based on spanning the major seas with telegraphic cables. These endeavors required systematic surveys of the seafloor, and the survey vessels on occasion took on board a naturalist.

One such naturalist was a rather isolated and controversial figure, G.C. Wallich (1815–99), an ex-India Army surgeon who sailed in 1862 across the North Atlantic aboard the survey vessel H.M.S. *Bulldog.* In his account of the cruise, Wallich strongly attacked the Azoic hypothesis. His most striking observation was the recovery of 13 starfish from a sounding line that had lain on the seabed at 1260 fathoms.[17]

But the Azoic hypothesis had such firm hold of the scientific imagination that some zoologists still trotted out the argument

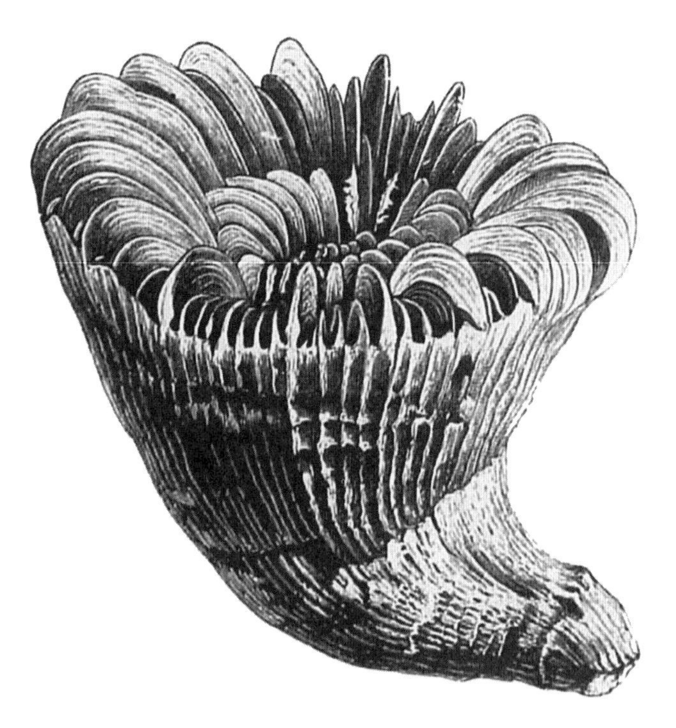

The solitary deepwater coral, *Caryophyllia borealis*, found in 1858 to be growing on a broken telegraph cable at the bottom of the Mediterranean Sea at about 1200 fathoms. (From C.W. Thomson, 1874, *The Depths of the Sea.* © Dr. David C. Bossard)

used against Sir John Ross, that the starfish might have swum into the line and taken hold of it while it was being raised to the surface. Such arguments could not be made, however, when some 15 species were found on a broken telegraphic cable brought up for repair from 2300 meters (1260 fathoms) in the Mediterranean. These included a solitary deepwater stony coral, *Caryophyllia*, encrusted and growing on the cable.

Still, the paradigm persisted until finally overturned by the determined and systematic sampling programs of Wyville Thomson, in collaboration with William Carpenter aboard the *Lightning* in 1868, the *Porcupine* in 1869 and 1870, and then the *Challenger* expedition of 1872–76.

Wyville Thomson (1830–82), who began his career in deep-sea biology dredging with Edward Forbes, for a long time accepted Forbes' view that life was limited to the upper depths of the ocean. By the 1860s, however, the mounting evidence led him to revise his view.

In 1866 Thomson, then Professor of Natural History at Queen's College, Belfast, visited the Norwegian marine biologist G.O. Sars, to examine specimens that his son Michael, a fisheries inspector, had dredged from Norwegian waters at depths to 825 meters (450 fathoms). Thomson had a particular interest in the crinoids, a class of echinoderms that flourished in the Paleozoic. Among Sars' specimens was a new species, *Rhizocrinus lofotensis*, belonging to a group until then believed extinct since the Jurassic. Several years before, another group of echinoderms known previously only from the fossil record, the brisingid starfish (p. 20), had been dredged from deep water. As Thomson wrote to his influential friend and colleague, William Carpenter, vice-president of the Royal Society, these discoveries indicated not only that there was life in the deep sea but 'that the forms at these great depths differ greatly from those met with in ordinary dredgings, and that, at all events in some cases, these animals are closely allied to, and would seem to be directly descended from, the fauna of the early Tertiaries.'[18]

Thomson's enthusiasm reflected a scientific preoccupation arising from the 1859 publication of Darwin's *Origin of Species*. Darwin had grappled with the extreme scarcity of intermediate forms linking one species with another or linking fossil and contemporary faunas. But at times, Darwin noted, one encountered so-called 'living fossils [which] have endured to the present day, from having inhabited a confined area, and from having been exposed to less varied, and therefore less severe, competition.'[19] And Darwin observed that 'the productions of the land seem to have changed at a quicker rate than those of the sea.'[20] When Thomson heard of the Sars' discoveries, he and Carpenter seized upon what seemed an immense opportunity: to discover a fauna perhaps unchanged since the so-called 'Age of the Chalk,' the Miocene fauna fossilized in the chalk beds of Britain. Was there the prospect of encountering the marine invertebrate equivalent of a Jurassic Park fauna, still extant in the deep sea?

The stalked crinoid, *Rhizocrinus lofotensis*, discovered and described by Michael Sars, a relict of the Mesozoic and earlier periods, when these echinoderms dominated the seafloor. (From C.W. Thomson, 1874, *The Depths of the Sea*. © Dr. David C. Bossard)

The transition to Big Science

Thomson returned to Ireland and planned a deep-sea dredging expedition with Carpenter. They realized that deep-sea dredging, with its need for several kilometers of 5 centimeter (2.5 inch) hempen rope, was beyond the resources and capabilities of any individual: the era of the amateur naturalist was at an end. Thomson included in his popular account of the early voyages of the *Lightning* and *Porcupine* the correspondence that he and Carpenter skillfully passed back and forth to obtain, first, the endorsement of the Royal Society and thence the support of the British Admiralty. With Thomson and Carpenter, we clearly pass into the modern era of oceanography as Big Science – forever more dependent on government funding and an attendant mix of skills in networking, politicking and manipulation of the popular imagination, as

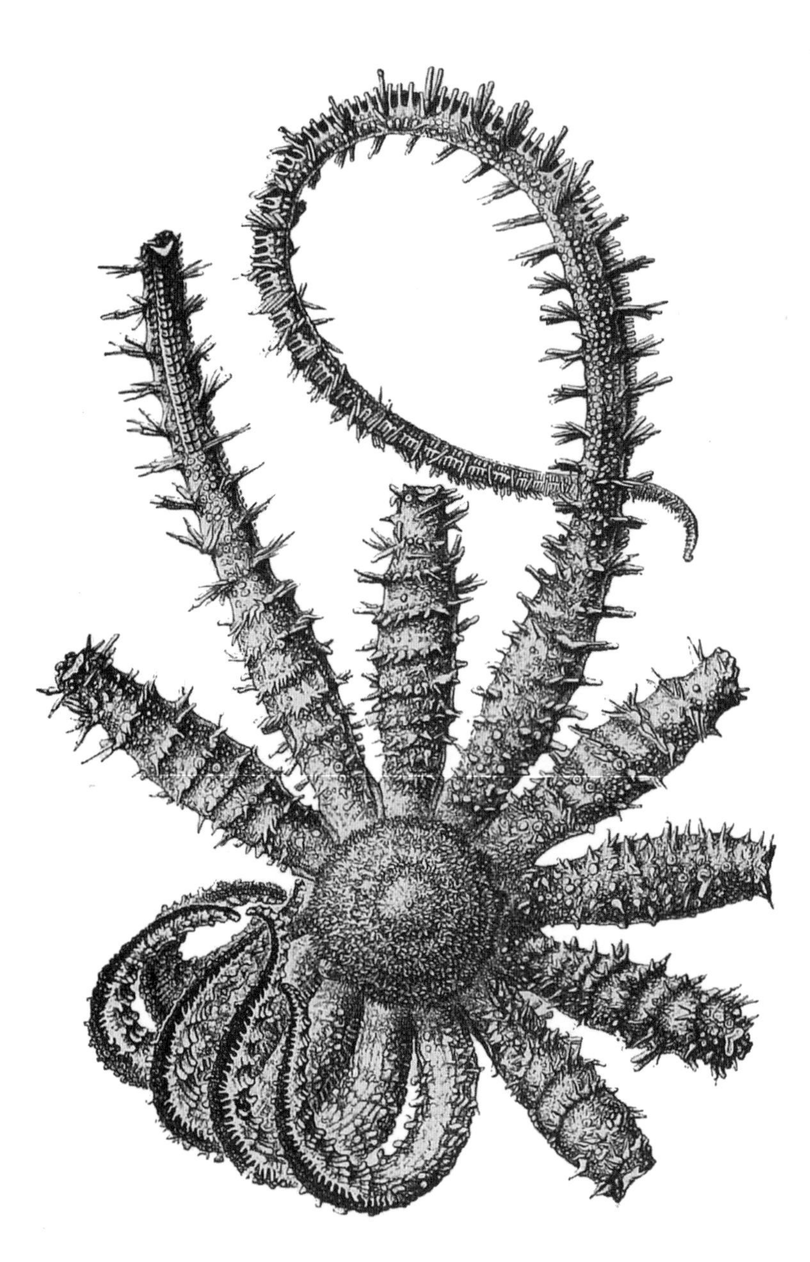

Brisinga coronata, described by G.O. Sars, a fragile, primitive starfish discovered during the early history of deep-sea dredging. (From C.W. Thomson, 1874, *The Depths of the Sea*. © Dr. David C. Bossard)

well as in the science itself.

The Admiralty granted use of a 'cranky' paddle steamer, as Thomson described it, H.M.S. *Lightning*, the oldest in the navy. In the late summer of 1868, the *Lightning* sailed from the north of Scotland and sounded and dredged along several tracks extending northeast into the deep waters of the North Atlantic and across the channel between the Shetlands and Faroes. But the weather is chancy at these latitudes (generally between 59° and 62° N), and poor weather plagued the voyage. The vessel leaked and twice almost came to grief as the rigging came undone, and it narrowly escaped being dismasted. In the end, they were only able to dredge for ten days of their six weeks at sea, and on only four days did they dredge in waters over 1000 meters (500–600 fathoms).

Although they obtained relatively few samples from deep waters, these few proved exceptionally interesting. Their northernmost transect between the Orkneys and Faroe fishing banks was dominated by Arctic outflows and North Atlantic Deepwater, a major deepwater mass formed by extreme winter cooling, with temperatures around 0°C (32°F) at bottom depths of approximately 800–1100 meters (450–590 fathoms). The sediments were sand, gravel and various rocks, probably glacially derived from Scotland and the Faroes. On their next track, around 60 nautical miles to the south, bottom temperatures jumped to around 6.4°C (43.5°F), and the bottom was covered with what Thomson referred to as 'Atlantic ooze,' a calcareous mud of largely marine origin. Abundant and varied life was found in both regions; not surprisingly, however, the species changed dramatically. As later soundings would confirm, the cause of this disjunction was a ridge, today named the Wyville Thomson Ridge, which separates the cold deepwater from warmer Atlantic water, flowing with the Gulf Stream into the eastern North Atlantic.

Their most interesting finding in the cold-water region was a new species of brisingid starfish, the group that had only been discovered extant 15 years previously in a Norwegian fjord at 100–200 fathoms. Unlike most starfish, which feed on bottom-dwelling organisms, the brisingids, which appear intermediate between a starfish and a brittle star, have long flexible arms that they extend into the water, from which they extract particles and small organisms. The specimen was described by Thomson as 'a rich crimson colour, passing into orange-scarlet ... They were certainly the most striking objects we met with. One was sufficient to give a glorious dash of colour to a whole dredgeful.' The initial discoverer of the extant group, P. Chr. Absjornsen, named them for a brilliant jewel (Brising) of the Scandinavian goddess, Freya.[21] In the warm Atlantic water, they also encountered new species of primitive and previously little-known siliceous hexactinellid and other sponges, as well as brittle stars, sea pens, sea urchins, crinoids, crustaceans, and other creatures. As Carpenter concluded in writing up the results of the voyage, the biological communities at these depths were not 'a degraded or starved out *residuum* of Animal life, but a rich and varied Fauna.'[22]

Thomson and Carpenter had taken the first steps toward discovering a new fauna adapted to the unique conditions of the deep sea.

Based on these promising, albeit limited, results, Thomson and Carpenter were able to obtain a much improved vessel, the *Porcupine*, to continue their investigations of the deep waters of European seas over the following two summers. On each successive cruise they worked further southward: first to the west of Ireland (51°–57.5° N latitude), where they dredged successfully to as deep as 2700 meters (1476 fathoms); southward from there on the second leg of their cruise in 1869 to the waters west of southern England and northern France, where they extended their dredging to 4450 meters (2435 fathoms) and obtained representatives of all major invertebrate groups: mollusks, crustaceans, echinoderms – including a new stalked crinoid – sponges, and others. Their deepest dredge sampling

The glass or hexactinellid sponge, *Hyalonema lusitanicum*. This primitive sponge, along with the stalked crinoids and brisingid starfish, raised hopes that the deep sea might contain a vast relict fauna from Mesozoic times. (From C.W. Thomson, 1874, *The Depths of the Sea*. © Dr. David C. Bossard)

took about seven hours to complete, but generally they encountered no significant technical challenges in going to ever greater depths. And most important, they found no depth limitation to the marine fauna.

The following summer, Thomson and Carpenter extended their sampling to the waters west of Spain and explored the deep waters of the Mediterranean, working with John Gwyn Jeffreys, a former close colleague of Forbes. It was here that Carpenter discovered the deep outflow of high-salinity Mediterranean water, produced by the excess of evaporation over precipitation in the Mediterranean region, through the Straits of Gibraltar into the Atlantic. Carpenter hypothesized that the global deepwater circulation, primarily the flow of cold water from the poles to low latitudes, might be similarly driven by differences in density between deep and near-surface water masses – a key hypothesis underlying the subsequent *Challenger* expedition.

During the cruises of the *Lightning* and *Porcupine*, Thomson, Carpenter and their colleagues succeeded in exploring the waters off Europe down to approximately 2000 meters – and in a few instances to twice that depth. They uncovered a diverse deepwater fauna containing several distinctive groups: the hexactinellid sponges; various stony corals, including extensive banks of *Lophelia pertusa*; stalked crinoids; and several primitive sea urchins, like the crinoids previously known only from the fossil record.

The discovery of a new fauna, and of 'living fossils' in particular, considerably whetted Thomson and Carpenter's appetites for deep-sea exploration. Carpenter sought confirmation, as well, for what Thomson called his 'magnificent generalization' regarding the mechanism underlying the oceans' deepwater circulation. What was to follow, however – the *Challenger* expedition – was more than a next step, a logical progression of the science. It was at once the culmination of a century of British global exploration and scientific enterprise – a century that had included the voyages of James Cook and Darwin – and, as the first and the greatest oceanographic expedition of all time, it marked the beginning of modern oceanography.

CHAPTER 2

On the Shoulders of Giants: The *Challenger* Expedition

If I have seen farther it is by standing on the shoulders of giants.

Sir Isaac Newton
in a letter to Robert Hooke

The *Challenger* expedition of 1872–76 marks the transition from Victorian to modern science: from the world of the gentleman naturalist to that of Big Science, with its requisite institutional, collaborative and multidisciplinary framework and national funding support. The pages of the 50 royal quarto volumes of the *Challenger* expedition report are graced with fanciful creations – mermaids drawing plankton nets through the deep – and elegant depictions of the naturalists aboard the vessel, properly attired in jacket, tie, and hat, overseeing the sailors as they bring in the nets and sort the catch. But there is nothing amateur about the monographs that comprise the report, a monumental effort that laid the foundations for modern oceanography.

The *Challenger* expedition was conceived and undertaken on the grand scale befitting the pre-eminent maritime empire of the time. Its objectives were no less than to map the geology and life of the deep seafloor, the chemistry of the world's oceans, the global pattern of deepwater circulation, and much else besides. And all this was to be accomplished by a scientific complement of four 'naturalists' in addition to Wyville Thomson and his combination secretary and expedition artist, John James Wild! The grand design of the *Challenger* expedition and the relatively few who conceived it –

primarily Thomson and his colleague, William Carpenter – speak for the breadth of the Victorian scientist at the dawn of the modern scientific age. These were men grounded in the Greek and Latin classics, whose interests spanned broad scientific fields: scientists in transition between the Aristotelian model of the philosopher and the modern specialist.

In undertaking to lead the *Challenger* expedition, Thomson resigned as Professor of Natural History at the University of Edinburgh, a post previously held by Forbes. Both men had started at university studying medicine – as solid a prospective career then as it is today – but had preferred natural history and eventually left university without ever completing degrees. But this seems hardly to have scotched their career prospects, both pursuing their scientific interests so avidly that each was to hold professorships of geology, botany, and zoology, as well as of natural history. And all this achieved at remarkably young ages: at 39, the age at which Forbes died, Thomson set out to sea on the *Porcupine* and was elected a Fellow of the Royal Society. Contrary to the image of the stuffy Victorian, these men carried off their learning and pioneering scientific endeavors with cheer and humor, as well as considerable aplomb. Picture Forbes leading fellow enthusiasts at dinners of the 'Red Lion Clubbe' (founded at the meeting of the British Association when Forbes' 'Dredging Committee' was established) in his 'Song of the Dredge':

Hurrah for the dredge, with its iron edge,
  And its mystical triangle.
And its hided net with meshes set
  Odd fishes to entangle!
The ship may move thro' the waves above,
  'Mid scenes exciting wonder,
But braver sights the dredge delights

Mermaids pulling a plankton net and a naturalist overseeing the sifting of a benthic sample. (Drawings by J.J. Wild, from J. Murray, 1895, Report of the scientific results of the voyage of H.M.S. *Challenger*. © Dr. David C. Bossard. Reproduced from original documents in the library holdings of Dartmouth College.)

As it roves the waters under.
Chorus: Then a-dredging we will go, wise boys!
A-dredging we will go![1]

And so on. Not great poetry, but doubtless great fun as an after-dinner drinking song!

Expedition planning and preparations

But there is a substantial break between the early collegial enthusiasm of the Dredging Committee and the professionalism, the organizational acumen that characterized Thomson and Carpenter's approach to marine science. Continuing with the approach that had secured them the *Lightning* and *Porcupine*, Carpenter initiated correspondence with the Royal Society in August 1871 as a member of its inner circle: 'the time is now come for bringing before our own Government the importance of initiating a more complete and systematic course of research than we have yet had the means of prosecuting.'[2] He recommended that a committee be formed with the aim of drawing up a research plan to be submitted to parliament. The immediate circumstance was a brief report in the fledgling but already influential scientific journal *Nature* that Germany and Sweden were planning deep-sea expeditions into the North Atlantic and the Arctic. And the Americans under Louis Agassiz were planning an expedition in both the Atlantic and Pacific Oceans to sail around South America. *Nature* and Carpenter spoke of Britain losing its lead in marine science; national prestige was at stake! Carpenter wrote and received favorable assurances of support from the First Lord of the Admiralty, George Goschen. At their summer meeting, the British Association adopted a resolution authorizing its president and council to work with the Royal Society 'for the promotion of the circumnavigation expedition specially fitted out for carrying the physical and biological Exploration of the Deep-sea into all the great oceanic centres.'

Planning, approval and preparation for the expedition proceeded remarkably quickly and smoothly. The Royal Society formed its committee, which included Carpenter, Thomson, and Jeffreys; Joseph Hooker, who had sailed as naturalist on Sir James Ross' expedition with *Terror* and *Erebus* to the Antarctic and collected some of the earliest evidence of life at great depth; and Thomas Henry Huxley, the prominent biologist and supporter of Darwin's theory of natural selection. By the end of November 1871, the committee had drawn up a detailed and ambitious set of objectives for the proposed expedition:

1. To investigate the *Physical Conditions of the Deep Sea,* in the great Ocean-basins – the North and South Atlantic, the North and South Pacific, and the Southern Ocean (as far as the neighbourhood of the great ice-barrier); in regard to Depth, *Temperature*, Circulation, Specific Gravity, and Penetration of Light; the observations and experiments upon all these points being made at various ranges of depth from the surface to the bottom.
2. To determine the *Chemical Composition of Sea Water*, not merely at the surface and bottom, but at various intermediate depths; such determinations to include the Saline Constituents, the Gases, and the Organic Matter in *solution*, and the nature of any particles found in *suspension*.
3. To ascertain the *Physical and Chemical* characters of the *Deposits* everywhere in progress on the Sea bottom; and to trace, so

far as may be possible, the sources of these deposits.
4. To examine the Distribution of *Organic Life* throughout the areas traversed, especially in the deep Ocean-bottoms and at different depths; with especial reference to the Physical and Chemical conditions already referred to, and to the connection of present with the past condition of the Globe.

The committee concluded that 'the Expedition should leave this country in the latter half of the year 1872; and as its perfect organization will require much time and labour, it is desirable that suitable preparations should be commenced forthwith.'[3]

A week later, the President and Council of the Royal Society submitted a one-page letter to the Secretary of the Admiralty, formally proposing support for an expedition to achieve these objectives. In April 1872 the House of Commons approved funds. H.M.S *Challenger* was selected, a 226 foot naval corvette of 2306 tons displacement, a sailing vessel with 1234 horse auxiliary steam power. Its only other mechanized power was an 18 horsepower steam winch, which, combined with the efforts of the crew, would raise the gear from the deep seafloor. The vessel was quickly outfitted as a research vessel: cannons removed; biological and chemical laboratories built; the staff selected and hired; the necessary scientific and other stores ordered and stowed. The *Challenger* sailed out of England shortly before Christmas in 1872 into the face of wintry North Atlantic gales.

Much has been written about the factors underlying the apparent ease and rapidity with which the expedition was approved. Clearly more was required than political and organization skills. This was an era of extremely limited government support for scientific enterprise. Many other factors contributed, foremost being the scientific issues and sense of national enterprise. And it was more than simply a question of whether there was life in the deep sea – that question had already been answered in the affirmative. But what kind of life was to be found there?

The big questions

Several lines of evidence pointed to the deep sea as a vast refuge for life-forms that had been extinct in shallower seas since the Mesozoic, the time of the dinosaurs. Indeed there was speculation that life may have originated in the deep sea. Darwin's *On the Origin of Species* had been published in 1859 and still engendered considerable debate. Key issues were the 'missing' evolutionary links between primitive and modern species and, most fundamentally, the origin of life from inanimate matter – contentious issues to this day. Thomson and Carpenter, as well as others like Huxley, were indefatigable, reporting on their results at meetings of the British Association and the Royal Society, writing for *Nature* and other leading scientific journals, giving public lectures and writing accessible but credible scientific works: for example, Thomson's still-fascinating *The Depths of the Sea*, his account of the cruises on the *Lightning* and *Porcupine*. Thomson and Carpenter succeeded in capturing the imagination of both the scientific world and the educated public: deep-sea science seemed on the threshold of truly exciting discoveries.

Three strands of evidence linked the deep

sea to ancient environments. There were, first, the species that initially excited Thomson and Carpenter in deep-sea exploration: the stalked crinoids, brisingid starfishes and glass (or hexactinellid) sponges, clearly linked to the faunas that dominated Paleozoic and Mesozoic seas. As Thomson wrote:

> Both on account of their beauty and extreme rarity, and of the important part they have borne in the fauna of some of the past periods of the earth's history, the first order of the Echinoderms, the Crinoidea, has always had a special interest to naturalists; and, on the watch as we were for missing links which might connect the present with the past, we eagerly welcomed any indication of their presence.[4]

If such interesting relict animals were obtained in their limited forays off the European continental margin, what might be uncovered by a concerted effort to explore the deep sea of the world's oceans? As Thomson wrote virtually on the eve of the *Challenger* expedition:

> It is unwise to prophesy; but when we consider that the first few scrapes of the dredge at great depths have added two remarkable new species [of stalked crinoids] to the living representatives of this group, until now supposed to be on the verge of extinction, and that all the known species are from depths beyond the limit of ordinary dredging, we are led to anticipate that crinoids may probably form rather an important element in the abyssal fauna.[5]

Underlying Thomson's hope that he might find an essentially Mesozoic deep-sea fauna was his belief that the deep sea was not only so constant that evolution there might have come to a virtual halt, but that the environment itself might be literally unchanged since the Mesozoic. It struck Thomson that the calcareous ooze he collected from the deep seafloor off Britain appeared essentially no different from the well-known English chalk deposits – the white cliffs of Dover – laid down beneath Cretaceous seas. Thomson believed that these formations had their origins as deep-sea deposits, such that there was, as he termed it, a 'continuity of the Chalk' between Cretaceous and contemporary deep-sea environments: essentially the deep sea might still lie within the Cretaceous.[6] No wonder then that he anticipated a deep-sea fauna dominated by species long extinct elsewhere: the deep sea was potentially a vast Jurassic Park waiting to be brought to the surface in the naturalist's dredge. Thomson's assertion was hotly contested by both the leading geologists of the day, including Charles Lyell, and marine taxonomists, such as Jeffreys, who argued, correctly, on the basis of the fossilized fauna within the chalk deposits that they derived from shallow seas. But Thomson continued to speculate along these lines, although he backed down from his more extreme claims.

Another of the expectations of these pioneers of deep-sea exploration was even more fantastical – although we will find resonance subsequently (Chapter 5), with some particularly intriguing contemporary discoveries that point to the origins of life in the deep sea. As Thomson described it:

> If the mud [dredged from the deep seafloor] be shaken with weak spirit of wine, fine flakes separate like coagulated mucus; and if a little of the mud ... be placed in a

drop of seawater under the microscope, we can usually see, after a time, an irregular network of matter resembling white of egg, distinguishable by its maintaining its outline ... This network may be seen gradually altering in form, and entangled granules and foreign bodies change their relative positions. The gelatinous matter is therefore capable of a certain amount of movement, and there can be no doubt that it manifests the phenomena of a very simple form of life.[7]

One senses in Thomson's description the excitement of discovery: the naturalist peering through his microscope at a most primitive and hitherto unknown form of life. The eminent Darwinian zoologist, T.H. Huxley, discovered it upon re-examining some old sediment samples and named it *Bathybius haeckelii* after Ernst Haeckel, the prominent German developmental biologist and champion of evolutionary theory. *Bathybius* seemed just the answer to several conundrums. If the bottom of the sea was covered with a vast sheet of *Urschleim* (primeval slime), as Haeckel termed it, then the subsistence of the deep-sea fauna no longer posed a difficulty. Haeckel himself seized upon the discovery as the missing link between the chemical and biological: the most primitive form of life, without nuclear structure, from which more advanced forms of life evolved.

The *Challenger* expedition was the 19th century British equivalent of the American space program a century later. An enterprise with national imperial prestige at stake was mixed with simple, fundamental scientific issues, capable of exciting both the popular imagination and the scientific community – the existence or nature of life in a remote, hitherto inaccessible region of the universe. And there was precedent in Britain for support of expeditions deemed of national or singular scientific importance, from James Cook's 18th century voyages of discovery to more recent expeditions to observe the solar eclipses of 1870 and 1871. Moreover, relatively little additional capital outlay was required, since the navy was committed, regardless of the expedition, to maintain its vessel in operation.

The length of the expedition – three-and-a-half years – was set by the duration of a

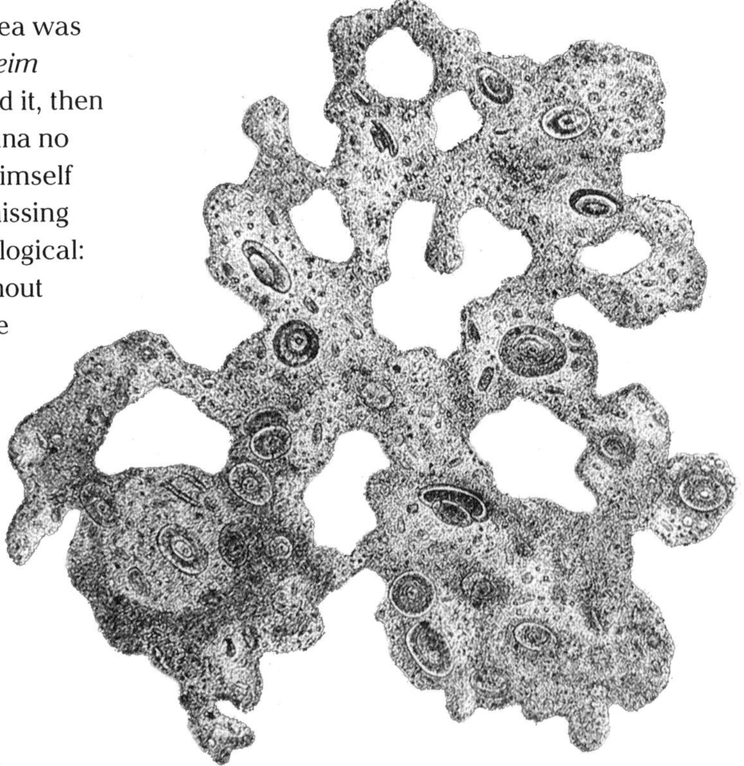

Bathybius haeckelii, the so-called *Urschleim*, originally believed to cover the bottom of the sea and provide a link between animate and inanimate nature. (From C.W. Thomson, 1874, *The Depths of the Sea*. © Dr. David C. Bossard)

survey vessel's typical commission in the navy. Carpenter, approaching 60, declared he would not go to sea again, so the leadership fell to Thomson, who temporarily resigned his professorship and received in compensation a stipend of £1000 per year. Interestingly, of the 240 men on board, only six civilians made up the scientific contingent, and the others were to receive a mere £200 for their annual stipends. Four were listed as 'naturalists': John Buchanan, a chemist; Rudolph von Willimoes-Suhm, a young German who died of erysipelas on the voyage; Henry Moseley, a keen broad-based naturalist who wrote a still-interesting account of the voyage, focussing on his times ashore (about 40 percent of the voyage was actually spent in port for repairs, recoaling and revictualing, as well as for designated faunal and floristic collections on oceanic islands); and most important was John Murray. He was to succeed Thomson in 1882 as director of the Challenger Expedition Commission, which oversaw the dissemination of the collections to a wide and international group of specialists, and the subsequent publication of the reports. In addition J.J. Wild combined the services of secretary for Thomson with those of artist, a vital role prior to the photographic era. Wild was to render many of the fine and accurate drawings that later graced the pages of the reports: sampling arrangements on the vessel, the various gears, the naturalists' laboratories, scenes from the cruise, detailed scientific drawings of new species while they remained fresh, and even the occasional fanciful mermaids. (Little wonder on a three-and-a-half-year voyage!)

Several aspects of the *Challenger* expedition appear remarkable to a modern oceanographer besides its duration (cruises today are measured in weeks and days rather than years and months). One is the implicit trust in the scientists. In contrast to the endless round today of pre-proposals, compendious proposals, milestone and other reporting requirements, the expedition proposal to the Admiralty was a single page. The Navy did not ask for a cruise plan until August 1872, a mere four months before the *Challenger* was to sail. Even more remarkable was the breadth of these naturalists. Whereas a major oceanographic project today might enlist several specialists to sample the plankton alone, Thomson set out in his reply to the Admiralty a sampling program for his small band of naturalists that would map out the broad outline of physical, chemical, geological, and biological oceanographic science for the next century. And as if this were not enough, Thomson added a program of sampling oceanic island faunas and floras, fully aware then of issues that are often considered distinctly modern concerns: the potential impact of introduced weeds and pests on these previously isolated environments, the extinction of native species, and the preservation of traditional knowledge:

> These [islands] are, in many cases, the last positions held by floras of great antiquity; and, as in the case of St. Helens, they are liable to speedily become exterminated, and therefore to pass into irremediable oblivion when the islands become occupied.
>
> Attention should be given to the esculent and medicinal substances used in various places ... As far as practicable, trustworthy information should be recorded as to the date and circumstances of the introduction of foreign species.[8]

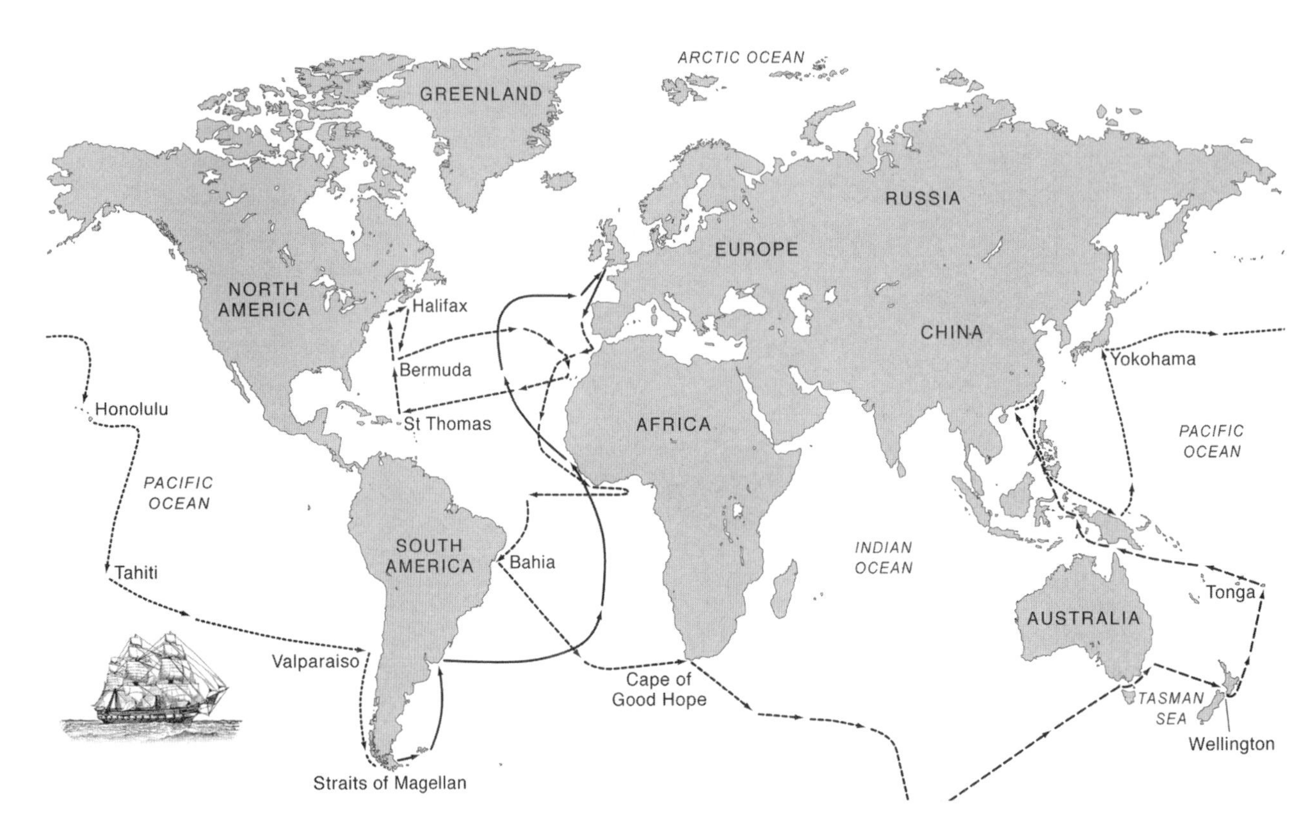

Track of the *Challenger* expedition. The different line types correspond to the different years of the voyage, and some of the waypoints are shown.

The expedition: tedium, passion, and the opportunity for great exploits

All told, the *Challenger* was to log about 69 000 miles over the almost three-and-a-half-year expedition, circumnavigating the globe from England to the edge of the Antarctic, crisscrossing the Atlantic and Pacific. The crew sounded, dredged, and trawled at some 362 stations. Moseley sums up the cruise experience most frankly in his account of the voyage, which, except for the last chapter, focussed on his excursions ashore:

> The vastness of the depth of the Ocean was constantly brought home to us on board the 'Challenger' by the tedious length of time required for the operations of sounding and dredging in it. When the heavy sounding weight is dropped overboard, with the line attached, it takes about … thirty-five minutes to reach the bottom in the average depth of 2500 fathoms.
>
> The winding in of the line again is a much slower process. It used to take us all day to dredge or trawl in any considerable depth, and the net usually was got in only at nightfall, which was a serious inconvenience, since we could not then, in the absence of daylight, make with success the necessary examinations of the structure of perishable animals.
>
> The ship, when deep-sea operations were going on, used to lie rolling about all day, drifting along with the wind, and dragging the dredge over the bottom. From daybreak

to night the winding-in engine was heard grinding away with a painful noise, as the sounding-line and thermometers were being reeled in.

At last, in the afternoon, the dredge-rope was placed on the drum, and wound in for three or four hours, sometimes longer. Often the rope or net, heavily weighted with mud, hung on the bottom, and there was great excitement as the strain gradually increased on the line. On several occasions the rope broke, and the end disappeared overboard; three or four miles of rope and the dredge being thus lost.

At first, when the dredge came up, every man and boy in the ship who could possibly slip away, crowded round it, to see what had been fished up. Gradually, as the novelty of the thing wore off, the crowd became smaller and smaller, until at last only the scientific staff ... and perhaps one or two other officers besides the one on duty, awaited the arrival of the net on the dredging bridge; and as the same tedious animals kept appearing from the depths in all parts of the world, the ardour of the scientific staff even abated somewhat, and on some occasions the members were not all present at the critical moment, especially when this occurred in the middle of dinner-time, as it had an unfortunate propensity of doing. It is possible even for a naturalist to get weary of deep-sea dredging. Sir Wyville Thomson's enthusiasm never flagged, and I do not think he ever missed the arrival of the net at the surface.[9]

Having spent all told several years at sea as an oceanographer, albeit spread over some 30 years, I can attest that Moseley's account has the ring of authenticity: the tedium, the constant noise, the frustration, the lack of comfort – and the fascination that kept Thomson returning to inspect each haul, as avid as a gambler peering to inspect the latest

The *Challenger* under sail. (From J. Murray, 1895, 'Report of the scientific results of the voyage of H.M.S. *Challenger*. © Dr. David C. Bossard. Reproduced from original documents in the library holdings of Dartmouth College.)

roll of the dice. Forbes' *Song of the Dredge* is a song of youth and innocence, to be sung in the club when the cruise is well behind; Moseley's tale follows upon long experience.

Murray was considerably more circumspect in his Summary of the Scientific Results, 'From beginning to end the history of the *Challenger* Expedition is simply a record of continuous and diligent work. There were few opportunities for any brilliant exploits during the voyage.'[10] Murray wrote this at the conclusion of almost 20 years work following the cruise: the time required to oversee distribution of the collections to various experts in Britain and abroad, their analysis, write-up and publication in the 50 volumes of the expedition report.

Contributing greatly to Moseley's disappointment was the naturalists' frustration that some of their fondest hopes and speculations were dashed, one after another, as the *Challenger* sailed across the oceans. The first speculative bubble to burst was Thomson's hypothesis of the 'continuity

The dredge as deployed from the stern 'derrick' of the *Porcupine*, showing the method of coiling and stowing the rope and the 'accumulator,' which indicated the strain on the line and took up tension from the pitching of the vessel. (From C.W. Thomson, 1874, *The Depths of the Sea.* © Dr. David C. Bossard)

of the chalk,' which was pricked first by the geology of the deep but ultimately by the nature of the deep-sea fauna itself. The *Challenger* had not sailed far into the Atlantic before the 'chalk' sediment began to disappear – the chalk that Thomson had believed would underlie the entire deep ocean, providing a contemporary link to the terrestrial chalk deposits dating from the Mesozoic.

As the tiny shells of foraminifera (protozoans ubiquitous in the oceans, related to amoebae but with shells; 'forams' for short), mostly *Globigerina*, that contributed calcium carbonate to the sediments became scarcer and eventually disappeared, the sediment took on a red clay character. Murray took upon himself the study of the deposits sampled from the soundings. With characteristic assiduity, he compared samples of the plankton collected at various depths through the water column with what he found in the sediments. Whereas Thomson had originally thought the deposits were the remains of benthic (bottom-living) organisms, Murray saw that they were composed predominantly of the tests (hard external coverings) of planktonic organisms. By detailed comparison of the calcareous oozes with chalks collected along the way, he eventually convinced Thomson that the deposits were in fact different: as Jeffreys had argued, terrestrial marine deposits were laid down in comparatively shallow water. The disappearance of the calcareous ooze over the deeper portions of the seafloor, as Murray and the chemist Buchanan surmised, was due to the dissolution of calcium carbonate. Calcite has the rather remarkable chemical property of becoming more soluble in seawater with decreased temperature as well as with increased pressure, so near-surface waters are always saturated, whereas deep waters are undersaturated. As the calcite tests of foraminifera and tiny phytoplankton cells known as coccolithophores dissolve, what remains is what Murray called 'red clay' (in fact it is usually brown) at depths below about 4500 meters in the Atlantic and 4000 meters in the Pacific.

What were the sources of red clay in the central ocean basins, far from land? Murray dried it, examined it under the microscope, subjected it to chemical analysis, tasted it, and passed magnets across it. He found occasionally the fossilized teeth of sharks and bones of whales, often from species known to have been extinct for millions of years. Clearly the accumulation rate of these sediments was incredibly slow: about a millimeter per 1000 years, according to modern estimates![11] Murray correctly surmised that much of the deep-sea clay was of terrestrial origin, either volcanic material or dust carried out to sea in dust storms. Small amounts he identified as being of cosmic origin, from the slow rain of meteoric fragments onto the Earth.

Following the *Challenger* expedition, Murray established himself as a repository for rock and deep-sea sediment samples collected worldwide by British and other survey vessels, eventually collecting some thousands of samples. His report on the *Challenger* samples effectively mapped the various sediment types throughout the world's oceans, showing how the distribution of red clays, of calcareous oozes from pteropods (a form of calcium carbonate known as aragonite), forams and coccoliths (calcite), and of siliceous or opaline oozes from diatoms and radiolarians corresponded with zones of varying surface productivity, depth, and chemical dissolution.

Some of this work was to have interesting economic implications. The *Challenger* found extensive portions of the deep ocean, particularly in the Pacific, paved with slowly accreting nodules, sometimes the size of potatoes, composed largely of manganese but also iron, cobalt, copper, and other metals. A century later, international mining consortia and the world's poorest nations were to vie for control of this potentially vast source of mineral wealth (see Chapter 8).

Of more immediate worth were the phosphatic rock deposits sent by the British survey vessel *Egeria* to Murray from Christmas Island. Murray realized their value as fertilizer, convinced the British government to annex the uninhabited island and himself formed a company to mine the deposits. The taxes paid to the British government from this operation reputedly exceeded the cost of the *Challenger* expedition; some of Murray's own royalties helped pay for publication of the final volumes of the expedition reports and a subsequent oceanographic expedition aboard the *Michael Sars* in 1910.[12]

The next speculative bubble to burst was that of *Bathybius haeckelii*. The naturalists spent many hours peering down the barrel of their microscopes at preserved samples of sediment, observing the primitive *Urschleim*, before Buchanan demonstrated that it was an artifact of preservation. Addition of alcohol to the samples caused calcium sulfate (gypsum) to precipitate out. This was no doubt most pronounced with samples of calcareous ooze, where this rather gelatinous substance agglutinated particles of sediment, causing the pseudo-protoplasmic behavior that so captured the imagination of Darwinian zoologists. Thomson wrote to Huxley of the finding, and he in turn, to his credit, immediately wrote to *Nature* to retract his earlier report.

Most disappointing, however, was the deep-sea fauna itself. Some thousands of new and interesting genera and species were discovered – sufficient to fill most of the 50 volumes of the expedition reports – but the expedition failed to open a door into the early evolutionary history of life on Earth. As Moseley wrote:

> In many respects, the zoological results of the deep-sea dredgings were rather disappointing. Most enthusiastic expectations were held by many naturalists ... who had hopes of finding almost all important fossil forms existing in life and vigour at great depths. Such hopes were doomed to disappointment but even to the last every Cuttlefish which came up in our deep-sea net was squeezed to see if it had a Belemnite's bone in its back, and Trilobites were eagerly looked out for. [The Belemnites are an extinct group of cephalopods that flourished in the Mesozoic, and were characterized by an internal shell or 'bone' often found fossilized in marine rocks of that period.] ... We picked up no missing links to fill up the gaps in the great zoological family tree.[13]

Such hopes die hard. When Australia's premier marine research agency, the Commonwealth Scientific and Industrial

Research Organization (CSIRO) Division of Marine Research undertook deep-sea trawling and fishery research in the late 1980s, its Chief, Roy Harden Jones, was long convinced that he had found a trilobite in a trawl. Alas, it turned out to be an asellote isopod, a group with superficial resemblance to the long-extinct trilobites.

In the end it fell once again to Murray to develop an understanding of evolutionary patterns in the deep sea, one that has largely withstood the tests of time. In his usual sober manner he states in his Summary of the Scientific Results, 'It is usual to assume that there is a universal and peculiar fauna of great antiquity in the deep sea. An examination of the statistics given in the foregoing pages does not seem to confirm this opinion.'[14] Rather, the deep-sea fauna appeared to have evolved from the continental shelf and slope benthic fauna in comparatively recent times. As he surmised, during the warm Mesozoic (the Age of the Dinosaurs) when tropical and sub-tropical conditions extended to the poles, waters in the deep sea were no longer renewed by overturning of the water column in high latitudes. As a result, deep waters stagnated, oxygen levels dropped, and life in the deep sea largely perished. With the onset of glaciation in the Cenozoic (our own modern era), deepwater circulation was re-established, the deep ocean was recolonized from shallower waters, and the deep-sea specialist forms evolved anew. The evolutionary history of deep-sea faunas remains an active research topic, but current views accord remarkably closely with Murray's original synthesis.[15] (See Chapters 4 and 9 for further discussion.)

Aftermath of the expedition: the *Challenger* report and its legacy

In a sense, the expedition's work had only just begun when the *Challenger* sailed back into port in May 1876. Thomson initially considered that about five years would be required to publish the results of the expedition – the time it had taken him to publish *The Depths of the Sea*, his account of the earlier *Lightning* and *Porcupine* expeditions. In fact it was almost a year before a small government department was established, the Challenger Expedition Commission, with Thomson as its head and Murray his assistant, charged with examining the expedition collections and publishing the results. Some 76 scientists were eventually engaged to work up the *Challenger*'s collections. Thomson's death at only 53 in 1882 was widely attributed to the stress and overwork required to maintain control of the expedition's collections (the British Museum attempted to garner them all immediately upon the vessel's return), to obtain continuing funds to properly work them up and to maintain accounts for all the myriad expenses of collaborators and artists. One battle that Thomson fought with particularly far-reaching consequences was to disseminate the *Challenger* collections to the best available scientists worldwide, a decision that enraged narrowly nationalistic interests but that established a framework of international collaboration in oceanographic research that has been maintained to varying degrees ever since. But in 1879 when he was just 50, a year before publication of the first volume of the reports, Thomson suffered a paralyzing stroke and never fully recovered. Murray succeeded him upon his death in 1882. No one foresaw that a further 13 years would elapse before he published volume 50, the last volume of the *Challenger* report, in 1895.

The *Challenger* reports, extending to some 29 500 pages and 3000 lithographic and copper plates, charts, maps and diagrams, established the framework for all major branches of oceanographic science, in keeping with the original design of the expedition. Murray, working with a Belgian geologist, Alphonse Renard, took it upon himself to synthesize the data from samples of deep-sea sediment from the *Challenger* and other vessels; to map the global distribution of seafloor sediments and develop an understanding of their origin. Buchanan ran afoul of the British Treasury and was barred from further work on the project, so further analysis of the seawater samples passed to the German chemist William Dittmar, then teaching in Glasgow. The resulting classic 'Report on the Specific Gravity of Samples of Ocean Water' demonstrated that the chemical constituents of seawater at all depths and in all oceans were in constant proportion to each other, so measurement of a single constituent, such as chloride, could serve to establish the concentration of all other dissolved ions.

Probably the least satisfactory result was in the physics of ocean circulation. Understanding the physics driving the deep circulation of the global oceans was a key objective of the expedition, particularly for Carpenter, but the project lacked a good physicist. Precise measurements of temperature and Buchanan's salinity determinations served as tracers to map the distribution of water masses throughout the world's oceans, from which deepwater circulation could be inferred.[16] However a theoretical understanding of deepwater circulation was to await the work of German oceanographers early in the 20th century.

Most effort went into the zoological studies. From knowing nothing about life in the deep sea, the reports provided definitive treatments by specialists on all major groups, describing the species, their distributions throughout the oceans, and their depth range, so far as could be known at that time. The reports still provide important reference material today. The final volume of the *Challenger* report, written by Murray and published a full 19 years after the naturalists returned to England, summarizes the results of the specialist reports.

In characteristic fashion – up-front yet tentative – Murray summarized the expedition's major finding in his opening sentence: 'The *Challenger* observations seem to prove conclusively that life is distributed all over the floor of the ocean.'[17]

Beyond this, Murray drew relatively few conclusions. He was an empiricist with few statistical or analytical tools at his command; unlike Thomson he was not given to sweeping generalization beyond what could be clearly demonstrated from the data. The biological data were enormously complex, far more difficult to interpret than the data on sedimentary deposits, governed simply by what rained down from above, and the chemistry and depth of the overlying waters. The expedition's several hundred trawl and dredge stations were distributed over all the world's ocean basins, over all depths and sedimentary environments. Thomson and others who had first sampled the deep sea had speculated that species there would have a highly cosmopolitan distribution, due to the uniformity of conditions – the same enormous pressure, lack of light, and temperature just a few degrees above freezing prevailing everywhere – combined with the lack of obvious barriers to migration. Hadn't Sir James Clark Ross noted a high degree of similarity in the deep-sea faunas of the Arctic and Antarctic – notwithstanding that his specimens were no longer available to be examined?

But Murray, poring over his lists of species, was struck by the lack of a cosmopolitan deepwater fauna. Only 5–7 percent of the species obtained at depths greater than 1500 meters were obtained at low and high latitudes in both hemispheres; only 3–4 percent of high latitude species had a bi-polar distribution, being found at high latitudes in the two hemispheres but missing from the tropics. Of 68 species obtained from *Globigerina* ooze from two stations at mid-equatorial latitudes in the Atlantic and Pacific, only one species was held in common. Indeed, species' distributions appeared potentially to be quite restricted. Comparing two stations in the North Pacific at similar latitude (~35°) and depth (~3700 meters) and separated by 3300 kilometers, Murray noted that of the 57 species obtained at one and 48 at the second, only five species were shared.

But why isn't there a 'universal fauna' in the deep sea? Is the distributional range of most species in the deep sea really so limited? Or are they simply so uncommon as to be inconsistently sampled? Both factors appeared to be at work. Going back to the species lists from the two stations in the North Pacific, 79 percent of the species were new to science (that is, they had never previously been sampled and described) and 40–46 percent were obtained nowhere else but at those single stations. The deep sea fauna is characterized by an extremely high proportion of rare species. As Murray noted:

> As a rule, in the *Challenger* trawlings deeper than 1000 fathoms, it was unusual to take more than four or five specimens of any one species in a single haul ... [whereas] In depths less than 1000 fathoms, and especially in depths less than 500 fathoms, immense numbers of individuals, all belonging to the same species, may be captured in a single haul.[18]

How many species are there in the deep sea? How widely distributed are they? These questions remain unresolved and engender enormous controversy to the present day, largely due to the combination of limited sampling and the high diversity of the fauna.

As Murray pointed out, the more mobile species – fishes and crabs, for example – were most commonly obtained over a wide range. He noted, further, that most species in the deep sea develop directly from the egg to the juvenile stage, foregoing the dispersive larval stage characteristic of the shallow benthic fauna, particularly in the tropics. Victorian evolutionists emphasized the role of natural selection working on populations isolated by geographic barriers, whereas modern theory has highlighted how genetic drift may lead small, long-isolated populations to diverge and speciate. Murray seemed to be pointing in this direction, although he lacked the theoretical framework for his observations. Unlike Thomson, however, he realized that the depths of the oceans during the warm Mesozoic, no longer renewed by cold polar waters, grew anoxic from the continued rain and decomposition of organic matter, leading to the extinction of higher life-forms. They must then have been successively recolonized by species from shallower water, thereby accounting for the relatively large number of genera in the deep sea.

One of the most influential of Murray's conclusions, albeit one that was not terribly exciting at first glance, was his observation based upon a summary table he drew up (Table 1), that 'all the intermediate zones show ... a progressive diminution of specimens and species with increase of depth.' Although the *Challenger* sampled with the dredge and trawl for several hours at a time over the bottom, on average only about 25 specimens were obtained per haul on the abyssal plain in depths greater than 2000 fathoms (3658 meters), compared with 150 specimens per haul on the continental slope at depths between 100 and 500 fathoms (200–900 meters). The *Challenger*'s bottom sampling was not strictly quantitative – the time that the dredge or trawl was actually on the bottom and the area it sampled was never known accurately – but the pattern was unmistakable. This new paradigm of the Depauperate Deep succeeded the Azoic Hypothesis as the dominant paradigm of deep-sea ecology for much of the next century.

However, Murray's analysis of species distributions was less successful than his analysis of deep-sea sediments, which established to this day the overall outline of the global distribution of sediment types and the framework for marine sediment geology. His analysis of species depth distributions combined topographically

Table 1 Summary of zoological data by depth zone (in fathoms; 1 fathom = 1.83 meters). (From Murray, 1913.)[19]

Depth zone (fathoms)	No. of stations	No. of specimens	No. of benthic species	No. of new species (%)	Species per 100 specimens	Species per genus
> 2500	25	600	153	137 (89.5)	25.50	1.29
2000–2500	32	820	247	229 (92.7)	30.12	1.44
1500–2000	32	1250	394	335 (85.0)	31.52	1.54
1000–1500	25	2000	493	358 (72.6)	24.65	1.58
500–1000	23	2000	616	493 (80.0)	30.80	1.75
100–500	40	6000	1887	1293 (68.5)	31.45	2.45
< 100	70	—	4248	1525 (35.9)	—	2.95

meaningful depth limits – 100 fathoms (about 180 meters) marks the approximate limit of the continental shelf, and 500 and 1000 fathoms (approximately 900 and 1800 meters), the limits of the upper continental slope and its base, respectively – with arbitrary 500 fathom (900 meter) depth zones. While he noted that the deeper species were found across a wider range of depth zones, he did not combine them into a single abyssal depth zone. This failure is understandable, given the limited statistical and computational tools available to him, the complexity of the faunal data and its limitations: the relatively few samples within any one ocean basin, the high proportion of rare species, and the relatively few with broad distributional ranges. One hundred years on, faunal boundaries in the deep sea – indeed, their very existence – remain a contentious issue.[20]

In completing the long series of *Challenger* reports and noting the lack of opportunity for 'brilliant exploits,' Murray seemed to share some of Moseley's disappointment. Following the massive effort that went into sorting and classifying the many thousands of specimens and describing literally thousands of new species, Murray bequeathed to deep-sea biology a great mass of detailed findings; but in way of overview little more than a vision of a sparsely populated and rather depauperate deep-sea environment. Following on from the *Challenger* expedition, the United States and various European countries mounted their own deep-sea expeditions, but this view remained virtually unchanged for almost 100 years.

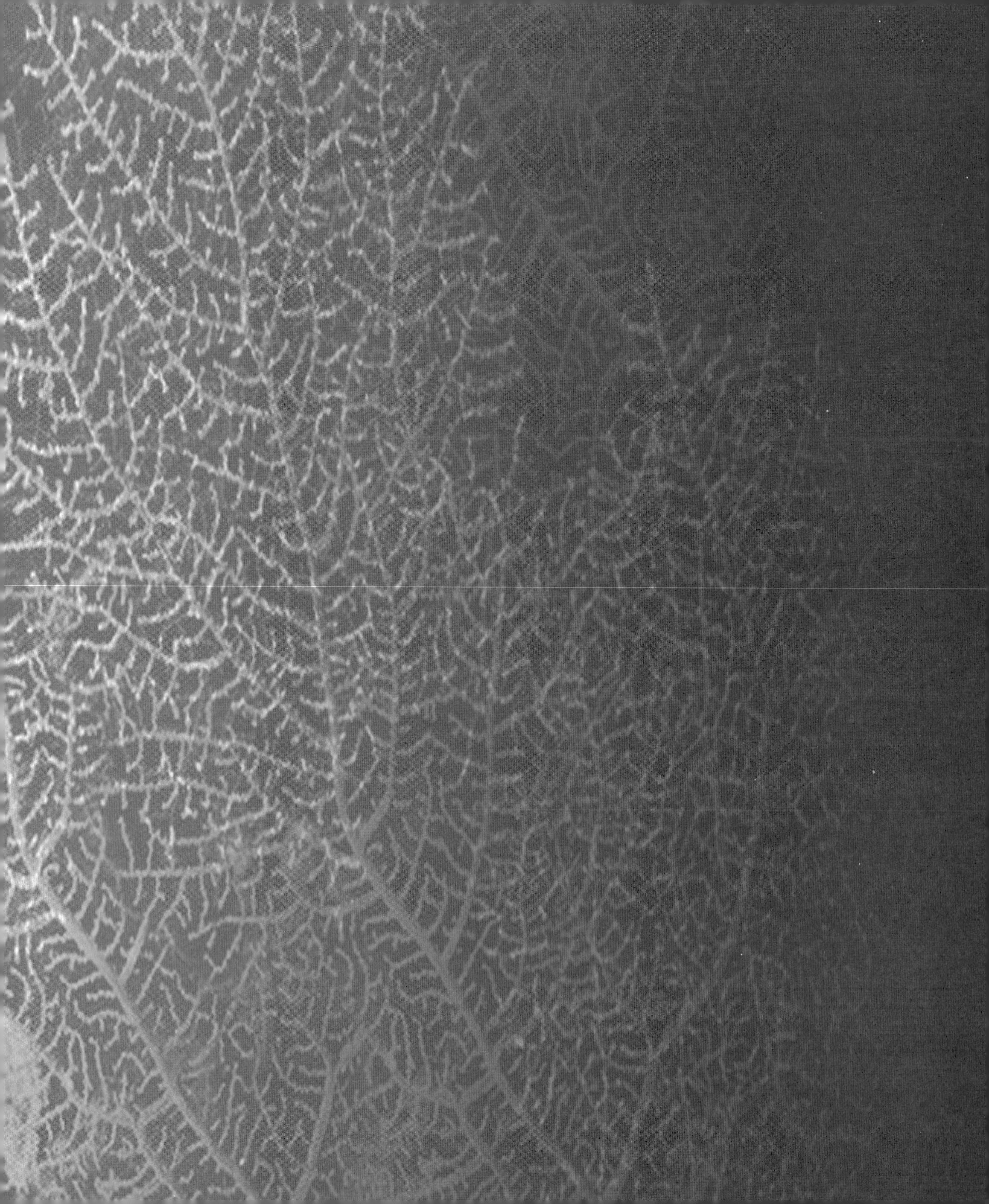

Part 2

The Ecology of the Deep Sea

The *Challenger* expedition ushered oceanography onto the scientific stage on the scale in which it has viewed itself ever since: a science of global scope, requiring national funding and collaboration across disciplines. For a period that lasted until after the Second World War, virtually every nation with any pretension of maritime significance, from the Netherlands and Denmark on up, mounted its own global oceanographic expedition. Yet oceanography entered a period of relative quiescence over this period. This is not to say that there were not significant achievements and discoveries, controversies waged and resolved; but in Thomas Kuhn's words, this was a period of 'normal' science. The sampling tools – the dredges, trawls and other nets – used aboard the *Challenger* continued to be used, and they brought onto the decks of successive expedition vessels much the same fauna. The paradigm of a depauperate deep, established by Murray and his *Challenger* collaborators, was elaborated and ramified but not significantly altered over that time.

The Second World War and the Cold War that soon followed opened up a new world for the sciences generally and for oceanography in particular. Developments in electronics and other technologies provided new ways to see into and sample the oceans. Science budgets expanded dramatically. Marine science, which outside the occasional national expedition had up to then been a low-key summer pastime, carried out at biological stations in picturesque seaside locations, became the province of government and university oceanographic institutions, which today are found in all coastal nations and, within the United States, in virtually every state having a coastline, from Rhode Island to California. Each carries out millions of dollars of research annually.

Since World War II, each decade has seen new worlds discovered within the deep sea: the astonishing life at mid-ocean depths; hitherto unimagined diversity on the vast abyssal plains of the open ocean; entirely new ecosystems thriving on chemical energy welling up from beneath the seafloor; and most recently, the extraordinary diversity of deepwater coral reefs on seamounts and elsewhere. In the following section, we explore these new worlds: how they were discovered, their ecology, and the limits of what is and isn't known. If most of us must stand on the shore, let us do so on the shoulders of giants – to look out and to see as far and as deep beneath the surface as is possible at this time.

CHAPTER 3

Life in the Twilight Zone and Beyond

There is every reason to believe that the fauna of deep water is confined principally to two belts, one at and near the surface and the other on and near the bottom, leaving an intermediate zone in which larger animals, vertebrate and invertebrate, are nearly or entirely absent.

Charles Wyville Thomson[1]

The recent discovery that a living cloud of some unknown creatures is spread over much of the ocean at a depth of several hundred fathoms below the surface is the most exciting thing that has been learned about the ocean for many years.

Rachel Carson, 1951, *The Sea Around Us*[2]

The deep interior of the oceans – below about 200 meters depth, the lower limit of the near-surface or *epipelagic* zone (from the Greek 'epi' meaning on top and 'pelagic' referring to the sea) – is the largest habitat on earth. These are the waters beyond the continental shelf, at depths where there is insufficient light for phytoplankton – or any other plant – to grow. To appreciate the extent of this habitat, consider the volume of the oceans, an estimated 1.37 billion cubic kilometers, some 11 times the volume of the land above sea level. Over 90 percent consists of waters below 200 meters.[3] And unlike the land, inhabited by higher life-forms only within a shallow surface layer, this vast habitat is inhabited throughout, except in those few regions where oxygen is depleted.

This inner core of the ocean, over the

continental slopes and abyssal plains, is commonly divided into a *mesopelagic* ('meso' meaning middle) or twilight zone between 200 meters and about 1000 meters depth, and an eternally sunless *bathypelagic* zone ('bathy' meaning deep) that extends to about 2500 meters. Below that is the *abyssopelagic* zone, which extends to within about 100 meters of the seafloor. The *benthopelagic* fauna, a distinct community of fishes, squids, and crustaceans, inhabits the waters directly above the bottom from slope to abyssal depths. The boundary between the meso- and bathypelagic faunas corresponds with the utmost detection limits for sunlight, about 1200 meters in the clearest waters.[4] This is remarkably close to the depth at which mesopelagic fishes can perceive daylight (about 1000 meters), whereas humans report the last traces of discernible light at about 600 meters.

The meso- and bathypelagic faunas (Plates 1–3), fascinate scientists and laypeople alike. These seemingly bizarre and alien life-forms continue to grace the pages of *Nature* and *Scientific American*, as well as *National Geographic* and other popular media, more than 100 years after their discovery by the *Challenger* and other early oceanographic expeditions: eel-like viper fishes with over-sized jaws and protruding fangs, bulbous angler fishes with ornate bioluminescent lures. It's as if evolution had taken a whimsical, macabre turn inspired by *New Yorker* cartoonist Charles Addams, creator of *The Addams Family*. The denizens of these realms range from the most abundant fish in the ocean, the diminutive *Cyclothone*, with some species only 2.5–5 centimeters (1–2 inches) in length, to the largest invertebrates on earth, the giant squid (*Architeuthis* spp.) and colossal squid (*Mesonychoteuthis*

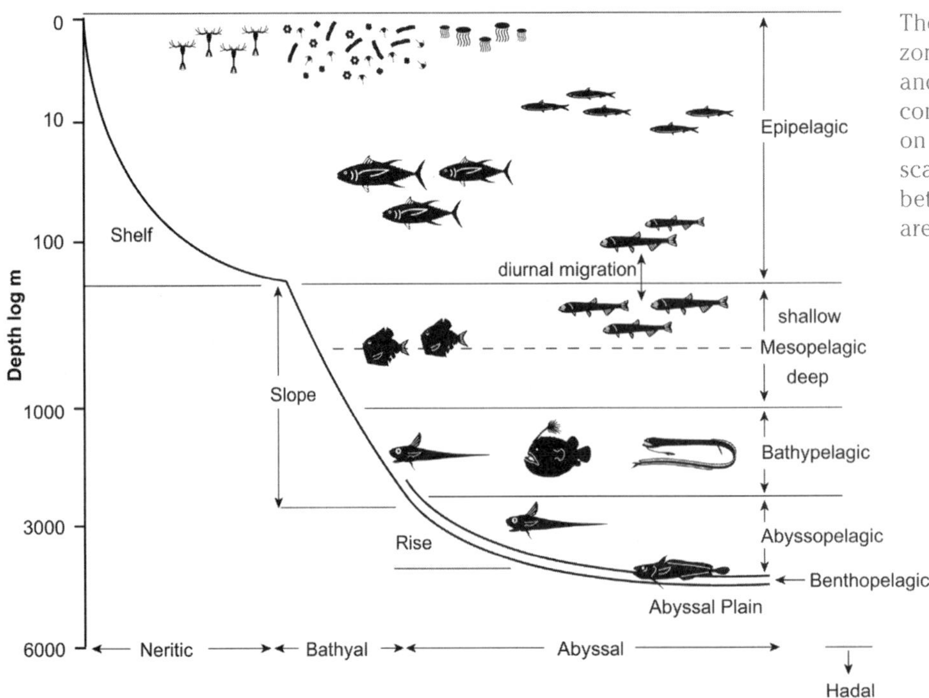

The main ecological zones in the ocean and their characteristic communities. Depth is on a quasi-logarithmic scale. The boundaries between depth zones are approximate.[a]

hamiltoni), estimated to range up to 10 and 12 meters long respectively, and the so-called 'living drift nets,' the 30–50-meter-long colonial siphonophore jellyfishes, *Praya dubia* and *Apolemia* spp.[5]

This deepwater pelagic fauna, so different from the fauna living on or over the seafloor, has excited the popular imagination since Jules Verne's *Twenty Thousand Leagues under the Sea*. However, the deep pelagic realm remains remarkably poorly explored. It is the only habitat on earth where some of the very largest animals have still not been seen or captured. The colossal squid *Mesonychoteuthis hamiltoni* made a media splash in April 2003 when it was recognized as the largest known squid on the basis of only the second specimen ever caught, a sub-adult female estimated to be half to three-quarters grown.[6] This awesome predator, with rotatable hooks set within suckers along its eight arms and two tentacles, the largest eyes in the animal kingdom, and two crescent-shaped photophores (light-producing organs) to help it find its prey in the deep,[7] is believed to feed on Patagonian toothfish (*Dissostichus eleginoides*), which grow to 2 meters in length. The colossal squid is not rare; in the Southern Ocean, it comprises some three-quarters of the diet of sperm whales and is a major prey item of the giant (3.9 meter) sleeper shark (*Somniosus* cf. *microcephalus*).[8] It simply eludes human capture.

In a remarkable study, British marine mammologist and squid expert Malcolm R. Clarke compared the size and species composition of squids obtained from the stomachs of commercially harvested sperm whales with those sampled with midwater trawl nets used for research and in commercial fisheries. Sperm whales feed predominantly on squid, and some 2000 squid beaks, representing 1–2.5 days of feeding, were found on average in a whale's stomach. To catch a similar number of squid required about ten days fishing with the commercial trawl – and 50 days with the smaller research trawl! Even more remarkable was the difference in size between the squid fed upon by the whales and those caught in the trawls. The beaks are species-specific and the size of a squid can be readily estimated from the size of its beak, so Clarke could determine the species and size composition of squid in the whales' diet on the basis of the beaks alone. Amazingly, for most families of squids the largest specimens from the research trawls were smaller than the very smallest found in the sperm whales' stomachs![9]

Other giants of the deep, such as the living drift-net siphonophores (Plate 2), colonial jellyfish distantly related to the Portuguese man-of-war, are too fragile to be collected with nets. Although they are the dominant midwater predators in areas such as Monterey canyon off central California, their role in midwater ecology was only realized when submersibles and remotely operated vehicles capable of operating in midwater were developed. From the dawn of deep-sea science, midwater ecosystems have proven the most difficult to study. Indeed, the very the existence of life at mid-ocean depths remained contentious long after the *Challenger* expedition had firmly established that there was life on the seafloor throughout the world's oceans.

Early exploration of midwater depths

The early controversy surrounding the issue of life in the deep pelagic realm bears an uncanny resemblance to the controversy over Forbes' initial Azoic hypothesis (see Chapter 1). Toward the end of the *Challenger* voyage, John Murray sampled the deepwater plankton with a plankton net, which he lowered to depths of 500 and 1000 fathoms (approximately 900 and 1800 meters) and hauled vertically to the surface. Obtaining species in these tows not found in near-surface tows, he inferred that there was a distinct deepwater pelagic fauna. But as the quote at the head of the chapter indicates, no less a figure than Wyville Thomson, leader of the expedition, remained unconvinced of the existence of significant life at mid-ocean depths. The issue remained in some doubt until nets were invented that could be both opened and closed at depth, thereby eliminating the possibility of contamination from shallower life-forms.

Alexander Agassiz was largely responsible for prolonging the controversy. Son of the eminent zoologist Louis Agassiz (founder of the Harvard Museum of Comparative Zoology) and himself a collaborator with Thomson and Murray on the *Challenger* reports and the leading American oceanographer of his day, Agassiz first used opening/closing samplers to collect plankton at midwater depths in 1880 – and came up with nothing! Agassiz *fils* was a man of considerable energy, enterprise, and ingenuity, who had the misfortune to become mired later in his career on the wrong side of two long-standing controversies: first on the question of life at mid-depths and later against Darwin's theory on the formation of coral atolls. Alexander Agassiz turned to oceanography following a highly successful career as a mining engineer. Enormously innovative, he was the first to replace the bulky hemp ropes then employed for deep-sea trawling with wire ropes like those used in mining – and used to this day in fisheries and oceanography.

Agassiz' lack of success in finding evidence of life below a few hundred meters stood in stark contrast to the success of others, most notably the German zoologist Carl Chun, who first deployed opening/closing nets at midwater depths in the Mediterranean in 1886–87 and subsequently led the German response to the *Challenger* expedition, a deep-sea oceanographic expedition aboard the *Valdivia* in 1898–99, in which more than 100 closing-net samples were collected of pelagic fauna, down to depths of 5000 meters.

Several factors contributed to Agassiz' failure. The sampling device he used initially was ingenious: a brass tube attached to the sounding wire that could be tripped, opened, and released with a brass 'messenger' sent down the wire. It then sampled the plankton as it descended a set distance, before coming to a stop against a mechanism that closed the trap. Unfortunately it was too small (the mouth opening was only 22 centimeters in diameter) to adequately sample the sparse midwater zooplankton, particularly with the coarse mesh he employed (0.42 millimeters).

Following Chun's success in sampling midwater plankton in the Mediterranean, convinced it was due to sampling near land and in the relatively enclosed Mediterranean, Agassiz carried out his most rigorous tests with an opening/closing net in the eastern Pacific between Ecuador and the Galápagos. Again Agassiz failed to find animals below about 300 fathoms (550 meters), and he concluded that 'in the open sea, even when close to the land, the surface pelagic fauna does not descend far beyond a depth of 200 fathoms, and there is not an intermediate pelagic fauna living between that depth and the bottom.'[10] Nothing was to shake him from this conclusion, so reminiscent of Forbes' conclusion regarding life on the seafloor. Like Forbes, he had the misfortune to sample in the wrong place: the mid-depth waters of the eastern tropical Pacific are unusual due to their extremely low oxygen levels; almost nothing but microbes can live there.

Adaptations of the midwater fauna

The most cursory examination of the deepwater pelagic fish fauna suffices to show how different it is from the familiar fauna of shallow coastal waters, where the race goes to the swift and the strong, with flashing shoals of silvery clupeoids (herrings, pilchards, and anchovies) pursued by tunas, sharks, dolphins, cods, and other predators. When midwater fishes were first encountered in the early days of deep-sea exploration, they were regarded as ecological freaks that had taken up refuge within the dark interior of the ocean to escape the Darwinian struggle. As one 19th century naturalist put it, these were 'out-of-date forms of life which can no longer compete with the vigorous shore-dwelling races [and are therefore] compelled to retreat to the ... deep-sea.'[11]

Of course, survival of the fittest holds in the deep sea just as it does near the surface, on a coral reef, or within a rainforest. But to understand the rules – the environmental constraints – that the deepwater fauna plays by, the deep-sea ecologist must abstract himself from the vantage point of more familiar life-forms. Over the past 30 years, studies of the morphology, biochemistry, physiology, behavior and life histories of meso- and bathypelagic organisms have revealed how this apparently strange fauna is as finely attuned to its distinct physical and biological constraints as is the epipelagic fauna (the dolphins, tunas, and silvery shoals).

From the early days of deep-sea exploration, scientists recognized that low temperature, high pressure, low food availability, and lack of light were the chief factors that shaped life in the deep sea. The first two factors – the extreme pressure and a constant temperature just a few degrees above freezing – most impressed early observers but it was soon recognized that they were in fact the most readily adapted to. Thus, although poikilotherms ('cold-blooded' animals) are highly sensitive to temperature change – their metabolism varies by approximately a factor of two for a 10°C change in temperature within a tolerable range (the so-called Q_{10} rule) – species' enzyme systems evolve to optimize activity levels within particular environments. Thus the metabolic rates of mesopelagic fishes off Antarctica and California are quite similar despite the 5–10°C difference in ambient temperature between the two regions.[12]

Pressure in the ocean is equal to an additional atmosphere for each 10 meters depth, so the pressure at 1000 meters depth

Bathypelagic viper fish *Chauliodus sloani* (length: 20 centimeters). Its protruding teeth, too large to allow it to close its jaws, enable it to seize large prey. (Karen Gowlett-Holmes/CSIRO Marine & Atmospheric Research.)

is equivalent to 100 atmospheres or about 100 kilograms per square centimeter. A styrofoam coffee cup lowered to this depth returns to the surface only slightly larger than a thimble – a common memento of deep-sea expeditions that graphically illustrates these pressure effects. But these pressures have less impact on deep-sea creatures than might be expected, because water is relatively incompressible and the pressure from within an organism is the same as that exerted from without. Thus a deep-sea creature is no more aware of the enormous pressure of the overlying ocean than we are inconvenienced by the approximately 10 tonnes-per-square-meter pressure of the atmosphere above us.

But deep-sea pressures can subtly influence metabolic enzymes, whose performance depends on their three-dimensional configuration. Fish biologists had often puzzled over sibling species – closely related species, virtually identical morphologically and genetically – that occupied somewhat different depth zones. How had they diverged through evolution, and how did they now coexist? What were their key adaptations to the different depth zones? Two biochemists at Scripps Institution of Oceanography in La Jolla, California, Joe Siebenaller and George Somero, first showed in the late 1970s that a key difference between two sibling species of thornyheads (a group related to the rock fishes common off the west coast of North America) lay in the pressure kinetics of a metabolic enzyme, lactate dehydrogenase (LDH). The LDH of the shallower species, *Sebastolobus alascanus*, which lived mostly between 180 and 440 meters, performed best overall, but only at pressures corresponding to shallower depths: at higher pressure its performance became reduced, indicating a pressure-induced change in protein structure. The LDH from the deeper-living (550–1300 meters) *S. altivelis*, on the other hand, though it performed less well at relatively low pressures, continued to perform stably at the pressures corresponding to depths of 680 to 4750 meters.[13] Further studies found that the LDH in species across four different

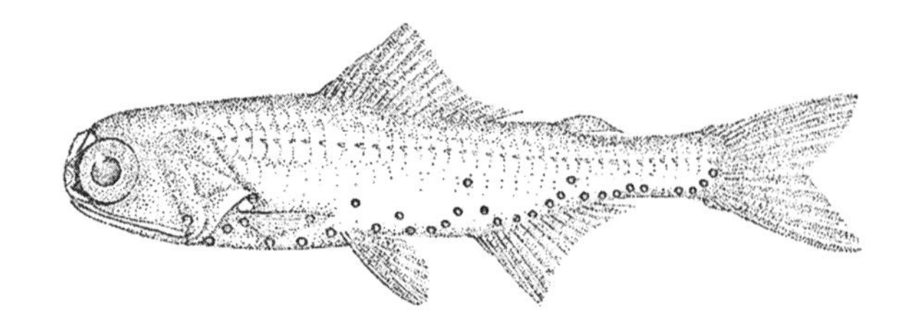

The headlight lantern fish *Diaphus effulgens* (length: to 15 centimeters) with a large photophore on its forehead to search for prey, as well as smaller photophores along its ventral surface that serve as counter-shading. The lantern fish is a myctophid, one of the most common and diverse families of mesopelagic fishes. (Courtesy of Sears Foundation for Marine Research.)[b]

The gonostomatid *Cyclothone acclinidens* (length: 3.6–6.5 centimeters). The diminutive fishes of this genus are the most abundant in the world's oceans. (F. Welter-Schultes; from original drawing in Brauer, 1906.)[c]

families of deep-sea fishes exhibited similar pressure-related performance, indicating that the evolution of stable pressure-adapted enzyme systems was likely a key element in the evolution of species adapted to living in the deep sea.[14] The boundary between the so-called bathy- and abyssopelagic zones at about 2500–2700 meters, where fish suddenly decline in abundance, may in fact be a pressure barrier. Organisms from depths less than 2500 meters (but not deeper) appear to survive being brought to the surface; conversely, surface-living organisms are able to survive submersion to only about 2400 meters.[15]

Until recently, biologists were virtually unanimous in considering low food availability to be the key factor shaping life in the deep sea. From the time of Murray, Agassiz and Chun, it was apparent that life became increasingly scarce below the euphotic zone, where there is sufficient light for photosynthesis (about 100 meters in the open ocean). As Soviet scientists demonstrated when they first adopted quantitative sampling methods in a global program of oceanographic research almost a century after the *Challenger* expedition, the biomass of plankton within the water column and of benthos on the seafloor declines exponentially with increasing depth, such that there is only 10 percent as much biomass at 1000 meters as in near-surface waters and a further 10-fold decline in biomass by 2000 meters near the base of the mesopelagic zone. The waters of the open ocean have been described as an exceedingly thin broth, but that is in large measure because the myriad life-forms from microbes and protozoans to crustaceans, salps, jellyfishes, and fish serve as an extremely fine filter: little of the material produced in the upper waters falls through the living mesh of jaws, tentacles and feeding appendages to the deeper waters and seafloor below.

Scarcity of food appears to underlie – or is at least consistent with – many of the interesting and bizarre adaptations of the deep sea. Albert Gunther of the British Museum, who worked up the first systematic collections of deepwater fishes from the *Challenger* expedition, was the first to note a number of features that distinguished deep and shallow-living fish, including 'the diminished amount of earthy matter' in the skeleton, along with reduced muscle in the trunk and tail.[16] More modern analyses show that even what remains of the muscle of deepwater fishes tends to be weak, with a high water content and reduced protein and lipid levels.

These factors, along with the general body plan of many deepwater fishes, imply reduced activity, apparently a strategy to conserve energy in the face of scarce food resources. Fishes like the mackerel or tuna that live in open, near-

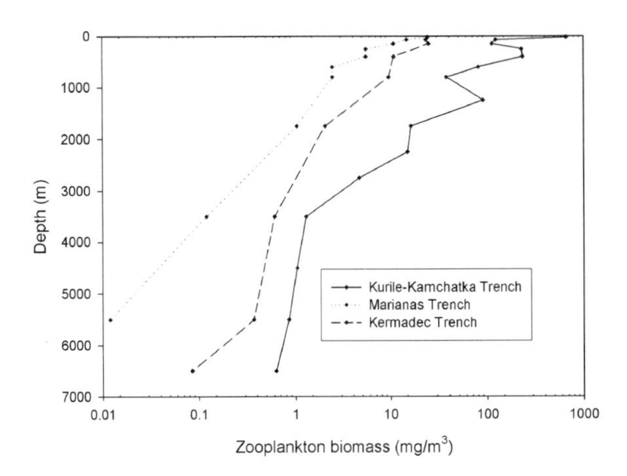

The decline of zooplankton biomass with depth over three trenches in the Pacific Ocean, plotted on a logarithmic scale. The Kurile-Kamchatka trench is in the northwest Pacific (45° N latitude, 149°–153° E longitude); the Marianas trench is in the equatorial North Pacific (11°–14° N latitude), where there is lower productivity; and the Kermadec trench is in the temperate southwest Pacific (30° S latitude, 177° W longitude). (Data from Vinogradov, 1968.)[d]

surface waters are clearly adapted to optimize rapid swimming performance: the crescent-shaped, rapidly beating tail or caudal fin for propulsion; well-oxygenated red muscle for sustained performance; the sleek torpedo-like body shape, even fitted with grooves to allow the pectoral fins to fold down flush to the body.[17]

Compare this to the design of typical mid-water predators: the eel-like dragon and viper fishes that swim sinuously and stealthily in search of their prey; or the angler fishes, ambush predators with bioluminescent lures to attract prey within gulping range (Plates 1, 3). Clearly they are not adapted to maximize swimming speed, so what are these fishes adapted to do?

Deepwater fishes like angler and viper fishes are stealth predators adapted to make a meal of as large a range of potential prey as possible. Open-water predators – whether herring feeding on copepods (crustaceans in the order of a millimeter to a centimeter in length) or cod preying on the herring – generally chase down prey a small fraction (about 1–10 percent) of their own size. Meso- and bathypelagic predators, on the other hand, often have distensible stomachs and either very large or unhinged jaws that enable them to swallow prey that are commonly 25 percent their own length or greater; sometimes, remarkably, even larger than themselves! A meal presents itself so infrequently that they can not afford to lose any opportunity that presents itself (Plate 1).

A further implication of scarce food resources is low population density. Finding a mate in the vast and featureless deep ocean is sufficiently challenging to have led to the evolution of an extreme form of reversed gender inequality. Among some anglerfish species, the male never leaves the female once she is found. Attaching himself to his mate, he lives out his life as a reduced vestigial parasite, his body fusing with hers and serving only to supply her with gametes.

The role of light and vertical migrations

Interestingly, light appears to be at least as critical as food availability – perhaps even more so – in shaping life in the deep sea. In the open ocean, where there is no refuge other than in darkness, light and vision play a critical role for both predators and prey. Just as many animals of field and forest live within burrows during the day and emerge at night to forage, while others hibernate in caves or underground during unproductive times of the year, much of the oceanic fauna descends into the refuge of deep water daily or seasonally.

But the deep ocean is vast and the extinction of sunlight is a gradual phenomenon down to about 1000 meters, which has enabled a diverse fauna to evolve within it, remarkably adapted to its unique and graded features. These adaptations extend to the physiology and biochemistry of deepwater creatures, as well as their morphology, internal structures, behaviors, life histories, and ecology. All differ dramatically from one depth zone to another, largely related to changes in the light field.

The mesopelagic fauna is a transitional fauna, its ecology closely linked to near-surface waters. Most species of mesopelagic fishes and crustaceans only spend their daytime hours in the twilight zone between about 150–200 and 1000 meters depth, swimming into near-surface waters at night to feed. This daily upward movement at sunset – and their subsequent descent around dawn – is the most massive animal migration on the planet, involving some hundreds of millions of tonnes of animals each day.[18] It is witnessed on ship's echo sounders everywhere in the world's oceans as the rising of 'deep scattering layers' (so named for their scattering of the sounders' pulses).

These daily vertical migrations are all the more impressive when one considers the diminutive size of most of the fauna: krill only a centimeter or so in length migrate several hundred meters each way; lantern fishes 2.5–7 centimeters long carry out migrations of 500 meters and more. These migrations, on the order of 10 000 body lengths each way, are the equivalent of a 15–20 kilometer hike or swim on the human scale.

Some migrations to midwater depths occur on a seasonal basis. Species of *Calanus*, the dominant copepod grazer on the phytoplankton over much of the world's oceans, are generally considered epipelagic; at the end of the most productive season, however, they descend as deep as 1000 meters and enter a diapause, remaining there in a state of torpor for approximately half the year.[19]

These daily and seasonal migrations are a key means by which biological production near the sea surface is transferred to the deep sea. The migrators provide prey to deepwater species that remain at depth, and they fuel the detrital base of the deepwater food chain. The primary source of organic material to the deep seafloor is the sedimentation of particles – senescent phytoplankton, fecal pellets and other detritus – from near-surface waters; about half as much is from the plankton and micronekton (from the Greek, meaning 'small swimmers'): larger pelagic crustaceans, gelatinous organisms, and smaller fishes and squids that feed in the upper waters and then digest and defecate after returning to depth.[20]

In the low-productivity waters of the central oceans, most primary production is from phytoplankton only a micron or a few microns in size, with exceedingly slow sinking rates. These are grazed upon primarily by microzooplankton, mostly various protozoan groups less than 200 microns in size, whose excretory products are rapidly taken up again by the phytoplankton. This highly efficient recycling, known as the 'microbial loop', is broken when the protozoans and larger phytoplankton are grazed by copepods and euphausiids (the crustaceans commonly known as krill), whose fecal pellets sink 100 meters or more per day, thereby creating a so-called 'biological pump' that rapidly transports organic matter out of the surface mixed layer into the interior of the ocean and to the seabed below.[21]

Although epipelagic and mesopelagic fishes for the most part feed on the same near-surface zooplankton, no one could mistake them. Their daytime habitat – their exposure to light and susceptibility to visually orienting predators – is clearly the dominant factor shaping the evolution of epipelagic fishes. Speed, schooling, and coloration are an epipelagic plankton-feeding fish's main defenses from predation. The silvery schooling fishes – anchovies, pilchards, herring, mackerels, and so on – are blue or green on their dorsal surface to blend in with the water when viewed from above; silvery underneath to blend in with the sky when viewed from below; and silver-sided both to reflect back the ambient light field when viewed side-on and to signal changes in direction when schooling.

In the upper mesopelagic zone, down to about 600 meters, many fishes are still silvery-sided, like the highly reflective hatchet fish, which causes them to blend into the background light when viewed from the side. However, they are typically dark-colored (brown or black) on their dorsal surface, enabling them to merge with the darkness below when viewed from above, and they have a row of photophores along their ventral surface. Many mesopelagic predators hunt by looking upward, seeking prey silhouetted against the faint light coming down from above: indeed the eyes of hatchet fishes are permanently directed upward, as is their jaw (Plate 1). The ventral photophores provide counter-shading: their faint glow enables midwater creatures to blend in with the downwelling light.

Photophores are often highly evolved structures, consisting not only of a light-generating cell, but also a reflector, lens, and color filter. Most photophores emit light in the blue end of the spectrum (around 475 nanometers) to match the wavelength of downwelling light, since red light is absorbed first within the water column. Mesopelagic crustaceans are typically red, which appears blackish at these depths because red light is rapidly absorbed, and they often have ventral photophores as well. The midwater shrimp, *Sergestes similis*, varies the intensity of light emitted by its ventral photophores to match the ambient downwelling light field.[22]

Not surprisingly, light organs, typically positioned around the eye, have also evolved as 'flashlights.' Many groups have these flashlights, including the colossal squid (*Mesonychoteuthis hamiltoni*), which has a large crescent-shaped light organ beside its dinner-plate-sized eyes. A particularly curious flashlight organ that emits red light is found in several genera of deepwater dragon fishes (*Malacosteus*, *Pachystomias*, and *Aristostomias*). Because there is no red light in the deep sea, the eyes of virtually all deepwater creatures – except those of these dragon fishes – are insensitive to red. Whereas most deepwater fishes are darkly pigmented (brown or black) to conceal themselves, deepwater crustaceans are often deep red, which, at depth, appears dark – except to the dragon fishes. In the spy versus spy world of the deep sea, they have evolved a stealth technology: a red beam to illuminate deepwater crustaceans that only the fish themselves can see. Their red beam also enables them to communicate secretly with each other – in particular to signal to potential mates, a key issue in the dark and sparsely populated world of the deep sea.[23]

Bathypelagic fishes, living deeper than any light penetrates, generally lack ventral photophores, and their eyes are much reduced. The eyes of mesopelagic fishes, on the other hand, are typically enlarged and contain various adaptations to enhance their sensitivity under low-light conditions: dense pigmentation, densely packed rods – the most light-sensitive receptors – sometimes in several layers, and in some fishes, a tapetum or reflecting layer behind the retina. It is estimated that the retina of midwater fish is 15–30 times more sensitive to blue light than the human retina.[24]

Whereas schools of epipelagic fish typically appear as tight dark marks on an echo sounder, the deep scattering layers formed by midwater fishes are diffuse but may extend for vast distances. Living at depth, there is not sufficient light to form tight schools – the fish cannot see each other adequately – but nor is there the need, since schooling is primarily an adaptation to avoiding visually orienting predators. However, they can remain loosely aggregated by using their 'lateral line' to maintain contact with their neighbors: this is a series of sensory cells along their side that can detect movement next to them.

Midwater fishes generally have large eyes to enable them to feed at low light levels, but also to communicate with each other in their loosely aggregated schools or to find mates, utilizing species-specific patterns of photophores arranged along their sides. These photophore arrangements have proven a boon to taxonomists as well, enabling them to distinguish some closely related and otherwise very similar species. This is particularly true among the lantern fishes (myctophids), the dominant and most diverse family of midwater plankton feeders, with well over 200 species described.

Exploration of the ocean's interior

In the early 1930s, William Beebe, Director of the Tropical Research Department of the New York Zoological Society, made the first manned descents to midwater depths in the waters off Bermuda, where oceanic depths can be conveniently found just a few kilometers from the Society's laboratory on New Nonsuch Island. Beebe descended in a simple device that he named the bathysphere, a steel sphere less than 1.5 meters (4 feet 9 inches) in diameter and weighing some two-and-a-half tonnes: just big enough for two people, with three fused-quartz viewing ports, 15 centimeters (6 inches) in diameter and 7.5 centimeters (3 inches) thick.

Unlike later research submersibles, which were designed to sink and rise again from the seafloor or maintain position in midwater through buoyancy control, the bathysphere was held suspended at midwater depths by a steel cable and was winched to depth and back from a barge that a tugboat towed out to sea and back. The bathysphere was remarkably crude by contemporary standards. Its atmosphere was maintained by two tanks with sufficient oxygen to keep two bathynauts alive for eight hours; moisture and carbon dioxide were absorbed by open trays of calcium chloride and soda lime. Beebe and his partner periodically circulated air over the chemicals with palm fans, a method eventually superseded by electric blowers. When conditions at the surface got rough, the chemicals spilled from the trays and the two men inside held on for dear life to avoid being thrown against the walls of the chamber. Beebe describes his initial dive, carried out after a test lowering of the bathysphere showed that it leaked 'only a quart of water [which] was collected at the bottom.'[25] Beebe and his dive partner, tumbled into the bathysphere through the 35 centimeter-diameter (fourteen-inch) port, euphemistically referred to as the 'door,' which was shut off with a 180 kilogram (400 pound) steel plate, affixed with ten massive brass nuts and a wing-nut that were tightened with a great wrench and sledge hammer, creating a deafening clang that the bathynauts endured from the inside.

Although the National Geographic Society sponsored Beebe's research, the bathysphere had neither an external camera nor any means to sample what Beebe saw. As a result, the many creatures unknown to science that Beebe sighted during his dives could only be reconstructed later by drawing upon memory. Beebe carried out some 35 dives, in the end going to 923 meters (3000 feet), at which point only a few turns of cable remained on the winch drum. However, despite the fascinating record that Beebe left behind in his classic work *Half Mile Down*, without more concrete evidence – photographs or specimens – considerable skepticism accrued about Beebe's more fantastical-seeming sightings, and his dives, which ended after 1934, had more influence on the popular imagination than on oceanographic science.

The exploration of mid-ocean depths began in earnest in the 1950s, as a by-product of the Cold War development of submarine-based military technology. The desire to conceal its submarines at sea and to find the enemy's turned the US Office of Naval Research into a major patron of American oceanography; the deep scattering layer, bioluminescence, and marine acoustics were all deemed to be of military as well as purely scientific interest and readily attracted research funding.

Because the ocean is virtually opaque to light, but highly transparent to a range of sound frequencies, ship sounders provide a detailed 'view' through the water column. A sounder transmits calibrated sound pulses into the water and records the echoes reflected back from density discontinuities; the stronger the difference in density between the seawater and an object, the greater the echo return. These discontinuities may be physical (for example, the seafloor or a strong thermocline) or biological. Echo sounding is the sensory modality used by marine mammals to locate their prey, enabling a sperm whale to hunt squid at depths of 1000 meters or so. But how does the whale know what it is 'seeing'? How

William Beebe sitting astride the bathysphere. (© Wildlife Conservation Society)

does it distinguish between a squid and a jellyfish? Marine organisms, particularly squid, jellyfishes, krill, and other creatures that are mostly water, lacking hard parts or air bladders, present weak acoustic targets. On the other hand, an air bubble presents a strong density gradient with water, so some 90 percent of the acoustic reflectance from fishes with air-filled swimbladders comes from these small structures, which are used to attain neutral buoyancy, although they represent only a few percent of a fish's volume. As a result, my fisheries acoustics group found it difficult, in its acoustic surveys of the Tasmanian orange roughy population, to distinguish 35–40 centimeter, 1.5 kilogram orange roughy from 10 centimeter, 100 gram lantern fish – only the lantern fishes had air-filled swimbladders, so they presented a similar echo return at one of our key frequencies.[26] There is little question that our signal processing capability was quite primitive compared to that of a dolphin or whale. To interpret what we so readily see on our sounders is one of the greatest challenges in marine acoustics.

When deep scattering layers and their daily movements were first observed, biological oceanographers anticipated that they were caused by lantern fishes with air-filled swimbladders, which were known to feed in near-surface waters at night. And many were. But other scattering layers were unexpectedly associated with physonect siphonophores, colonial jellyfishes typically recovered from midwater trawls as only scattered fragments. Gelatinous organisms generally provide extremely low acoustic reflectance, but this group contains individuals specialized to produce and retain bubbles of carbon monoxide to provide buoyancy, while other individuals have swimming bells or physonectae (from the Greek, 'physo' meaning bladder and 'nect' swimming, as in nekton), which enable them to move through the water column (Plates 2,3).

As Eric Barham, a biologist working for the US Navy Electronics Laboratory in San Diego, observed off California from a two-person 'diving saucer,' the siphonophore *Nanomia*, about 60–70 centimeters in length, migrated more rapidly than the lantern fish in its daily excursions to the surface layers and back.[27] Although these siphonophores do not have proper eyes, they have light-sensitive cells and a simple nervous system that enables the swimming bells to beat synchronously for efficient directional propulsion, or asynchronously in order to shift direction, as Barham observed when they veered to avoid his diving saucer's lights.

Despite Barham's observations in the early 1960s, the diversity and importance of these midwater gelatinous predators was not fully appreciated until the 1990s, when Bruce Robison developed a sustained program of midwater observations at the Monterey Bay Aquarium Research Institute, or MBARI (pronounced 'em-bar'-ee'): the research institute that David Packard, the billionaire co-founder of Hewlett-Packard, founded in 1987. Packard believed that deepwater research could be best achieved from remotely operated vehicles (ROVs) rather than manned submersibles, for reasons of safety and endurance. No more need to worry about oxygen supply and caustic scrubber chemicals; the time an ROV can spend at depth is essentially limitless. Maneuvered with precision by operators working from a surface vessel, the MBARI ROV *Ventana* was able to hover in midwater at depths down to a kilometer, to sense with side-scan sonar what might be around it, to approach creatures, and to zoom in and out with its video system to resolve structures as small as a centimeter or extending to over 100 meters.[28]

Like Beebe, Robison had ocean depths of several thousand meters virtually on the doorstep of his research institute, within the Monterey Canyon. But unlike Beebe, Robison and his colleagues enjoyed generous sustained funding and the fruits of modern technology. Within the canyon they observed a variety of jellyfish predators, including the so-called 'curtain of death': the siphonophore *Praya dubia*, which trails its feeding zooids on long tentacles from a central chain extending up to 40 or 50 meters (Plate 2).[29] Robison and his co-workers confirmed the most important of Beebe's observations and extended them considerably. Perhaps their chief conclusion was the disparity between what is observed directly at mesopelagic depths and what is sampled with nets. Beebe collected over 1500 net samples from the area where he carried out his bathysphere dives. As John Tee-Van, an ichthyologist, wrote after accompanying Beebe on his 35th dive:

> What I had seen out of the windows was unexpected, even after all of the reports I had heard coming over the telephone wires. For years I had watched our deep-sea nets arrive at the surface after having been towed in the ocean for four or five hours, each net containing a scant pint or so of minute animal life plus occasional larger fish or shrimps. From these results I had visualized the depths of the ocean as beautifully transparent and rather sparsely populated. But through the bathysphere's windows was evidence that our nets, which are the best of oceanographic nets, gave a totally false picture of the abundance of life in this part of the ocean. Discounting the larger fish and shrimps that constantly advertised themselves by flashing lights, we passed through vast numbers of small and moderate-sized organisms, distributed through a wide range of the phyla of the animal kingdom. Nothing that our nets had produced had prepared me for as much life as I had seen.[30]

Bioluminescence

The aspect of life at midwater depths most remarked upon by both Beebe and Robison, however, was its bioluminescence. This is a relatively rare phenomenon on land, exhibited by fireflies, glow-worms and a few other creatures, but at sea and particularly at midwater depths, it is virtually ubiquitous. Creatures ranging from bacteria to sightless jellyfishes and filter-feeding gelatinous *Pyrosoma* (literally, fire-body) as well as pelagic copepods, shrimps, squids, and fishes, all display brilliant bioluminescence, such that Beebe made many of his observations with the bathysphere's searchlight turned off. But what is the purpose of these deep-sea light displays?

In fact, what has proved most fascinating about light production in the ocean is its variety of purposes, some still somewhat mysterious. Mention has already been made of the blue photophores along the underside of fishes and shrimps that serve to camouflage them from predators that seek prey silhouetted against the downwelling light. However, Beebe noted that the photophores along the sides of some fishes emit a yellowish light; used to communicate position to their near neighbors, the yellow light is absorbed relatively quickly and thus does not betray the organism's presence to distant predators.

Many bioluminescent displays apparently serve to startle, blind, or present a decoy to a predator. Thus the mesopelagic copepod *Gaussia princeps* emits a brilliant bioluminescent flash when startled and darts away.[31] Other crustaceans, such as the giant scarlet mysid *Gnathophausia*, up to 35 centimeters in length, and a variety of mesopelagic prawns, emit bioluminescent fluids when startled, before scooting away.[32] The well-known escape response of epipelagic squids – emitting a dark cloud of ink just prior to propelling themselves from a predator – is modified among some mesopelagic squids, such that the ink is bioluminescent. Some jellyfishes jettison bioluminescing tentacles before making their escape, much as a lizard may let go its still-writhing tail to occupy its predator while effecting its escape.[33]

Bioluminescent displays may also serve to warn organisms of a predator's presence – or alternately of a potential predatory opportunity. Thus Robison describes how his submersible, once it disturbed an organism and initiated a bioluminescent display, often set off cascading effects as other organisms moved away and set off their own disturbances.[34] On the other hand, the bioluminescent lures of angler fishes and other deep-sea fishes, which seem to mimic the display of a small organism, perhaps a copepod, serve to attract potential predators that then fall prey themselves. The bioluminescence of certain bacteria that colonize detritus and parasitized or dead organisms may have evolved to advertise the availability of prey and hence enhance the likelihood of their quick passage into the gut of a fish, their preferred habitat.[35]

But were the brilliant bioluminescent displays that Beebe observed in fact 'natural'? On recent dives with a neutrally buoyant, motionless submersible in Monterey submarine canyon, scientists observed virtually no spontaneously emitted bioluminescence, leading them to conclude that within the dark, still mesopelagic realm, there is little advantage for an organism to advertise its presence. Many organisms, however, emitted bioluminescent flashes when struck by the submersible.[36] It seems likely that many of Beebe's observations were set in train by the constant bobbing motion of the bathysphere itself.

Biochemical and physiological adaptations

The behavioral and morphological adaptations of the deepwater fauna are endlessly fascinating, providing continued insight into the challenges of life in the deep and the myriad ways in which organisms have adapted to meet them. But deepwater creatures have evolved a wide range of biochemical and physiological adaptations as well; though these are subtler and more difficult to discern, they also offer remarkable insights into the life of the deep.

Starting in the early 1970s, a group of researchers led by Jim Childress of the University of California in Los Angeles began to explore such adaptations within the mesopelagic and bathypelagic faunas. An often surprising story has emerged, helping us better understand the factors driving the evolution of deepwater animals as they fashioned virtue out of necessity.

Childress and his collaborators began working with the deepwater fauna off the coast of southern California, but later extended their studies to the exceptionally clear waters off Hawaii and the uniformly cold waters off Antarctica to test hypotheses related to the influence of light and temperature. Through a research program spanning some 30 years, they studied the depth adaptations of gelatinous plankton, cephalopods, and crustaceans, as well as fish, to tease out whether the adaptations were specific to only a single group, to all organisms, or only to those that were visually orienting.

Childress and his co-workers first compared the composition of deep and shallow-water faunas: the relative proportion of protein, lipid (fats and oils), and water within their bodies. Some striking patterns emerged from this seemingly prosaic study – a good sign that they were onto something. First, the water content of open-water fishes markedly increased in relation to their depth of occurrence, accompanied by a decline in protein, lipid and overall energy content (the caloric value of the flesh, expressed as the kilocalories per 100 grams wet weight). The water content of fishes that fed near the surface was generally about 70 percent, but increased to 80–90 percent for fishes permanently residing at about 1000 meters depth.[37] Interestingly, this pattern held for the fishes' minimum rather than maximum depth of occurrence; in other words, vertical migrators had flesh characteristics more similar to near-surface fishes than to deepwater fishes.

As Childress pointed out, variation in body composition can serve a variety of purposes. Buoyancy control is one. Bone and muscle are negatively buoyant, so to avoid sinking, a fish must either expend energy by swimming upward, maintain a gas-filled swimbladder, build up reserves of some other buoyant substance such as lipid, or take the low-energy strategy of reducing muscle and bone and increasing water content, which is taken to its extreme by gelatinous plankton. In fact, each of these strategies is followed by different groups of fishes, consistent with their depth distribution and migratory habits.

Fishes that maintain their depth by actively swimming and those with swimbladders tend either to live near the surface or to migrate to the surface from relatively shallow depths; they have relatively high protein content and low water and lipid contents. Fishes with gas-filled swimbladders are restricted to relatively shallow migrations and generally reside within the upper few hundred meters, because the gas expands as the fish ascends through the water column. The fish therefore needs to release the gas to avoid an uncontrolled ascent and then to re-inflate the bladder when descending. This is energetically costly in the deep sea, given the enormous pressures that the bladder must be pumped up against and the lack of readily available gas. Fish use oxygen, often available at low concentrations in the deep ocean, which they extract from their blood and excrete into the bladder.

As Barham noted, another group of midwater fishes, predominantly lantern fish, undertake more extensive vertical migrations and enter a state of apparent torpor when they return to depth. These lantern-fish species generally have high lipid levels (about 15–20 percent) to maintain buoyancy, as well as relatively high protein levels (greater than 10 percent) to sustain their extensive migrations and nocturnal feeding activity. On the other hand the deeper, non-migratory bathypelagic fishes tend to take the low-energy route, reducing their protein, lipid and bone content by approximately half and increasing their water content to about 85 percent of their bodily composition. As Childress noted, the substantially higher water content of the bathypelagic fauna enables it to achieve larger size and reduces the energy required for growth.

The differences in bodily composition between migratory and non-migratory deepwater fishes implied considerable differences in energetics, but no one had yet measured the metabolism of deepwater fishes. Such measurements are routinely made with shallow-water fishes by placing them in a respirometer, a chamber in which the oxygen uptake from the water can be measured. The problems in carrying out such an experiment with deepwater fishes are obvious – how to capture the fish and bring them to the surface without damaging them.

Together with Robison and others, Childress developed improved methods to capture and maintain midwater fishes, including the use of large, thermally insulated cod ends (the buckets at the end of nets that hold the catch).[38] Carrying out the first ship-board measurements of these fishes' metabolism, they discovered that oxygen consumption dropped off dramatically in relation to the minimum depth that the fish inhabited: a fish living permanently at 1000 meters depth or below requires only 1 percent of the energy a surface-dwelling fish needs to maintain itself. Non-migrators required 73 percent less energy than species living at the same depth that migrated daily into near-surface waters.[39] This reduction in metabolism was far greater than could be accounted for by the decreased temperatures at depth or the higher water content of bathypelagic fishes. In fact similar differences in metabolic levels between epipelagic and bathypelagic fishes were observed among crustaceans and fishes off Antarctica, where temperatures were much the same throughout the water column.[40]

Interestingly, the metabolism of visually orienting midwater shrimps was similarly reduced,[41] but not the metabolism of deepwater invertebrates without well-developed eyes, such as various pelagic worms (chaetognaths or arrow worms, polychaetes, and others) and jellyfishes.[42] Childress reasoned that if reduced metabolism was a response to scarce food resources, as he and others had previously believed, it would be found universally across animal groups. That it was found only in visually orienting animals indicated that it was because they no longer needed to avoid visually orienting predators by constantly swimming away from them.[43]

Childress' hypothesis turned the commonly held view of the deepwater fauna on its head. Perhaps the watery flesh and weak musculature of deepwater fishes wasn't a consequence of the poverty of living conditions in the deep sea; maybe the strong musculature and high metabolism – all the sustained swimming and darting to and fro – of the silvery fishes living in near-surface water was simply expendable in the deep sea, due to the reduced risk from visually orienting predators. Strong swimming performance is costly but necessary in near-surface waters, where maintaining distance is the only way to avoid being eaten, and closing that gap the only way to obtain a meal. But in the bathypelagic zone, a body capable of strong burst swimming performance is unnecessary and requires considerable energy to maintain. For the evolution of fauna in deep water, where little energetic expenditure is required and little energy is available, fitness meant energetic efficiency.

With this new perspective on the deep sea, Childress and his group examined patterns of growth in fishes living across a range of depths. Interestingly, they found that bathypelagic fishes actually grew more rapidly, as well as more efficiently, than their mesopelagic counterparts. Growth is the difference between food intake and metabolic costs and is readily studied by recording the size (length or weight) and age of specimens. (Age is readily determined by counting the growth rings laid down in the ear bone – the otolith.)

Although bathypelagic fishes eat less, their reduced metabolism combined with the reduced energy content of their flesh more than compensates, enabling them to grow more quickly. Comparing two species from the lantern fish genus *Lampanyctus*, one which vertically migrates (*L. ritteri*) and another (*L. regalis*) that lives permanently at 500 meters and below, the deep-living *L. regalis* was 70 percent larger (17 centimeters versus 10 centimeters) than its migratory congener at four-and-a-half years of age. So the race did not necessarily go to the swift or the strong. Life in the slow lane at depth actually held evolutionary advantages in terms of enhanced growth, as well as reduced mortality.[44] Enhanced growth presumably enables bathypelagic fishes to avoid certain predators, as well as to capture certain larger prey; fecundity (the number of eggs that a female produces) is also closely related to body size.

The mesopelagic migratory fishes achieve lower growth rates and smaller adult size than their bathypelagic or epipelagic counterparts, largely because they tend to use lipid to achieve neutral buoyancy, so their flesh has the highest energy density, higher even than epipelagic planktivores such as the pilchard.[45] The generally small size of mesopelagic fishes led the great Norwegian fishery biologist Johann Hjort, who first described the vertical zonation of this fauna based on the *Michael Sars* expedition of 1910 (which incidentally was privately funded by John Murray), to term these fishes the 'Lilliputian fauna' in his classic book *The Depths of the Ocean*, published in 1912.[46] (Despite the utterly spartan set-up within the bathysphere – and the lack of light! – this volume was Beebe's constant companion on his dives.) Hjort's use of the term is not the last time we will find deep-sea biologists referring to communities in the deep sea as Lilliputian.

However, this decrease in size runs counter to a general trend toward gigantism in much of the deep-sea midwater fauna: the great jellyfishes and giant squids, the giant mysid crustacean *Gnathophausia* (up to 35 centimeters in length compared with 2–3 centimeters for most mysids),[47] and the giant larvacean, *Bathochordaeus*.[48] Larvaceans are planktonic creatures at the base of the chordate line that leads to the vertebrates; they spin a mucous house around themselves, complete with flow channels, coarse and fine mesh filters and an escape hatch. By beating their tail, which is typically several times larger than their body, they set up water flow through the filters and live on the retained plankton. Most larvaceans live within mucous houses a few centimeters in diameter, but *Bathochordaeus*' is football-sized!

Why does large size seem to be favored in the deep-sea pelagic environment? The giant squids appear to be associated with seamounts and other topographic features that harbor massive aggregations of orange roughy, Patagonian toothfish and other fishes, the subject of a subsequent chapter. Midwater depths, however, appear to be dominated by large gelatinous organisms, a fauna that was virtually unknown and vastly underestimated prior to the advent of submersibles and ROVs. Gigantism might seem an anomalous evolutionary outcome for this food-poor region. However, as J.L. Acuña of the University of Oviedo in Spain recently pointed out, gelatinous organisms are uniquely adapted to food-poor regions. Having diluted their organic content through a large gelatinous matrix, they are able to search or filter large volumes of water with minimal energetic cost: a salp or larvacean may have a thousand times less organic content than a crustacean relative to its size.[49] Furthermore, at depth these large but slow and potentially highly vulnerable organisms find sanctuary from predation, enabling them to evolve to massive proportions.

On the other hand, to return to Hjort's Lilliputian fauna, the small size and less efficient growth of lantern fishes and other mesopelagic vertical migrators does not imply less success on evolutionary time scales. It may seem energetically inefficient to commute daily, when only 5–10 centimeters in length, some hundreds of meters through the water column in order to feed. However the risk of mortality is reduced, and the energetic cost of vertical migration must be balanced against the cost of avoiding predators through the day.

In fact, the evolutionary evidence indicates that there are considerably more ecological niches available to midwater fishes. Of the estimated 1250 pelagic fish species living beyond the limits of the continental shelf, only about 250 are epipelagic, mostly open-water sharks, scombroids (tunas, mackerels, and billfishes), and flying fishes. The mesopelagic fish fauna contains more than three times as many species – some 850 – mostly lantern fishes (Myctophidae), hatchet fishes (Sternoptychidae), bristle-mouths (Gonostomatidae) and several families of deep-sea predators: the viper fish (*Chauliodus*) and several families of dragon fishes, including *Malacosteus* with its red light organ. Only an estimated 150 species are bathypelagic, mostly predatory angler fishes and gulper eels.[50]

The largest ecosystem on earth, the mid-depths of the ocean, remains one of its most poorly known. We are still, literally, gaining our first glimpses of some of its largest and most elusive fauna; it is still too early to say a great deal about their ecology. But nature's own verdict appears to be that this is an environment with more diverse ecological opportunities than the near-surface ocean with its more familiar fauna – the tunas, dolphins, and flying fishes – which break the surface and so engage our interest and delight.

CHAPTER 4

Bugs in the Mud: The Lilliputian Fauna of the Deep Seafloor

As Robert Frost noted in the poem that opens Chapter 1, despite our fascination with the oceans, we actually see very little when we look at it – neither out far nor in deep. No doubt this explains our fascination with whales and whale sharks, dolphins, and sea turtles: the sea creatures that command our attention as they rise to the surface of the seemingly limitless ocean. As terrestrial organisms, our senses are poorly adapted to aquatic environments. The wonder of oceanography is revealed to us through our various sampling tools; it is through these observational prisms that we are able to perceive the complexity, diversity, and patterns of the watery portion of our blue planet.

As a result, oceanography has been a highly empirical science, advances in the field driven more by technology – the development of new sampling tools – than by theory. This is often seen as a sign of the science's 'immaturity': that oceanography – and biological oceanography in particular – has not advanced to the level of such sciences as physics, where, based on a powerful theoretical framework, advances may be predicted well ahead of observation. But this 'physics envy,' most deeply felt by doctrinaire Popperians who believe there is only one correct way to do science, misses the point: ecology generally, and the ecology of the oceans in particular, is intrinsically more complex than physics. The beauty and delight of studying marine ecology – and no small measure of its frustration – is found in its richness, its variety, its many layers;

its resistance to prediction from simple models. The ocean depths *haven't* yet been fully explored; many surprises remain in store for the investigator who conceives of a new way to peer into our vast, opaque and downright inhospitable global ocean environment.

Scientists probing the deep sea have been much like the proverbial blind men groping about an elephant: the one who has taken hold of the tail thinks it a vine, the one feeling his way along the rough, hairy flank believes he has come upon a great mossy rock, while another holding the trunk cries out that he has met up with a python. The technological limitations of particular sampling devices become the crude lens through which reality is filtered.

Simple dredges and trawls, similar to those used by Wyville Thomson and the other naturalists on the *Challenger*, remained for almost 100 years the primary tools employed to explore the deep sea. Over this period, the so-called *heroic age* of ocean exploration, successive national deep-sea expeditions – American, French, Swedish, Norwegian, German, Dutch, Danish, Soviet, and others – set off to fill in the picture that Murray and others initially sketched in the *Challenger* reports. Attention was paid to regions neglected or only briefly sampled by the *Challenger* voyage, such as the Indian Ocean and the Antarctic. As recounted in the previous chapter, the existence of unique meso- and bathypelagic faunas was confirmed and studied.

Exploration of the oceanic trenches

The last of the global expeditions – the Danish *Galathea* Expedition (1950–52) and successive Soviet cruises from 1949 into the 1960s aboard the *Vityaz* and a fleet of other research vessels – fittingly closed off this initial exploratory period by trawling the greatest ocean depths, within the ocean's trenches. These features are generally found around the rim of ocean basins, particularly in the Pacific, seaward of certain island arcs and continental margins where the earth's great tectonic plates collide, one slipping beneath the other to be consumed within the earth's mantle. Trawl samples obtained at depths of almost 11 kilometers – twice the maximum depth sampled by the *Challenger* on the oceans' abyssal plains – confirmed the existence of life at the ocean's greatest depths.

Trawling within the trenches was a significant technological achievement, not possible before echo sounders came into widespread use during and after the Second World War. Only the narrow, flat floor of the trenches, sometimes just a few kilometers wide, could be trawled; landing on the steep, canyon-like walls of the trench would almost certainly have resulted in losing the gear. Days of painstaking acoustic mapping of the trench bathymetry were required before actual sampling could commence. The trawl operation involved paying out some 12 kilometers of wire over 5–6 hours and maintaining the trawl against the vagaries of wind, sea, and current, some five kilometers behind the vessel, as it slowly, steadily threaded its way along the axis of the canyon.[1]

Anton Bruun, leader of the *Galathea* expedition, named the biological community within the trenches below about 6000 meters the 'hadal' fauna, from the Greek word for the underworld (Hades). The hadal fauna was clearly evolved from, yet altogether distinct from, the fauna of the abyssal plain. Fish became scarce at these depths, and certain groups, including the decapod crustaceans (shrimps, crabs, and lobsters), disappeared altogether. Approximately 400 species were identified from about 80 bottom trawls, of which approximately half to two-thirds were endemic to the trenches and often to a particular, isolated trench. Holothurians (sea cucumbers) were typically dominant, particularly at depths greater than 8000 meters, often comprising more than 90 percent of the biomass.[2] The trenches, due to their topography and proximity to land, served as natural sediment traps, and therefore had richer feeding conditions and greater densities of a few, relatively large creatures than the abyssal plain.

Within ten years of Danish and Soviet oceanographers trawling up the first biological samples from hadal depths, a curious international collaboration led to the first – and only – manned descent to the bottom of the deepest known trench in the ocean. Most improbably, the vision and driving force behind this achievement came from land-locked Switzerland. Auguste Piccard, a brilliant Swiss physicist and engineer, had pioneered manned exploration of the stratosphere through development of an airtight pressurized cabin (the principle used today in jet air travel) suspended from a large balloon. Always fascinated by the deep sea, he adopted similar principles in building the first free-diving deep submersible, the bathyscaph (Greek for 'deep ship'). Successor to William Beebe's bathysphere, the bathyscaph was essentially a small, watertight compartment suspended below a large 'balloon,' which was filled with gasoline rather than air, because of its lesser compressibility. Piccard initially collaborated with the French navy (including a young Captain Jacques-Yves Cousteau) to build the bathyscaph, but the collaboration foundered and the French continued the project alone. However, the indomitable Piccard, now joined by his son Jacques, obtained sponsorship to build another bathyscaph in Trieste. In a stroke of good fortune, the US Navy purchased the Piccards' bathyscaph, precipitating a race between the US and French navies to first achieve the world's deepest dive. This rivalry, only one facet of the postwar rivalry between the two countries, ended in triumph for the US. In January 1960, Jacques Piccard piloted the *Trieste* to the bottom of the Challenger Deep in the Mariana Trench off Guam in the western Pacific – a depth of 10 910 meters (35 800 feet).[3]

It is perhaps worth noting, before passing on, that the Challenger Deep, the deepest point in the world's oceans, is in fact not named after Thomson and Murray's vessel, but a British successor to the original, which discovered this eponymous deep in 1951. As for the balloon-based bathyscaph, it was soon succeeded by smaller, more maneuverable submersibles. And while France remained committed to deepwater research and to this day maintains its own deepwater submersibles, the lead in oceanographic research, as in virtually all other scientific fields, passed in the 1960s from the Old World to the New. As noted in the previous chapter, the oceans had become another theatre in the Cold War, and the US Navy a major patron of oceanographic science.

The advent of quantitative sampling

The Danish and Soviet expeditions provided a fitting transition to the modern oceanographic era by introducing the use of quantitative sampling methods. The Danish oceanographer E. Steeman Nielsen developed a new method to estimate primary production in the ocean, based on the uptake and fixation by phytoplankton of a radioactive isotope of carbon (^{14}C), a method still in use today. He obtained more than 700 such measurements during the course of the *Galathea* expedition,[4] leading to the first maps of global ocean productivity (Plate 3).

The Danish and Soviet expeditions were also the first to quantitatively sample the deep seafloor. Rather than simply dragging a trawl or dredge an indeterminate distance over the seabed, they deployed grabs and corers to remove a fixed area of sediment, so the abundance of organisms and their biomass could be determined per unit of area.

The global Soviet oceanographic program extended over several decades, throughout the period of rapidly expanding distant-water fishing fleets. The Soviets, leaders in this endeavor, were prospecting – quantitatively assessing the potential productivity of the world's oceans for fishing (see Chapter 6).[5]

Not surprisingly, the distribution of biomass on the seafloor of the world's oceans closely mirrors that of the primary production above. Primary production is highest over the continental shelves, where regenerated nutrients are readily mixed back into near-surface waters during periods of winter cooling and storm, and in regions of upwelling along the west coasts of continents, where there is an upward flux of nutrients. The large central gyres of each ocean basin, all regions of downwelling, comprise the world's marine deserts in terms of their productivity.[6] The close correspondence between global patterns of physical circulation, nutrient concentration, the abundance of phytoplankton and zooplankton at the base of the marine food chain, and of fish and fisheries at its apex became the key paradigm of modern biological oceanography, which linked the physics, chemistry, and ecology of the oceans from the base to the top of the food chain.[7] Yet another dimension was added to this global synthesis with the close match between the biomass on the seafloor and the primary productivity of near-surface waters, which confirmed Murray's early qualitative observations. The amount of food reaching the seafloor is primarily a function of the productivity of the overlying waters and the water depth: the greater the depth, the greater the proportion of material that is consumed by organisms within the intervening water column.

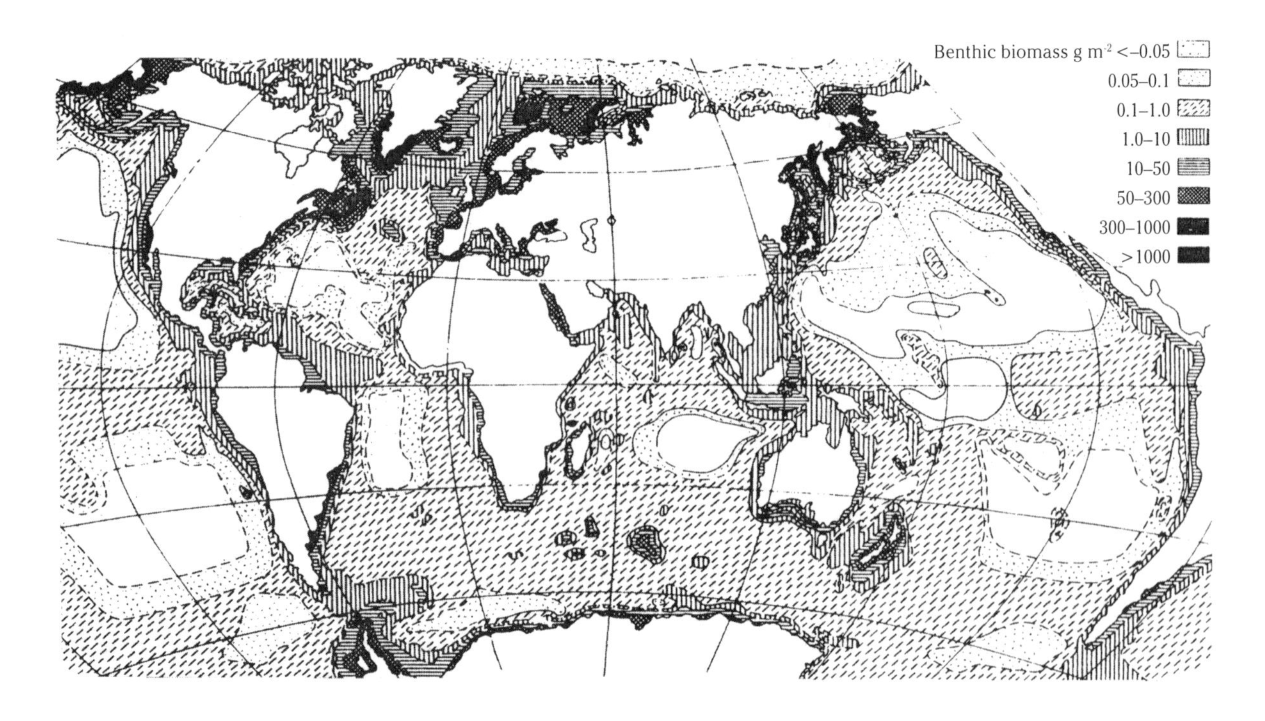

The distribution of benthic biomass in the world's oceans, based on Soviet quantitative sampling (from Belyaev et al. 1973).[e] Biomass on the seafloor is closely linked with surface productivity (Plate 3) and declines with depth as well.

The Big Picture – deep-sea global biogeography

Seventy-five years of ocean expeditions also made possible the emerging science of deep-sea biogeography: a synthesis of the distribution patterns of benthic species throughout the world's oceans – or at least of the megafauna, species large enough to be retained in the coarse meshes of most trawls and dredges. This synthesis, which had eluded Murray, capped the career of Sven Ekman, a Swedish benthic ecologist and professor of zoology at the University of Uppsala, who in 1953 – a year following the *Galathea* expedition – published the English edition of his classic *Zoogeography of the Sea*. Reviewing the vast store of data on the distribution of fishes, mollusks, echinoderms, corals, sponges, and other major faunal groups collected by oceanographic expeditions since the *Challenger*, Ekman delineated three primary depth zones – the shelf, the slope or bathyal zone, and the abyss – observable across all oceans, and two principal biogeographic provinces, the Atlantic and Indo-Pacific, as well as two minor provinces, the Arctic and the Southern Ocean.[8]

Continental shelf faunas differ dramatically from one climatic zone to another – between tropical, temperate, boreal, and arctic regions in the Northern Hemisphere, for example – as well as from one side of an ocean basin to the other and one ocean to another. Entirely different families of fishes and other major groups dominate the different climatic regions. Ekman showed that the deeper faunas were more broadly distributed, but even abyssal species were rarely cosmopolitan, that is, found within all major ocean basins.

At abyssal depths only two water masses – Antarctic Bottom Water, originating in the austral winter in the Weddell Sea region, and North Atlantic Deep Water, originating in winter in the Greenland/Norwegian Seas – extend over the seafloor of all the world's oceans. As a result, physical conditions are virtually uniform across the abyssal plains right around the world: temperatures range between about –1 to 4°C. (Because of its salt content, seawater freezes at around –2°C, rather than at 0°C.)

At bathyal depths, there is slightly greater physical diversity. One intermediate-depth water mass originates in the North Pacific, and there are two in the North Atlantic: one forms in winter from the convective overturning of the Labrador Sea; the second from the relatively warm but saline outflow at intermediate depths through the Straits of Gibraltar from the Mediterranean Sea. Yet another major water mass, Antarctic Intermediate Water, forms in winter around the Southern Ocean at 50°–55° S; it can subsequently be traced into the North Atlantic to about 20° N as a tongue of characteristically low-salinity water at about 1000 meters depth.

This uniformity of deepwater conditions, combined with the apparent lack of barriers to dispersal, led oceanographers from Thomson to Bruun to postulate a cosmopolitan distribution for the abyssal fauna. If abyssal species were mostly known from only a single ocean basin, this was because of limited sampling, not the species'

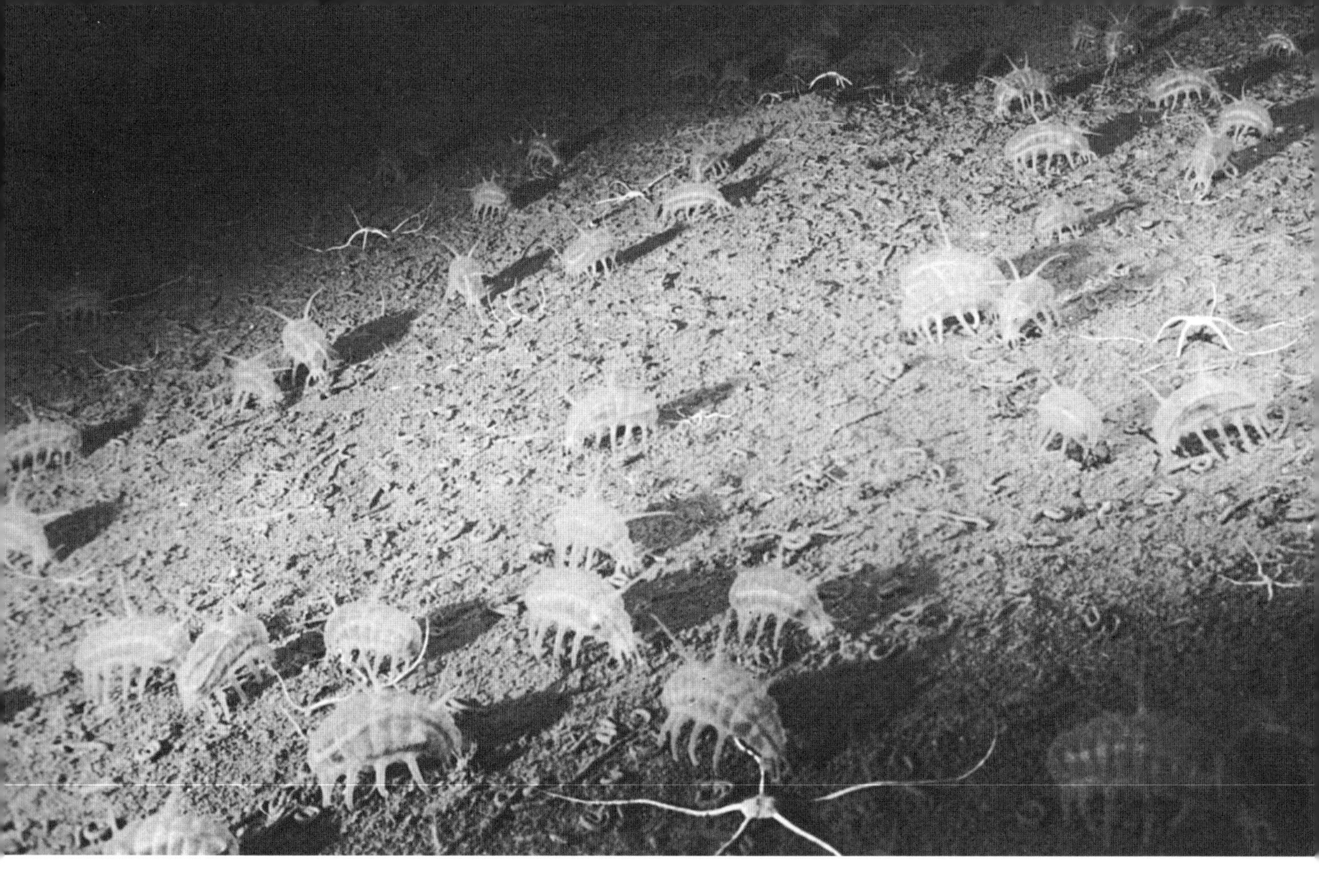

The deep seafloor showing elasipod holothurians and brittle stars on the continental slope. (From Heezen and Hollister, 1971.)[f]

limited distribution. But just as Murray rejected Thomson's speculation along these lines in the *Challenger* reports, Ekman rejected Bruun's position in his *Zoogeography*. As he showed, abyssal families and even genera were generally distributed globally, but species' distributions were generally limited to one or at most two ocean basins. Murray and Ekman's conclusions have stood the test of time, being confirmed by analysis of the extensive Soviet deep-sea studies from the 1950s, as well as more recent studies carried out around the world.[9]

Bruun appeared not to have been aware of developments in genetics and evolution, showing that random genetic drift leads to speciation between widely separated populations, if they have limited ability to disperse and mix. In 1950, Danish benthic ecologist Gunnar Thorson generalized that deep-sea species tend either to have limited periods as larvae in the plankton or forego the larval period altogether, being released directly as juveniles.[10] Currents at abyssal depths are generally only a few centimeters per second, or a few kilometers per day. Although the deep ocean is connected throughout the world, the continents effectively isolate the major basins from one another on evolutionary time scales for organisms with limited dispersal ability.

In all the major groups comprising the benthic megafauna of the deep seafloor – that is, the animals greater in size than a few centimeters, large enough to be identified in photographs of the seafloor and retained by trawls – there are families (or even orders or classes) characteristic of the deep sea, with genera found throughout the world's oceans. Thus, there are the primitive hexactinellid, or glass, sponges (so-called because the spicules that comprise their skeletal structure are made of silica) and the stalked crinoids that so fascinated Thomson and his contemporaries, previously known mostly from the fossil record (see Chapter 1). Echinoderms generally – brittle stars, sea urchins, sea cucumbers (or holothurians), and sea stars, in addition to crinoids – often dominate the benthic megafauna of the deep-sea, both in biomass and number of species. Some characteristically deep-sea groups include the brisingid sea stars that extend their arms up into the water to feed on prey wafted past in the currents, and the elasipodid holothurians, which feed on the sediment and are often found in dense aggregations (like herds moving across the deep plains) in more productive environments, such as deep trenches and areas where layers of phytoplankton detritus periodically settle from the surface. Many of these sea cucumbers can flap their bodies to swim into the water column to avoid predators or find better feeding grounds.

Among the arthropods, some of the most characteristic deep-sea forms are the giant

The giant scavenging amphipod, *Eurythenes gryllus*. Typically found at depths greater than 1000 meters, the species grows to 140 millimeters (5.6 inches). (Cédric d'Udekem d'Acoz, Tromsø Museum. Photo from Crustikon, Crustacean photographic website, University of Tromso, Norway.)

Benthogone rosea, an elasipodid holothurian 'walking' on paired tube-feet. (Tony Rice, Southampton Oceanography Centre.)

sea spiders (or pycnogonids), distantly related to terrestrial spiders, the largest having a half-meter leg-span (Plate 2), and giant scavenging amphipods, related to beach hoppers but growing to 14–32 centimeters. These giant amphipods, among the dominant deep-sea scavengers, were only discovered when time-lapse cameras were set on the seafloor above drums of bait to record the unique fauna attracted to large food-falls.[11]

Fishes are among the most characteristic deep-sea creatures. Some, such as the remarkable tripod fish (*Bathypterois*), adopt a sit-and-wait predatory strategy on the seafloor. Elongated rays of its pelvic and caudal (tail) fins enable it to perch just above the bottom, facing into the weak current. Like many deep-sea fishes, its eyes are reduced, but it fans in front of it the elongated rays of its pectoral fins as antennae to detect prey. Unquestionably, the most successful group of deep-sea fishes are the Gadiformes, the order of cod-like fishes; and within that order, the macrourid or grenadier family, which has some 38 genera and approximately 300 species.[12] Many of these species have, to a greater or lesser extent, an elongated spade-like rostrum for rooting around in the sediments to feed on the benthic infauna.

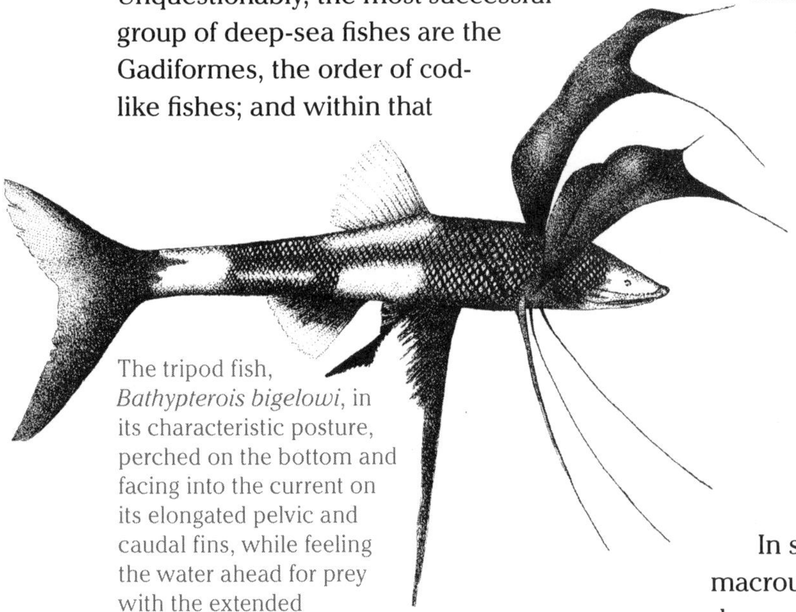

The tripod fish, *Bathypterois bigelowi*, in its characteristic posture, perched on the bottom and facing into the current on its elongated pelvic and caudal fins, while feeling the water ahead for prey with the extended antennae-like rays of its pectoral fins (Marshall 1979).[g]

Most near-bottom deepwater fish, however, inhabit the benthopelagic zone, the water column just above the seafloor. The dominant fishes in this habitat range across a number of families and orders that have mostly converged in their evolution upon an elongate body form, often with dorsal and anal fins running along much of the length of their bodies: adaptations for slow, sinuous swimming in the weak currents over the deep seafloor. Examples include the cutthroat (synaphobranchid) and notacanthid eels and halosaurs; and in the Gadiformes, the brotulids and cusk fishes, the morid cods and many macrourids, such as *Coryphaenoides*, which combines scavenging with feeding on small crustaceans and fishes over the bottom and in the water column. These fishes are most diverse and abundant over the continental slope, where food is most readily available: productivity is greatest near the continental margins, and the deep scattering layers of meso- and bathypelagic organisms impinge on the slope, providing zones of higher prey concentration.

In some regions, the abundance of macrourids, brotulids, cusk eels, and other deepwater species over the continental slope is sufficient to support commercial fisheries.

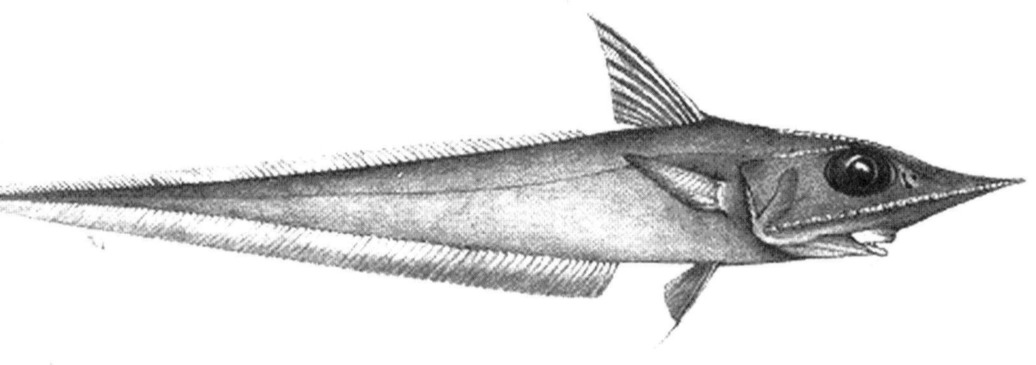

The shovelnose grenadier, *Caelorinchus braueri*, whose extended rostrum enables it to grub in the sediment for prey. (Courtesy FAO.)[h]

Many of these groups, however, extend to abyssal depths as well, dominating the fish fauna there, albeit at reduced densities.

Commercial fishing has stimulated research into the growth and productivity of deepwater fishes such as the macrourids. Most appear to live to about 50–70 years, considerably longer than the distantly related cods living over the continental shelf, which live to about 25 years.[13] (But one must wonder about the age of the monster cods – up to 1.5–1.8 meters in length (5 or 6 feet) and weighing around 90 kg (200 pounds) – caught in centuries past![14] Perhaps the difference is not so great.)

On the question of species diversity, Ekman again followed Murray in pointing to the severe decline in the number of species and their abundance on the abyssal seafloor. The long trawlings on board the *Challenger* at depths of 3600 meters or greater yielded only 25 specimens per trawl on average, compared with 150 specimens per trawl when deployed on the continental slope at depths of 180–900 meters (100–500 fathoms). Not surprisingly, the expedition's 25 hauls from depths greater than 4600 meters (2500 fathoms) yielded less than a tenth as many species as the 40 trawl samples from the slope at 180–900 meters (161 compared to 1893 species).[15]

However, when ecologists speak today of *species diversity*, they recognize that the concept has several aspects. The simplest is *species richness*: the total number of species within a community or ecosystem. This concept, while it corresponds closely to our commonsense notion of biodiversity, has the disadvantage that it is very difficult to ascertain, particularly for an environment like the deep sea, from which there are very few samples, each containing only a relatively few specimens. In speaking of species diversity,

The grenadier, *Coryphaenoides subserrulatus*, from mid-slope depths off southeast Australia. The genus is among the dominant fishes over the continental slope and abyss in much of the world's oceans. (CSIRO Marine & Atmospheric Research.)

ecologists therefore often invoke the concept of *evenness* or equitability, which is inversely related to the degree to which a biological community is dominated by one or only a few species. What is most striking, for example, about the diversity of a tropical rainforest is not only the species richness but the lack of dominance by particular species.

In a point missed by subsequent deep-sea ecologists until the 1960s, Murray noted not only the decline in overall species numbers (i.e. species richness) in his deeper samples but also their great variety or evenness: 'in striking contrast to the abundance of individuals in shallow water is the fact that in deep water beyond 1000 fathoms the "Challenger" rarely took more than three or four specimens of any one species in each haul.'[16] Murray took this no further, having neither the data, the speculative inclination, nor the conceptual tools to do so. Yet here was an example of apparent diversity existing under the most depauperate conditions, so different from communities typically associated with high diversity, such as coral reefs or tropical rainforests, where diversity flourishes amidst a profusion of life.

With Ekman's synthesis and the close of the age of global oceanographic expeditions, deep-sea oceanography appeared to enter a period of stasis. Deepwater science was slow, laborious, and expensive. What incentive was there to mount further expeditions? The monotony that Moseley had remarked upon in his cursory examination of the *Challenger* trawl samples continued for the next 75 years of large-scale oceanographic expeditions: all exhibited the same monotonically declining number of specimens and species with depth (see Table 1, p.38). Although oceanographers typically towed for an hour over the deep seafloor, their samples generally consisted of only a few dozen specimens and ten or so species. Only two samples from the deep abyssal plain – one from the *Challenger* and one from the *Galathea* expeditions – in that whole 75-year period of deep-sea exploration contained more than 100 species.

However, it was the perspicacious Professor Ekman who pointed the way forward. After summarizing Murray's data to show the dramatic decline in species obtained from abyssal depths, Ekman noted that the 'poverty of the deepest zones is not so very great if one considers that with the methods employed the microfauna was only caught to a very small extent.'[17]

About a decade after the publication in English of Ekman's classic monograph, Howard Sanders, leader of the benthic ecology laboratory at the Woods Hole Oceanographic Institute (WHOI, pronounced 'who-ee'), and his young post-doc Bob Hessler began their study of the smaller deep-sea benthic fauna along a transect of stations between Gay Head, Massachusetts and Bermuda.[18] In addition to using an exceedingly fine-meshed sieve – 0.4 millimeter rather than the standard 1 millimeter mesh – to separate the organisms from the sediment in their samples, Hessler and Sanders built two new sampling devices, both with fine-meshed collecting bags, designed to retain the sediment rather than allow it to be winnowed away. The first was an 'anchor dredge' that dug into the seafloor to a fixed depth and primarily sampled organisms burrowing within the sediment (the infauna); second was an epibenthic sled that was towed on runners along the seafloor to better sample organisms living on the surface of the sediment (the epifauna).

There was nothing 'high-tech' about Hessler and Sanders' samplers: the use of dredges dated back to Forbes and the early Victorian naturalists. However, the use of fine meshes provided an entirely new prism to view deep-sea benthic communities. The first five samples Sanders and Hessler obtained in 1964 with their epibenthic sled, at depths of between 1330 and 4680 meters, each contained from 3737 to 25 242 specimens and between 196 and 365 species. On average Hessler and Sanders' samples contained 15 times more species than those collected by the *Challenger* at depths greater than 915 meters! One of their epibenthic sled samples, obtained within Buzzards Bay near WHOI, contained 3845 individuals but only 90 species – less than half the number in their poorest deep-sea sample. Indeed, the diversity of their deep-sea samples was comparable to a shallow-water sample from the Bay of Bengal, a tropical site and one of the most diverse that they had previously encountered.[19] The deep seafloor had been transformed from one of the most depauperate to one of the most diverse environments on earth.

So what did Hessler and Sanders find in their samples? And why did it take 80 years of circum-global oceanographic expeditions by every Western nation with maritime aspirations to discover the amazing diversity of life on the seafloor? The simple answer is the technological one. As Hessler and Sanders pointed out, the deep-sea macrofauna was considerably smaller than the soft-sediment

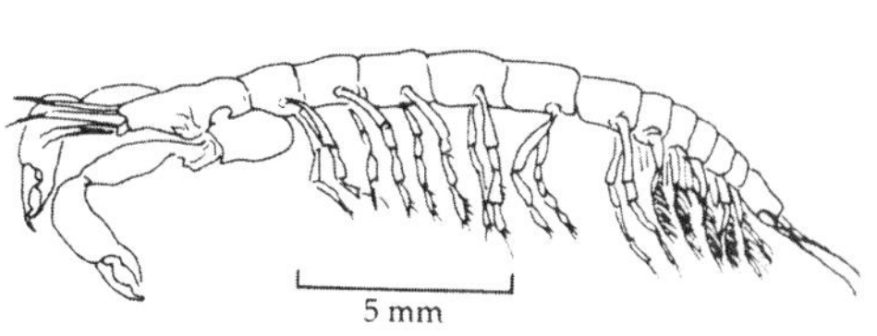

The tanaid crustacean *Neotanais serratispinosus hadalis*, from 8.2 kilometers depth within the Kermadec Trench in the South Pacific.[i]

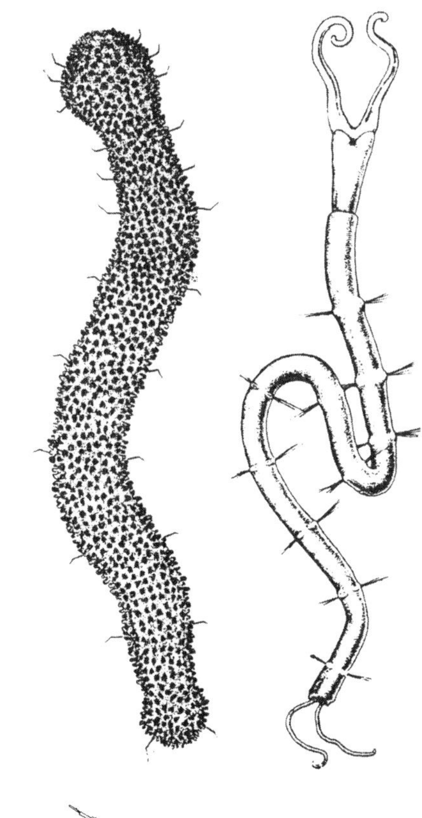

Two deposit-feeding polychaete worms from the Gay Head–Bermuda transect: (a) *Flabelligella papillata* (body length: 3.8 millimeters) and (b) *Myriowenia gosnoldi* (body length: 20 millimeters). (Courtesy of University of Southern California, on behalf of the USC Library Specialized Libraries and Archival Collections.)[j]

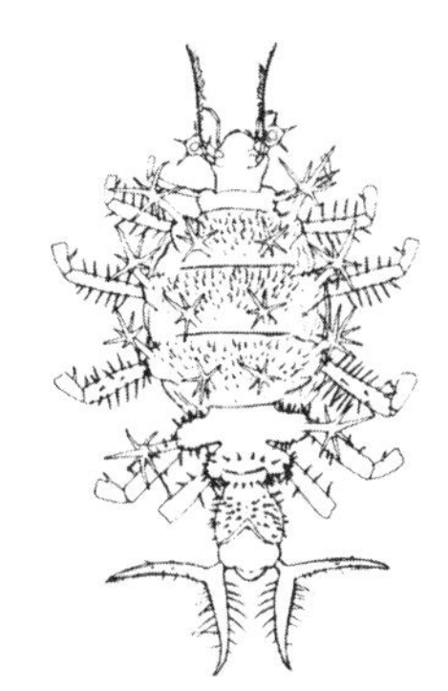

The deep-sea asellote isopod *Dendromunna compsa* (body length: 4.0 millimeters).[k]

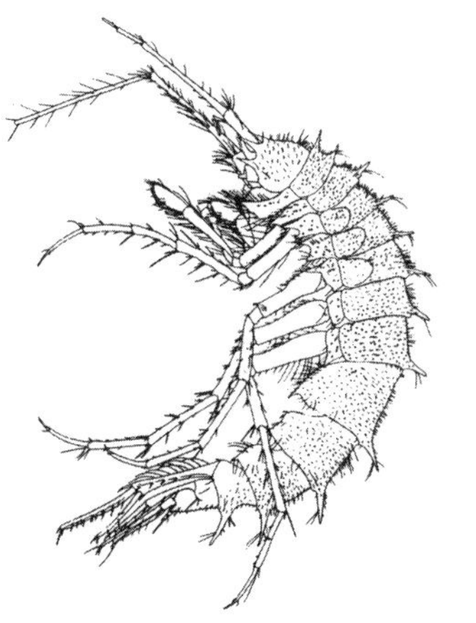

The eyeless, deposit-feeding amphipod *Lepechinella cura*, from the east Pacific.[l]

benthic macrofauna in shallower waters, though composed of the same general taxa – primarily polychaete worms, crustaceans (mostly amphipods, isopods and tanaids), and mollusks (mostly gastropods (snails) and bivalves). Most of these 'bugs,' diminutive but endlessly fascinating in their remarkable diversity, were simply winnowed away, along with the finer sediment, in the long (usually about 5-kilometer) haul of the crude trawls and dredges from the deep seafloor back to the vessel. And what was not winnowed away was lost on deck, washed through the coarse-meshed sieves used to separate the sediment from the organisms (see p. 24). The 1-millimeter mesh screens used on previous expeditions in fact represented the erstwhile accepted size cutoff, based on studies in shallow sediments, between the macrofauna and *meiofauna*. The meiofauna (derived from the Greek 'meio,' meaning smaller) is composed of yet another group of taxa, in the deep sea mostly nematode worms, small benthic copepods known as harpacticoids, and unicellular foraminiferans.[20] Hessler and Sanders decided to retain the term *macrofauna* for their crustaceans, mollusks and polychaetes, even though they were reduced in the deep sea to the size of shallow-living meiofauna. Indeed when Bob Hessler moved to Scripps Institution of Oceanography and began his own research program in the North Pacific, he further reduced his sieve mesh to 0.3 millimeters in order to retain even more of the Lilliputian macrofaunal taxa of the deep sea.

But there may be another reason why for 80 years successive deep-sea expeditions explored different oceans and developed gear to sample the trenches, the mesopelagic and bathypelagic faunas, without looking more closely within the deep-sea sediments. The paradigm of a depauperate deep-sea fauna was too ingrained; it was consistent with and fit comfortably with prevailing paradigms of species diversity. When little was obtained using the 1-millimeter mesh sieves, virtually no one – Ekman being a notable exception – gave it much attention.

How many species in the deep sea?

How many species are there in the deep sea? In posing such a question, ecologists don't ever expect to enumerate all the species and obtain a precise answer. They would like, however, to know the ecosystem well enough to be able to answer within an order of magnitude – that is, within a factor of ten. Not until 1992, almost 30 years after Hessler and Sanders discovered the remarkable diversity of the deep seafloor, did someone even attempt such a ballpark estimate for the number of deep-sea species.

Fred Grassle, successor to Bob Hessler as a post-doctoral fellow in Howard Sanders' WHOI laboratory, was Director of the Institute of Marine and Coastal Sciences at Rutgers University when he published a paper in 1992 with Nancy Maciolek of Batelle Ocean Sciences. It had a curiously ambivalent title, highly qualified yet still ambitiously addressing the big question: 'Deep-sea species richness: regional and local diversity estimates from quantitative bottom samples.'[21] The reasons for this ambivalence were clear. The study was based on the most extensive collection to date of quantitative deep-sea benthic samples from any region: some 233 box-core samples from 14 stations off the coasts of New Jersey and Delaware, mostly obtained along a 176 kilometer transect at a single depth (2100 meters), though with some covering the depth range from 1500–2500 meters. The study had modest origins as a contract to Batelle from the US Department of the Interior for a baseline survey for oil and gas exploration activities. However, the authors enlisted the cooperation of some 18 taxonomists at WHOI and Batelle, as well as specialists at other institutions, to assemble a formidable, albeit localized, data set.

The box-corer's sides were 0.5 meters, so it obtained sediment samples 0.25 square meters in area. However, only nine subsamples, each with an area of 0.01 square meters (10 x 10 centimeters), were examined for their species composition from each sample. Grassle and his colleagues are fond of pointing out that the total area of seafloor actually examined in this way amounts to some 21 square meters, the size of a modest living room. From these samples they found some 798 species from 171 families and 14 phyla, out of a total of just over 90 000 individuals. Almost half the species were polychaete worms, and the remainder mostly crustaceans and mollusks, much as Hessler and Sanders had found in their research across the northwest Atlantic. More than half of the species (58 percent) were new to science. Most were rare: 39 percent occurred only once or twice in the data set, and only 20 percent of the species were found at all ten stations at 2100 meters depth. The most abundant species accounted for only 7–8 percent of individuals in the samples. The deep-sea sediment samples thus contained a far higher proportion of rare species, and far less in the way of dominant species than samples from shallow water. By all measures – species richness and evenness – the macrofauna of their region was highly diverse.

But how far can such a data set – the 21 square meters of sediment obtained mostly from a single depth off the coast of New Jersey and Delaware – be extrapolated? Examining the rate of occurrence of new species along their transect, Grassle and Maciolek observed a rate of increase of about ten new species per 10 kilometers, or one species per kilometer. In the most controversial part of their study, they proceeded to extrapolate from this rate of discovery, which they generalized to one new species per *square* kilometer, to the 300 million square kilometers of seafloor at depths greater than 1000 meters in the world's oceans. Since the abundance of the macrofauna on the abyssal plain is more than an order of magnitude less than that on the slope, they scaled back their estimate of global deep-sea benthic diversity from about 100 million to about 10 million species.

The dangers of extrapolating from some 798 species obtained along a 176 kilometer (approximately 100 mile) transect off the coast of New Jersey to give an estimate of the number of species in all the world's oceans are obvious. For one, given the broad distribution of most deep-sea species, it is highly unlikely that the number of new species will continue to increase indefinitely at the same rate. In a commentary on the paper in *Nature*, the Oxford mathematical ecologist Robert May estimated that the number of benthic species was probably closer to half a million, about double the number of marine species that have been described to date, based on the relative proportion of species new to science in their study (just over half).[22]

I myself pointed out the fallacy of Grassle and Maciolek's approach, based on the results of an expedition I had recently led to survey for the first time the fish fauna along the continental slope (200–1400 meters depth) off the west coast of Australia. Our survey identified 310 demersal fish species from 65 trawl stations between Northwest Cape (at the northwest corner) down to Cape Leeuwin at the southwest tip of the continent. The demersal fish fauna, like the benthic megafauna, was highly diverse: more than half of the fish species (57 percent) occurred only once or twice in our samples. However, deep-sea fish are relatively well-studied, and only about 7 percent of the fish species were unequivocally new to science. Approximately 2650 species of deep-sea demersal fishes are known to date and John Paxton, ichthyologist at the Australian Museum, estimated that there are likely some 3000–4000 species in total: a reasonable figure, given that our survey of a large and previously unexplored region with a diverse fauna came up with less than 10 percent previously unknown species. Using Grassle and Maciolek's method, however – extrapolating from our rate of encounter of new species along the Western Australian slope to the expanse of the world's deep ocean – yields a global estimate of some 60 000 deepwater demersal fish species, which is at least 20 times higher than Paxton's more reasonable estimate.[23]

In their defense, Grassle and others have pointed out that some other well-cited estimates of global biodiversity – for example Terry Erwin's claim that there are some 30 million species of arthropods in the world[24] – are based on equally dubious statistics. Erwin's estimate was based on his finding some 955 species of beetles in a single species of tree in a single region of the Amazon; he proceeded to extrapolate this upward, based on the number of species of trees in the forest, the proportion of beetle species restricted to a particular host tree, the relative proportion of beetles to other insects, and so on. These exercises clearly tell us less about biodiversity itself than about how little is known about the diversity of life on our planet.

More important than how many species there are in the deep sea – whether half a million or 100 million species – is why the deep sea is so diverse, how such diversity arose, and how it is maintained. Clearly it is among the most diverse environments on earth, yet this runs counter to most of what we thought we knew about ecology. The extremely low productivity and monotony of the deep seafloor would, on the face of it, make it highly unlikely that it should be exceptionally diverse as well. These issues have occupied deep-sea ecologists ever since Hessler and Sanders' initial discovery.

Why is the deep sea so diverse?

Why are some biological communities highly diverse and others depauperate? This is a central question in ecology, and one of the most vexed. Various factors appear to contribute to diversity or its absence: productivity, predation pressure, physical disturbance versus stability, geological and evolutionary history, habitat complexity and the availability of ecological 'niches.' For some of these factors – for example, productivity, predation, and disturbance/stability – nature seems to follow the classic dictum, 'moderation in all things.' Too little productivity, and there are insufficient resources for a number of species to share. Highly productive ecosystems, on the other hand – sewage sludge ponds or upwelling systems, for example – are typically dominated by just a few species, those capable of taking up the available nutrients most quickly. Similarly, without predation or physical disturbance, a competitively dominant species may take over; with too much disturbance, on the other hand, only the fastest growing, most weed-like species are likely to thrive.

Based on these commonsense notions, the deep sea seems a most unlikely candidate for one of the world's most diverse ecosystems. To begin with, it is a biological desert, supported mainly by an exceedingly fine drizzle of detritus from near-surface waters. Sediment traps suspended through the water column over the abyssal plain in the Atlantic and Pacific oceans indicate that about 5 percent of near-surface primary production is transported to mesopelagic depths (200–1000 meters), and only about 1–3 percent reaches the abyssal seafloor.[25] The zooplankton are efficient grazers, consuming most of the detritus as well as the plankton produced in the upper waters. Still, the meager rain of detrital material is estimated to provide about 90 percent of the energetic input to the deep sea, with the remainder supplied by the occasional fall of large organisms – a whale or a fish carcass – to the deep seafloor.[26] This rain of detritus to the abyssal plain is almost unimaginably small, particularly over the central oceanic regions, which are the least productive oceanic environments to begin with, and characterized by the greatest recycling of primary production within the near-surface waters. Typical values for the input of organic carbon to the deep seafloor range from 1–10 grams per square meter per year.[27] To put this in perspective, a common fish-oil tablet contains one gram: imagine several spread over a square meter of seafloor and providing the entire year's food input for the biological community! And of this meager organic input to the deep sea, much is highly refractory and can only be metabolized by higher organisms after processing by bacteria. This extra step in the deep-sea food chain consumes a further 60–70 percent of the detrital organic content.[28] The abyssal seafloor is, without question, one of the poorest ecosystems on earth. One has to search hard – deep subterranean caves, for example – to find a less productive ecosystem.

At this time all one can say is that productivity appears to contribute far less to species diversity than our commonsense notions would indicate. In fact, a number of our most diverse ecosystems are also among the least productive. Lake Baikal, which contains 20 percent of the freshwater on the surface of the Earth, is noted for both its exceptional water clarity, due to low productivity, and its high diversity, having some 1800 plant and animal species, most of which are endemic.[29] Southwestern Australia was recently nominated as another of the world's biodiversity hotspots, based on the remarkably diverse heath-like plant communities that have evolved on its micronutrient-depleted soils.[30] Returning to the marine environment, zooplankton communities are notably more diverse in the least productive, central sub-tropical regions of the major ocean basins.[31]

It may be that food is less important than we, as warm-blooded creatures maintaining our high metabolism with three meals per day, seem to believe. Despite its low food input, conditions in the deep sea are not as harsh as, say, conditions in the Sahara. Organisms living on the deep seafloor appear to have adapted to the extremely limited food available to them much as bathypelagic organisms have, through a remarkable slowing of their metabolic processes. This was demonstrated most convincingly in an unintended experiment involving a bologna sandwich. In October 1968 the submersible *Alvin* was lost in about 1540 meters of water when a cable snapped as it was being lowered into the water. The crew escaped, leaving the hatch cover open, and the submersible sank to the bottom. When it was recovered more than ten months later, the greatest surprise of all was that the crew's lunch – six bologna sandwiches wrapped in waxed paper, two apples, and some meat broth in a thermos, all packed within a lunch box – was found unspoiled and (apparently) still edible. The sandwiches, when placed in a refrigerator at a similar temperature to that on the seafloor, spoiled within a few weeks. The implication, confirmed by subsequent experiments, was that microbial activity is much reduced in the deep sea, although there are bacteria, particularly in the sediment, adapted to deep-sea conditions.[32]

The metabolism of higher organisms is also dramatically reduced in the deep sea, as Ken Smith at Scripps Institution of Oceanography in La Jolla, California demonstrated, using respirometry chambers that he could bait and deploy with a submersible in order to capture fishes over the seafloor and measure their metabolism *in situ*. These measurements, made at about the time that Childress carried out his first ship-board measurements of the metabolism of midwater fishes, provided comparable results. The grenadier, *Coryphaenoides armatus*, and the thornyhead, *Sebastolobus altivelis*, exhibited metabolic rates about 100 times lower than those of fishes living over the continental shelf or in near-surface waters.[33]

Monotony and diversity in the deep sea

A more formidable challenge for ecological theory is to explain seafloor diversity in the face of the environment's homogeneity. Common sense and ecological theory both point to the need for moderate environmental heterogeneity to sustain high species diversity. Environmental heterogeneity provides ecological niches, and the more niches available, the more species an environment can support. Thus we intuitively comprehend the diversity of life within a coral reef or tropical rainforest.

Ecological niche theory – simply stated, the view that there needs to be a distinct niche for each species supported within an ecosystem at equilibrium – has always struggled to account for the diversity within aquatic ecosystems. How are some dozens of species of simple organisms, such as phytoplankton or copepods, with only a few simple ecological requirements – light and a few essential nutrients in the case of phytoplankton – able to co-exist over the millennia? Planktonic systems are also structurally simple – at first glance it's all just water. The pre-eminent limnologist G. Evelyn Hutchinson first referred to the diversity of planktonic ecosystems as 'the paradox of the plankton' in 1961.[34] However, he posed the puzzle only to dispel it, pointing to a variety of mechanisms that might enable competitors to co-exist: for example, predation, heterogeneous conditions within the water column, and changing seasonal conditions that prevent the system from coming to equilibrium, leading to a succession of species.

Typical scene from the abyssal equatorial Pacific (5300 meters depth) showing mounds from burrowers, as well as fecal 'loops' and tracks from fauna at the sediment surface. (From Heezen and Hollister, 1971.)[f]

But Hutchinson was primarily a limnologist, and the temperate lakes he studied generally only supported a few tens of planktonic species. The sense of paradox was only heightened when plankton ecologists turned to the central gyres of the North Atlantic and North Pacific oceans and found not a few tens of species, but on the order of 100 or more within a single region. As if to compound the paradox, the most diverse regions by far were these sub-tropical central gyres, the least productive and least seasonal planktonic systems of all.[35]

Still, if we view ecosystems along a continuum, although the central oceanic gyres were the least productive and the least seasonal of epipelagic oceanic ecosystems, they were a hundred times more productive and had far greater seasonal dynamics and physical heterogeneity than the underlying abyssal seafloor – and yet the deep seafloor contains vastly more species. The deep seafloor is simply one of the most monotonous environments on earth. Its physical characteristics – temperature and salinity – are virtually unchanging seasonally and from year to year. The composition of the sediment can vary little for hundreds of kilometers. Still, Grassle and Maciolek found 55–135 macrofaunal species within the 900 square centimeters of each box-core that they examined. Not only was there virtually nothing to eat in the deep-sea sediment, but it all appeared to be very much the same. What could all those species be doing there? How could they possibly co-exist? The communities of the deep-sea benthos pose the greatest challenge of any ecosystem on Earth to our understanding of the planet's biodiversity.

How are we to explain this 'riot of species in an environmental calm: the paradox of the species-rich deep-sea floor'? That this is the title of a recent review by two of the best of the next generation of deep-sea ecologists, Paul Snelgrove of Memorial University in Newfoundland and Craig Smith at the University of Hawaii, former students of Fred Grassle and Bob Hessler, respectively, indicates how deeply this question still vexes deep-sea ecologists. Snelgrove and Smith do not pretend to have resolved it.[36]

Sanders put forward the first explanation of deep-sea diversity in 1968, hypothesizing that it was a product primarily of the stability of the deep-sea environment over prolonged geological time. This hypothesis, which became known as the Stability–Time Hypothesis, assumed that relatively constant environments were regulated by biological interactions rather than physical stresses, thereby allowing, over long periods of constant conditions, the gradual diversification of species.[37]

Sanders' hypothesis has considerable intuitive appeal. The most diverse environments on earth are generally those with a long evolutionary history, regardless of their productivity: tropical rainforests and reefs, ancient lakes such as Lake Baikal in Siberia, the plankton communities of the central oceanic gyres, the heath communities growing in the nutrient-limited soils of southwestern Australia. But although these are all *reasonably* stable environments (a subjective assessment, to be sure), none exhibits the exceptional stability of the deep seafloor: the tropics experience wet and dry seasons, cyclones and storms; Lake

Baikal experiences Siberia's extremes; and southwestern Australia undergoes periodic drought. Ecological theory predicts that an exceptionally stable ecosystem, particularly one characterized by a monotonous habitat (endless expanses of mud) and virtually a single resource – a thin drizzle of detritus – should simplify, not diversify, over time as the ecosystem comes to equilibrium, much like the classic laboratory experiments that underlie ecological niche theory.[38] Moreover, there is little evidence that the deep-sea fauna has evolved and specialized as Sanders postulated – that it partitions the organic matter in the sediment the way coral reef fishes or tropical insects partition the myriad resources available to them.

Around the time Sanders put forward his hypothesis, ecologists working over a range of notably diverse ecosystems were also finding orthodox competitive exclusion theory too restrictive. Paradoxically, ecologists working with relatively complex organisms – vertebrates, particularly birds and lizards – were able to reconcile the diversity of their communities with its stringent requirements,[39] but ecologists working with simpler organisms found themselves unable to do so. As Joseph Connell, professor of biology at the University of California in Santa Barbara, wrote in *Science* in 1978, the hundreds of species of trees within a few hectares of tropical rainforest, or of corals within a tropical reef, cannot reasonably be imagined to be co-existing near equilibrium, such that each species has evolved a distinct niche. The requirements for a tree to flourish – light, nutrients, and water – are just too simple, as is the case for corals.[40]

Ecologists working in the field could see a variety of mechanisms operating in natural ecosystems that prevented them from coming to that equilibrial endpoint where competitive exclusion set in. But can any mechanism be invoked to explain the diversity of the deep seafloor, perhaps the most extreme natural environment in terms of its simplicity, homogeneity, and extremely muted seasonal and interannual variability?

One of the first mechanisms addressed by Connell and others was the role of disturbance in preventing ecosystems, such as tropical reefs and forests, from coming to equilibrium, and hence enabling them to evolve exceptional diversity. Hessler, who had moved to Scripps, collaborated with Paul Dayton, a Scripps ecologist noted for his studies of the role of predation in maintaining the diversity of shallow-water benthic communities, to speculate in 1972 that similar processes might be at work in the deep sea. In other words, cropping by predators might serve as a form of disturbance, maintaining the diversity of the deep-sea benthos by preventing populations from building up to levels at which they excluded each other through competition.[41]

Sanders and his then new post-doctoral fellow, Fred Grassle, wrote a rejoinder, pointing out that the life-history characteristics of the deep-sea fauna made it extremely unlikely that it was preyed upon heavily.[42] Heavily cropped populations display a fast turnover and are dominated by young recruits to the population (as ecologists term them) to replace those lost to predation. But it was already apparent – and the evidence would continue to accumulate for decades to come

– that deep-sea populations displayed just the opposite characteristics: slow growth, extreme longevity, highly sporadic recruitment, and dominance by old individuals. In marked contrast to meso- and bathypelagic fishes, whose life-history characteristics are not so different from fishes living in near-surface waters, fishes such as macrourids, which are dominant over the deep seafloor, live on the order of 50 or more years.[43] Even more surprising, minute benthic invertebrates have similar life spans. Yale geochemist Karl Turekian pioneered the use of radio-isotope dating methods to determine the age of such creatures; his initial results indicated that a deep-sea clam, *Tindaria callistiformis*, from 3800 meters depth, was about 100 years old, though only 8.4 millimeters (a third of an inch) in length. When it attained sexual maturity at a mere 1.4 millimeters, it was already 50–60 years old![44]

Organisms with such life histories clearly could not have evolved or persisted in the face of heavy predation. Still, some of the seafloor fauna are predators, and even modest predation may limit competition between prey populations. But are there other forms of disturbance at work in the deep sea? Could our view of the unchanging deep, like that of the depauperate deep, simply be the product of our preconceptions, and of not having looked with the right sampling tools?

The first evidence of heightened activity on the deep seafloor came from John Isaacs' 'Monster Camera.' A one-time fisherman, Isaacs lowered cans filled with bait to the deep ocean floor with a time-lapse camera set above them, just to see what might happen. The camera became known as the Monster Camera for the approximately 8-metre shark that it recorded: all that could be seen of it was the creature's head, which filled the frame.[45] At that time such monsters were not yet known from the deep sea. More typically, however, the camera recorded a succession of scavengers: macrourids and hagfish, followed by amphipods and, later, less mobile crabs and urchins, all of which appeared highly attuned to the scent emanating from food-falls. It thus became clear that there were at least two branches to the deep-sea food chain: one based on the fine rain of detritus, the other on occasional large food-falls from dead whales or fish.

In the 1980s, investigators followed up on Isaacs' simple experimental design by placing a variety of instruments in the deep sea and watching to see what happened: current meters, sediment traps, and unbaited time-lapse cameras. It might seem that nothing could be more boring and less promising than a series of time-lapse photographs of the deep seafloor, where nothing ever happens.

In fact, far more was happening than anyone had previously realized. One of the first surprises was the occurrence of benthic 'storms', observed beneath the large eddies that spin off the Gulf Stream and in areas where bottom water forms in winter, sufficient to raise and transport large clouds of sediment from the bottom.[46]

A camera placed over the seafloor in the Porcupine Seabight off the UK recorded thick falls of 'phytodetritus' at depths of 1370–4100 meters within weeks of the spring phytoplankton bloom. Some of the material had already been ingested but some appeared fresh, suggesting that the bloom had outpaced the ability of near-surface zooplankton to graze it. The benthic fauna responded quickly, with various worms extending from their burrows to sweep away the material around them. Within about a month, the green layer over the seafloor was gone. Further studies have now observed pulses of detritus to the deep seafloor following seasonal peaks in primary production across a range of deep-sea environments: beneath oligotrophic sub-tropical central oceanic gyres as well as underlying productive tropical waters or temperate ecosystems with a marked seasonal production cycle.[47]

Certain benthic species rapidly colonize enriched or disturbed areas, contributing to species richness.[48] In an extension of Isaacs' baited-camera experiments, Craig Smith used a submersible to place individual fish on the seafloor and observe how the feeding activity of scavenging fish disturbed the sediment, leading first to a decline in species diversity, followed by movement of colonizing species into the disturbed patch. Similar processes occur around mounds, on the order of 10 centimeters high and half a meter in diameter, that dot the deep seafloor. The mounds form as worms deposit their feces around their burrows. Diversity is initially reduced around the mounds, but an influx of colonizing species follows, leading to an increase in diversity that persists for almost two years.[49]

There are now a few sites in the North Atlantic and Pacific where researchers have observed the deep sea for approximately a decade. In both oceans, they have observed dramatic climate-related changes. Ken Smith, whose *in situ* measurements of the metabolic rates of deep-sea fish were noted earlier, also built respirometry chambers that could be placed over the seafloor to measure the metabolism of the benthic community as a whole. Working in the North Pacific, he first used these chambers to show that the metabolism of deep-sea benthic communities varied several-fold between summer and winter. Then, over a seven-year period from 1989 to 1996, he found that the supply of material sedimenting to the seafloor declined by more than 50 percent, such that it was no longer sufficient to meet benthic food requirements. These changes were consistent with an increase in sea surface temperature and a concomitant decline in nutrients, primary production, and zooplankton over the central North Pacific. More recently, these changes have reversed (see p. 188).[50]

In the Porcupine Abyssal Plain west of Ireland, European researchers observed over ten years (1989–1999) an astounding 100–1000-fold increase in the abundance of *Amperima*, a holothurian (sea cucumber). It went from being a minor element of the benthic community to comprising 70 percent of the megafauna biomass. Another holothurian also increased substantially. Since both feed on phytodetritus, the ecologists speculated that there was a large-scale change in surface productivity, but there are unfortunately no data to substantiate this.[51] The precise cause of what has come to be known as the '*Amperima* event' may be unclear, but it *is* clear that the deep sea – even the remote abyssal plain – can no longer be regarded as unchanging. Activity and metabolic levels are much reduced in the deep sea, and it is buffered from events occurring on the short-term time and space scales of the day-to-day weather at the sea surface. Still, seasonal cycles and their interannual variability are rapidly transmitted to the deep sea, which responds physiologically and ecologically, exhibiting changes in community composition and dynamics.

Far more dramatic changes occur on geological time scales. Long before direct evidence was available, Murray recognized why relict species from the Age of the Dinosaurs were not found in the deep sea: this environment must undergo massive change, resulting in waves of extinction and recolonization, as Earth's climate periodically warms and cools. The present deep-sea environment, filled with dense, cold water formed in winter in the Antarctic and far North Atlantic, is a product of the Earth's current glacial configuration. During the warm epochs in the Earth's history, when there were no glaciers, the densest water formed in low-to-mid-latitude regions of great warmth and evaporation, similar to the Mediterranean today. The deep sea then filled with warm salty water, oxygen levels dropped, and the cold-water deep-sea fauna largely went extinct. With remarkable insight, Murray interpreted the present diversity of the deep-sea megafauna, which, unlike the slope or bathyal fauna, is characterized by many genera but relatively few species per genus, as evidence that the deep sea has been recolonized, all around the world, from the bathyal zone.[52] Today, the study of paleoclimate and paleo-oceanography is a highly active research area, with critical lessons as we enter the 'Anthropocene,' the climate period dominated by human activities. But this is the subject of a later chapter (Chapter 9).

Deep-sea ecologists today accept the paradox of exceptionally high diversity in the deep-sea benthos, an environment characterized by extremely low productivity and little variability spatially or temporally. Certainly no single factor explains it, although a combination of factors appear to contribute:

- the extreme stability and age of the ecosystem;
- its vastness, combined with the limited dispersal of much of the fauna, which facilitates genetic differentiation and speciation;
- its lack of high productivity, which prevents a single species from achieving high abundance and competitively excluding others;
- the slight disturbances, seasonal and on longer time scales;
- the environmental heterogeneity induced by mound-producing worms, and by the fall of wood, seaweed, and the carcasses of fish and whales;
- and perhaps even a degree of predation, insofar as the preponderance of older organisms implies a relatively high mortality of the young stages.[53]

Debates over the number of species in the deep sea may seem rather academic, the modern scientific equivalent of the medieval scholastics' debate over the number of angels that can dance on the head of a pin. After all, does it really matter whether there are half a million macrofaunal species in the deep sea, or 5 million or 50 million? The number itself is of little consequence, but the issues underlying such an estimate are of critical importance if we are to be able to assess the potential impact of some of the major projects mankind has proposed for the deep sea.

In large measure, the quest for 'the answer' to the diversity of the deep sea led ecologists down a *cul de sac*, insofar as it caused them to narrow their focus to smaller and smaller scales of species distributions and the micro-heterogeneity of their environment. Key broad-scale issues were missed in the process, such as the biogeography of the 'bugs in the mud' that Sanders and Hessler discovered some 40 years ago. What are their patterns of distribution across the seafloor of the world's oceans? And what is the distributional range of these species? These are the issues that Sven Ekman and succeeding deep-sea ecologists addressed for the megafauna.[54] They are also some of the questions that pelagic ecologists working with the zooplankton, organisms of a similar size to the Lilliputian macrofauna of the deep sea, first sought to answer, thereby achieving a global overview of the distribution of zooplankton communities in relation to the ocean's main water masses and circulation features.[55] The deep sea is unquestionably more difficult to sample, and the communities more difficult to describe due to their greater diversity. But today that overview is sorely missed as ecologists consider the potential impacts of large-scale deep-sea activities by humans. This need has already stimulated new research activity; the abyssal seafloor remains a vital focus for deep-sea science.[56]

CHAPTER 5

Vents, Seeps, and Whale Falls: Alternative Energy Sources in the Deep Sea

Plate tectonics – the scientific revolution at the heart of the story

The next great advance in deep-sea ecology, after the discovery of the enormous diversity of life on the seafloor, followed in the wake of plate tectonics. The key to this scientific revolution lay in the deep sea, which virtually overnight transformed marine geology from a non-event – largely dismissed in a single sentence in Sverdrup, Johnson, and Fleming's classic 1942 synthesis of oceanography: 'from the oceanographic point of view the chief interest in the topography of the sea floor is that it forms the lower and lateral boundaries of water'[1] – into one of the most dynamic fields of oceanographic science.

The tectonic revolution was deep-rooted and its impacts continued for decades, eventually resolving many long-standing puzzles: the history of the seas and continents, the origins of mountains and mineral deposits, and the evolution and global distribution of plant and animal communities, past and present. Plate tectonics provided new insights into ocean chemistry, including the age-old question of why the sea is salt (and not saltier). But its greatest impact outside geology was on marine biology, where it led to the discovery of the first major ecosystems on Earth not ultimately based on the energy of the sun and photosynthesis. In the end, tectonics led back to one of the most fundamental riddles of all – the origin of life itself.

Prior to the tectonic revolution, laymen and professional geologists alike shared an essentially static view of the Earth. Ice sheets might advance and retreat, sea level and great mountain chains rise and fall, but the underlying configuration of the continents and oceans remained fixed: trilobites had scuttled across the floor of familiar seas, and the dinosaurs roamed the same continents that we do now. But with plate tectonics, the familiar oceans and continents came unstuck in geological time, with an Invisible Hand constantly moving them about like the pieces in a great jigsaw puzzle. And even the pieces were not set: plates grow at one end and are consumed at the other. Land masses split apart, creating new seas and oceans; other plates collide, closing old seaways and accreting new bits to old continents.

Today, the tenacity with which geologists clung to their static worldview seems nothing short of remarkable. Almost from the time that maps of the Old and New Worlds became available, keen observers, starting with Francis Bacon in 1620, speculated about their jigsaw-like fit.[2] Evidence that the continents had once been joined mounted through the 19th century, based on similarities in fossil plants and animals in coal deposits on the two sides of the North Atlantic and in South America, southern Africa, India, Antarctica, and Australia: continents now utterly disjunct and inhabited by different – albeit sometimes curiously linked – floras and faunas. In 1912 Alfred Wegener published the German edition of *On the Origin of Continents and Oceans*, in which he put together the evidence from similar geological formations spanning these regions, as well as the fossil record, to reconstruct how the continents were once joined in an ancient super-continent that he named Pangaea. However, Wegener envisaged the continents plowing through the ocean crust to arrive at their present positions; without a convincing mechanism for 'continental drift,' geologists as a whole remained unconvinced.[3]

The breakthrough was to come from study of the seafloor. The first clue goes back to the *Challenger* expedition, whose soundings revealed a mid-ocean ridge that extended the length of the North and South Atlantic – 'the most striking feature of the Atlantic Ocean,' as John Murray wrote.[4] As early as 1910, an American geologist, Frank Taylor, proposed that the New and Old Worlds broke apart along this ridge to form the Atlantic Ocean, anticipating plate tectonics by half a century.[5]

Harry Hess, a Princeton geologist and geophysicist, is often credited with the epiphany that heralded the dawn of the plate tectonic revolution, based on his 1962 publication *History of Ocean Basins* – coincidentally, this was also the year that Kuhn published his essay on the nature of scientific revolutions.[6]

Hess left Princeton to become a US naval commander during the Second World War. Echo sounders were then still relatively new instruments, and Hess was fascinated by the trenches, seamounts, and other seafloor features that he encountered and mapped as he steamed back and forth across the Pacific. Assiduously – some would say obsessively – he collected seafloor data and thought deeply about what he saw.

The seafloor, Hess came to recognize, must be far younger than the rest of the Earth. Sedimentation rates over much of the open ocean are on the order of a centimeter per thousand years – or 1000 meters per 100 million years – yet there was only a thin veneer of sediment over the mid-ocean ridges, and nowhere was there sufficient sediment to indicate that the seafloor was more than a few hundred million years old, compared with the approximately 4-billion-year age of the Earth. Similarly, there were far too few seamounts in the ocean – again by approximately a factor of 10 – given the Earth's rate of volcanic activity, if the seafloor was as ancient as the Earth itself.

The well-respected English geologist Arthur Holmes, who, based on rates of radioactive decay, had first correctly estimated the age of the Earth and developed the geological time-scale, first speculated around 1930 that mantle convection might underlie Wegener's continental drift hypothesis.[7] He reasoned that the heat generated by the decay of the heavy radioactive elements concentrated near the center of the Earth might drive convection cells within the Earth's mantle, much the way heating a kettle of water causes heated water to rise from the bottom and cooler water at the surface to sink.

Hess' genius was to pull together these (and a few other) disparate elements into a coherent and convincing hypothesis. And the time was ripe for the Kuhnian revolution: too many patterns could no longer be accommodated within the old paradigm, and the explanatory power of the new was overwhelming. In what he called an 'essay in geopoetry,' Hess conceived in his *History of Ocean Basins* that convection cells within the Earth's mantle drove upward to form new seafloor along the mid-ocean ridges and consumed the old in the ocean trenches, the

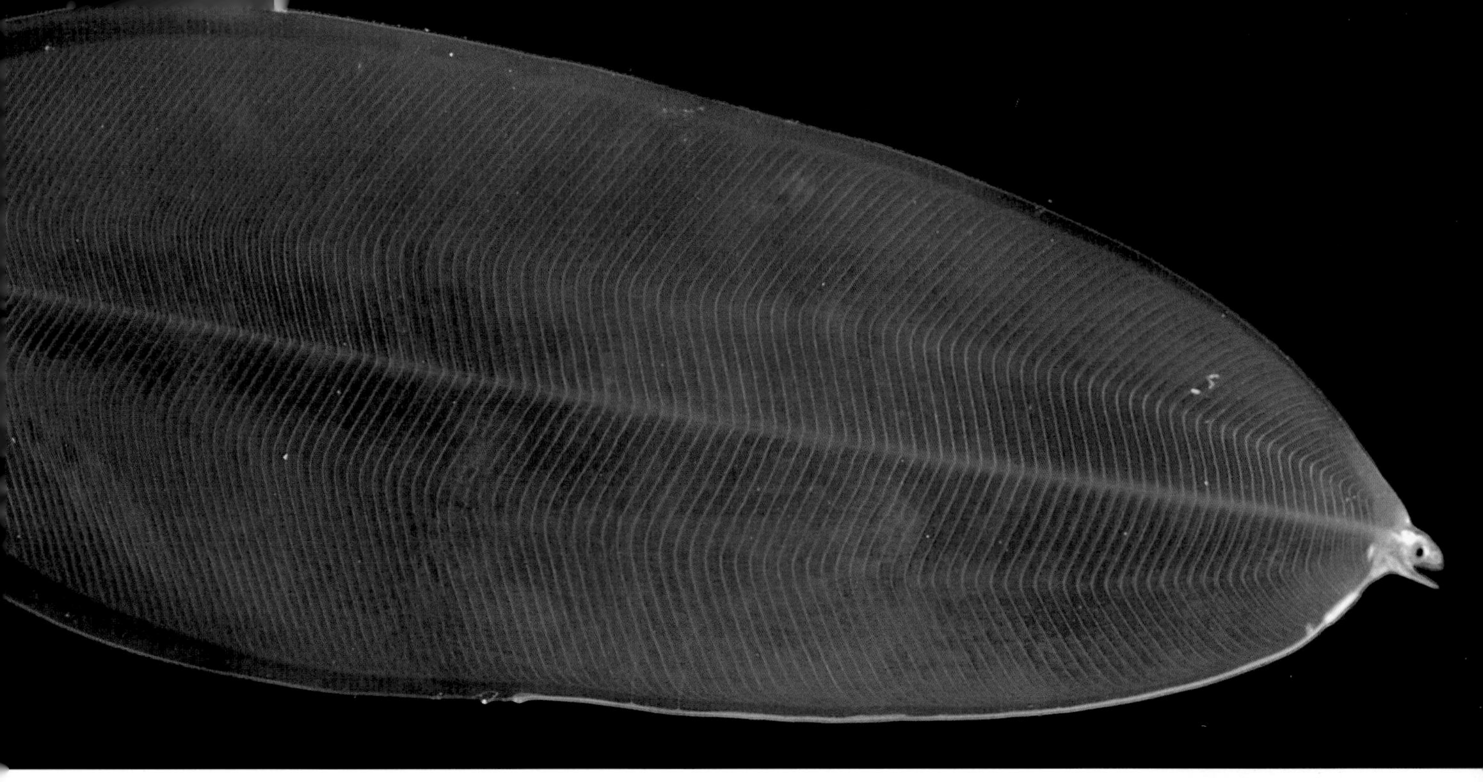

^ The giant leaf-shaped leptocephalus (meaning 'small head') larva (20 x 8 cm) of the deep-sea eel, *Thalassenchelys coheni*, collected from the North Pacific at 400 m depth. Eels are characterized by this larval form, but this species is known only from larval specimens. (Courtesy of Peter Herring/Imagequestmarine.com, National Oceanography Centre, Southampton.)

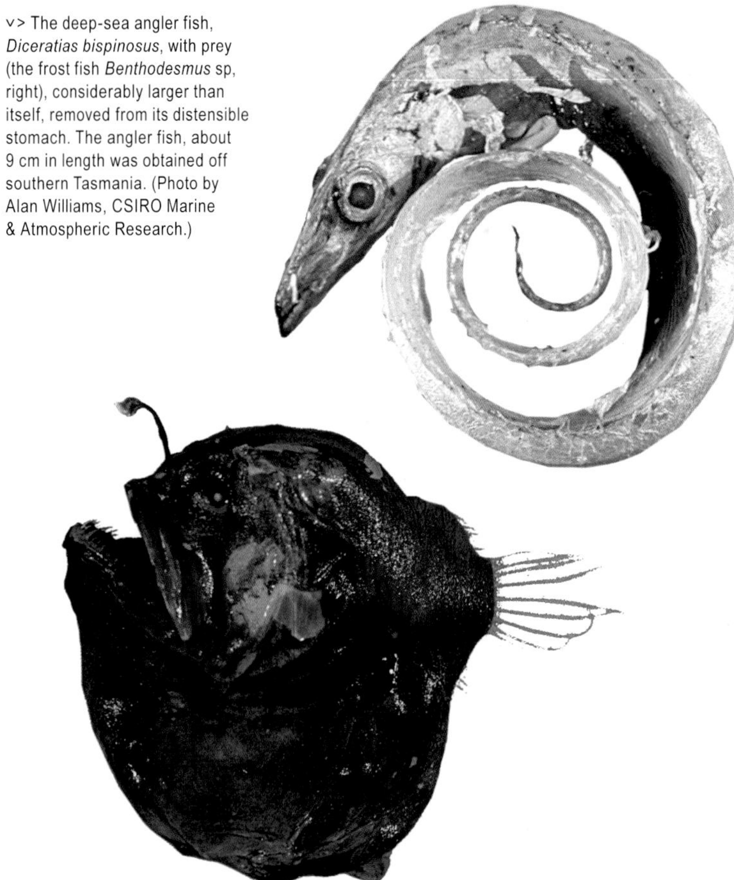

v> The deep-sea angler fish, *Diceratias bispinosus*, with prey (the frost fish *Benthodesmus* sp, right), considerably larger than itself, removed from its distensible stomach. The angler fish, about 9 cm in length was obtained off southern Tasmania. (Photo by Alan Williams, CSIRO Marine & Atmospheric Research.)

^ The Pacific hatchet fish, *Argyropelecus affinis* (maximum length: 8 cm). Its upwardly directed eyes and jaw are adaptations to prey on plankton silhouetted against the weakly downwelling light at mesopelagic depths. (Courtesy of F. Welter-Schultes, Zoologisches Institut, Berlin University, scanned from a drawing in Brauer 1906.[i])

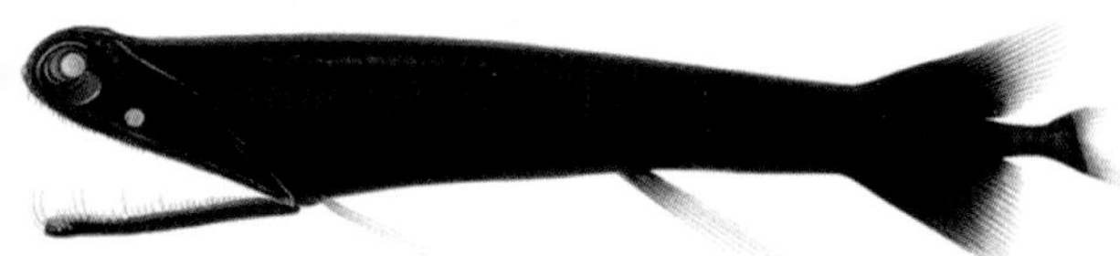

^ The dragon fish, *Malacosteus niger* (length: to 24 cm), with a large red photophore beneath its eye to search for prey and a highly extensible jaw which lacks a floor or back wall. (Courtesy of F. Welter-Schultes, Zoologisches Institut, Berlin University, scanned from a drawing in Brauer 1906.[ii])

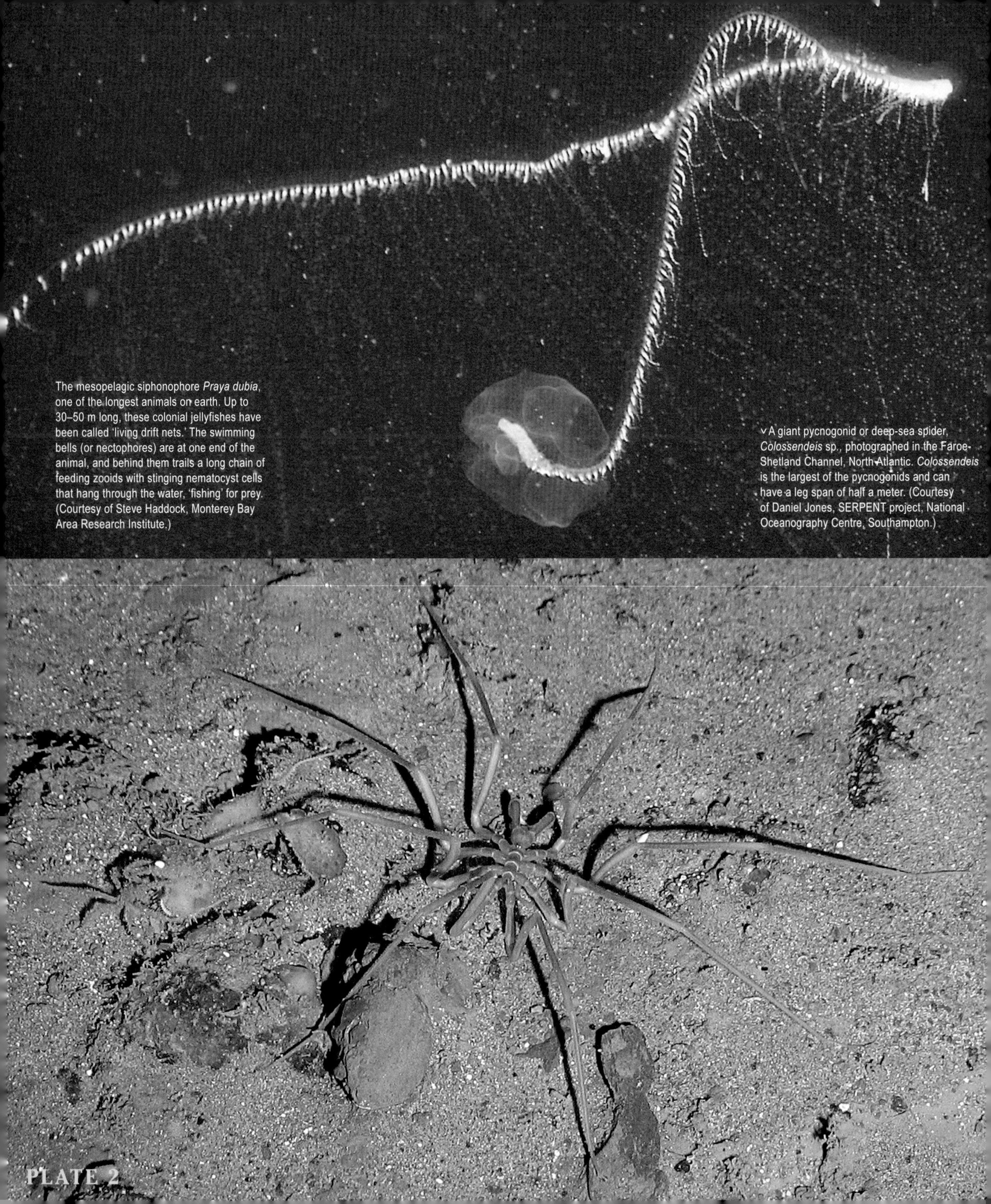
The mesopelagic siphonophore *Praya dubia*, one of the longest animals on earth. Up to 30–50 m long, these colonial jellyfishes have been called 'living drift nets.' The swimming bells (or nectophores) are at one end of the animal, and behind them trails a long chain of feeding zooids with stinging nematocyst cells that hang through the water, 'fishing' for prey. (Courtesy of Steve Haddock, Monterey Bay Area Research Institute.)

∨ A giant pycnogonid or deep-sea spider, *Colossendeis* sp., photographed in the Faroe-Shetland Channel, North Atlantic. *Colossendeis* is the largest of the pycnogonids and can have a leg span of half a meter. (Courtesy of Daniel Jones, SERPENT project, National Oceanography Centre, Southampton.)

PLATE 2

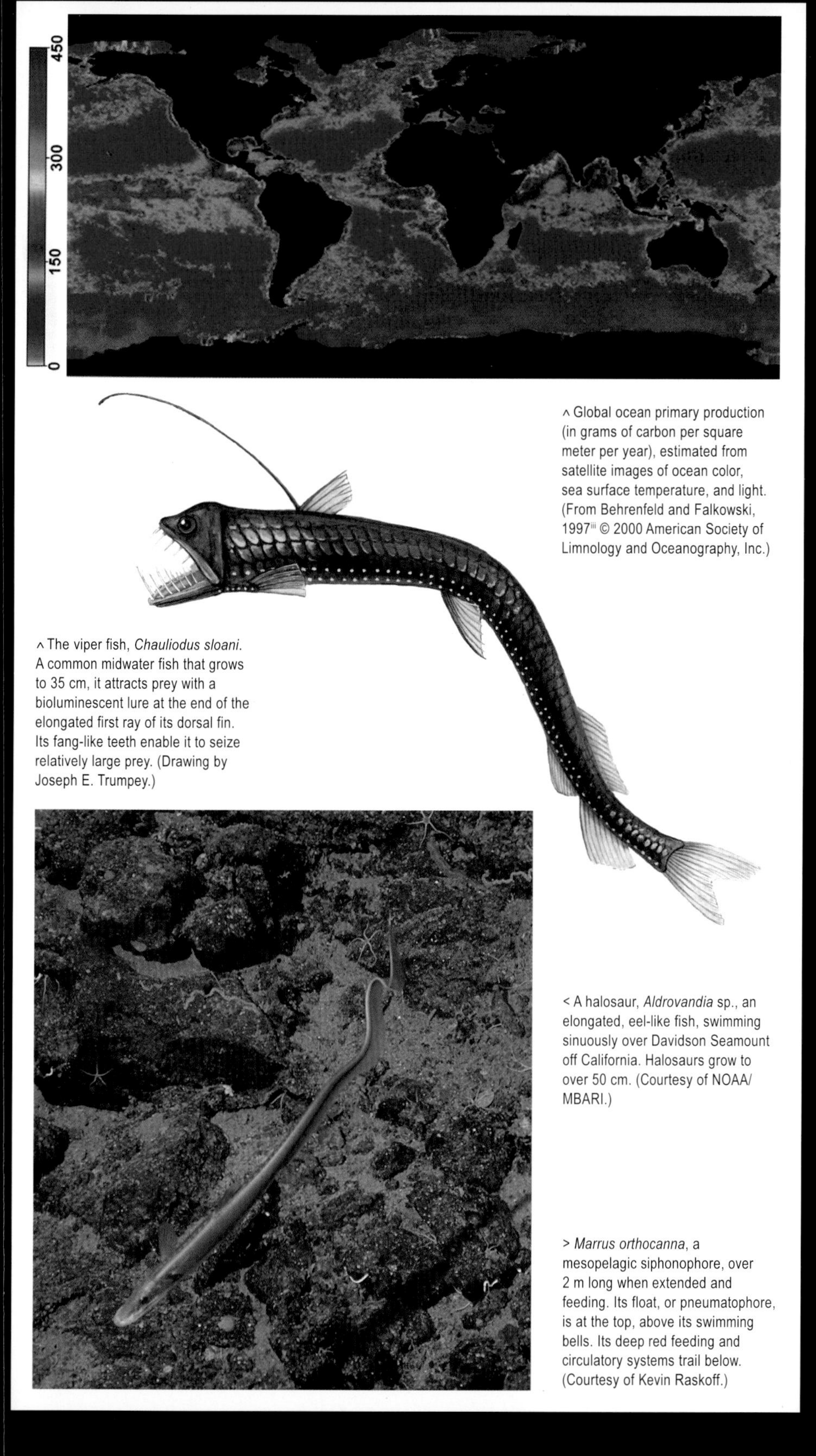

^ Global ocean primary production (in grams of carbon per square meter per year), estimated from satellite images of ocean color, sea surface temperature, and light. (From Behrenfeld and Falkowski, 1997[iii] © 2000 American Society of Limnology and Oceanography, Inc.)

^ The viper fish, *Chauliodus sloani*. A common midwater fish that grows to 35 cm, it attracts prey with a bioluminescent lure at the end of the elongated first ray of its dorsal fin. Its fang-like teeth enable it to seize relatively large prey. (Drawing by Joseph E. Trumpey.)

< A halosaur, *Aldrovandia* sp., an elongated, eel-like fish, swimming sinuously over Davidson Seamount off California. Halosaurs grow to over 50 cm. (Courtesy of NOAA/ MBARI.)

> *Marrus orthocanna*, a mesopelagic siphonophore, over 2 m long when extended and feeding. Its float, or pneumatophore, is at the top, above its swimming bells. Its deep red feeding and circulatory systems trail below. (Courtesy of Kevin Raskoff.)

v^ Images of the abyssal seafloor beneath the relatively productive equatorial North Pacific. Above: A burrowing urchin creating a furrow about 10 cm wide. The lump of sediment (far right, mid-photo) is a xenophyophore, a 'giant' single-celled foraminifera. Below: An echiuran worm trace and (far right) a holothurian. (Courtesy of Craig Smith, University of Hawaii.)

^ A close-up of the plumes or 'gills' of the hydrothermal vent tubeworm, *Ridgeia piscesae*, from the Juan de Fuca Ridge, northeast Pacific. The brilliant red color is due to its unique hemoglobin, which binds and transports hydrogen sulfide from the vent fluid, as well as oxygen from the ambient seawater. (Courtesy of Kim Juniper, University of Victoria.)

v A halosaur swimming over the abyssal plain near Hudson Canyon off New Jersey. Photo taken with the submersible *Alvin*. (Courtesy of Cindy Lee Van Dover/NOAA.)

^ Ophiuroids, or brittle stars, on a seafloor pockmarked with hundreds of worm tubes at 1100 m depth on the Pakistan Margin. The region is near the boundary with the oxygen minimum zone, which is characterized by high levels of organic matter. (Courtesy of The Deepseas Group, National Oceanography Centre, Southampton.)

< A dense aggregation of the vesicomyid clam, *Calyptogena soyoae*, from a cold seep at bathyal depths (~1200 m) in Sagami Bay, Japan. (Courtesy of and © Japan Agency for Marine-Earth Science and Technology (JAMSTEC).)

> A dense aggregation of the bathymodiolid mussel, *Bathymodiolus* sp. (probably *B. japonicus*) near a hydrothermal vent on Minami Ensei Knoll in the Okinawa Trough off Japan. (Courtesy of and © JAMSTEC.)

> A dense clump of the tube-worm *Riftia pachyptila* growing on the side of a black smoker vent on the East Pacific Rise. (Courtesy of Kim Juniper, University of Victoria.)

< A small polychaete worm, *Bathykurila guaymasensis* (maximum length: about 15 mm), obtained from a gray whale skeleton at 1600 m depth in the Santa Cruz basin off California. It feeds on mats of the giant sulphide-oxidizing bacteria, *Beggiatoa* (filaments to 120 μm wide and 10 mm long) that grow on the skeleton (Plate 7). (Courtesy of Adrian Glover, Natural History Museum, London.)

A scanning electron micrograph of the head of a polychaete worm, *Nereis* sp., from a hydrothermal vent. (Courtesy of Daniel Desbruyères, French Research Institute for the Exploitation of the Sea (IFREMER).)

The skeleton from a 35 tonne whale on the seafloor of Santa Cruz basin off California 18 months after death. Scavenging hagfish still swim and feed around the bones, which have now been largely picked clean. (Courtesy of Craig Smith and M. Degruy.)

< Vestimentiferan tubeworms and the remains of lucinid clams (*Mesolinga soliditesta*) at a cold seep at 300 m depth at Kanesu-no-Se Bank in the Ensu Sea off Japan. Lucinid bivalves, like the more common vesicomyids (e.g. *Calyptogena spp.*) host endosymbiotic bacteria that metabolize hydrogen sulfide. (Courtesy of and © JAMSTEC.)

> Vent shrimp (*Rimicaris kairei*) swarming around the first hydrothermal vents discovered in the Indian Ocean in 2000 at a depth of 2450 m near the triple junction of the African, Antarctic and Indo-Australian plates. *Rimicaris* vent shrimps, also found at vents in the Atlantic Ocean, host symbiotic sulfur-metabolizing bacteria from hairs on their carapace that cover them in a fuzzy white coat. Mussels, anemones, and crabs can also be seen. (Courtesy of and © JAMSTEC.)

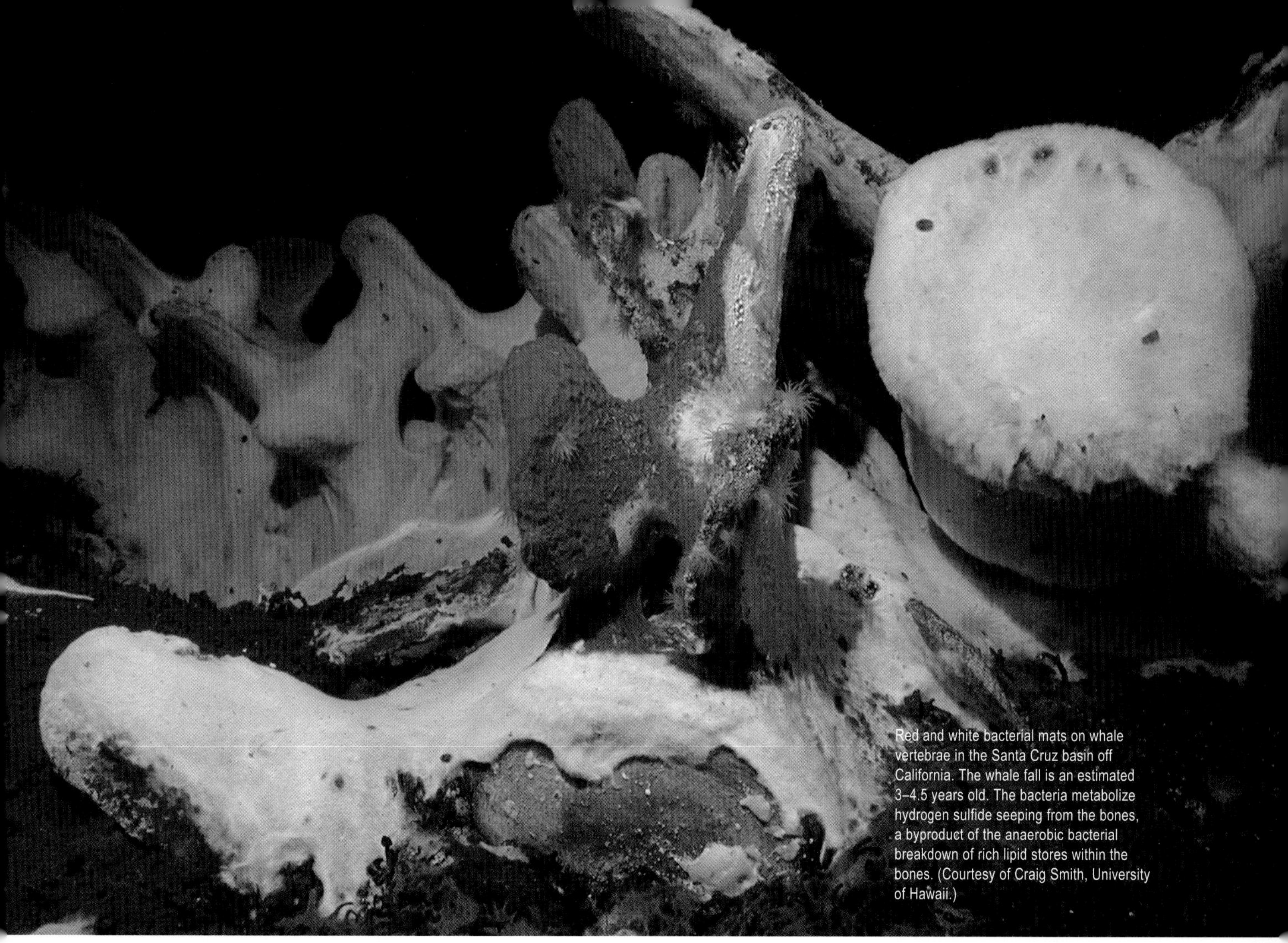

Red and white bacterial mats on whale vertebrae in the Santa Cruz basin off California. The whale fall is an estimated 3–4.5 years old. The bacteria metabolize hydrogen sulfide seeping from the bones, a byproduct of the anaerobic bacterial breakdown of rich lipid stores within the bones. (Courtesy of Craig Smith, University of Hawaii.)

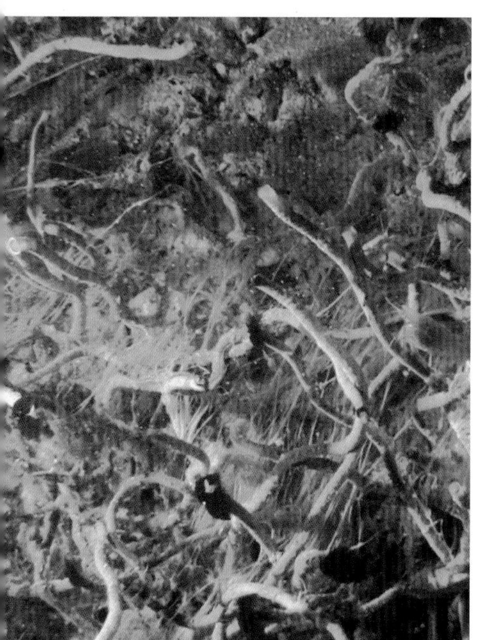

< Dense vestimentiferan tubeworms at a hydrothermal vent site in the Manus Basin off Papua New Guinea. The site is characterized by its extreme acidity (pH = 2). (Courtesy of and © JAMSTEC.)

v A 'living fossil,' the decapod crustacean *Neoglyphea neocaledonica* (length: 12 cm), discovered at 400 m depth on Capel Bank in the Coral Sea. Described as 'halfway between a shrimp and a mud lobster,' *N. neocaledonica* is the second species discovered from a family, the Glypheidae, that thrived in the Age of Dinosaurs but was once thought to have been extinct for about 50 million years, since the Eocene. (Courtesy of Bertrand Richer de Forges, IRD; photo by Joelle Lai.)

> A close-up of giant vestimentiferan tubeworms, *Riftia pachyptila*. With no mouth or gut, the worms, among the fastest growing invertebrates on the planet, depend for their nutrition on endosymbiotic bacteria that metabolize hydrogen sulfide. The worms absorb sulfide from the vent fluids with their red plume-like 'gills' and transport it in their blood to the bacteria. (Courtesy of NASA-JPL-Caltech.)

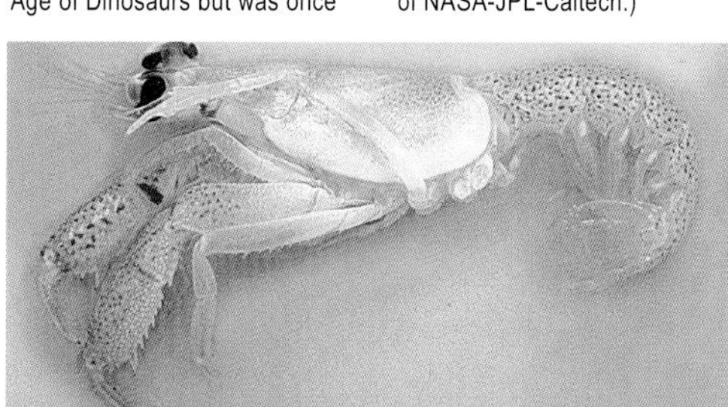

> Echogram of a seamount off the northeast coast of Tasmania, the spawning ground of the main orange roughy stock off southeast Australia. The seafloor is shown in red, with the peak at about 650 meters depth, and its base at 1000–1100 meters. The spawning aggregation (in blue) extends into the water column along its slope. A school of cardinal fish (*Epigonus telescopus*) hovers over the peak. The seamount is about two kilometers in diameter at its base. (Courtesy of CSIRO Marine and Atmospheric Research.)

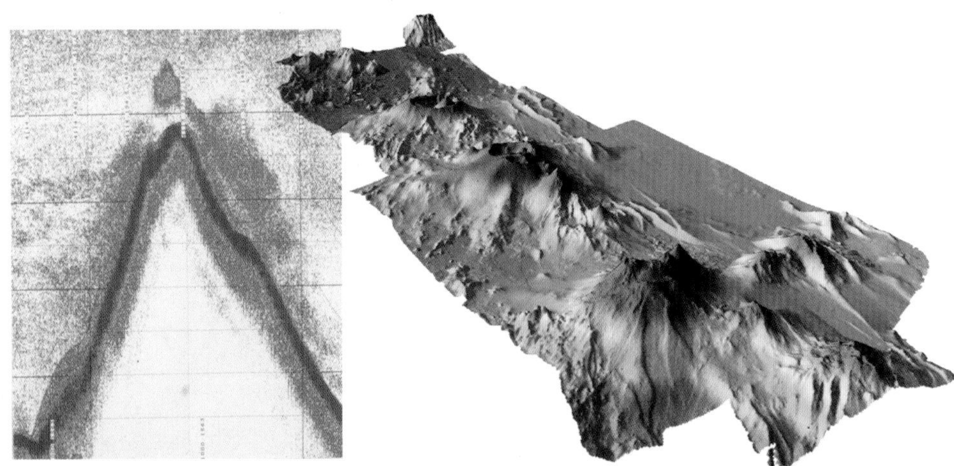

< The topography of the Brothers Seamount chain along the Kermadec Ridge, northeast of New Zealand. Brothers Seamount is 1300 m high and peaks at 1200 m depth. With massive sulfide deposits rich in zinc, copper and gold, it has been licensed for mineral exploration. The high-resolution three-dimensional topography is based on multi-beam sonar mapping. (Courtesy of NIWA New Zealand.)

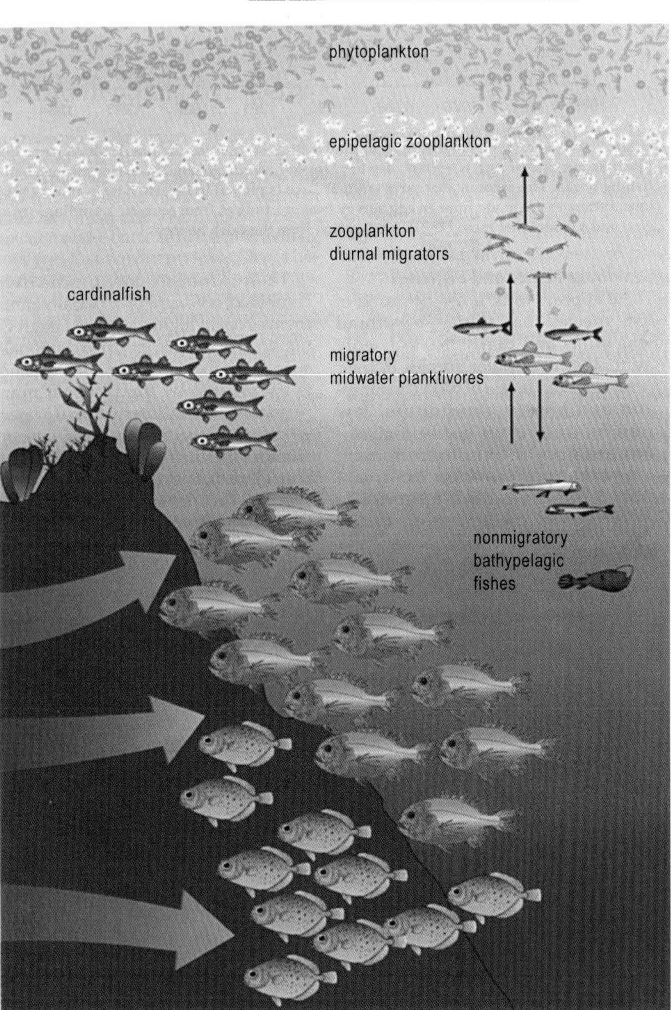

^ The seamount ecosystem, showing sessile benthic invertebrates (corals and sponges) on the seafloor near the crest and orange roughy, oreos, and cardinal fish in the water column feeding on vertical migrators and other prey intercepted and swept past in the amplified currents. (Courtesy of *American Scientist*, from Koslow 1997.)[iv]

> A block of *Solenosmilia variabilis*, the coral that forms the underlying matrix for the seamount reefs off Tasmania. The polyps grow off the stolons or branches, which continue to grow upward into the current, eventually overlying and starving the polyps underneath. The living coral is pinkish, whereas the dead coral turns dark brown or black as it is coated in precipitates of manganese and iron oxide. The interstices provide refuge and habitat for a variety of associated invertebrates. (Photo by Karen Gowlett-Holmes/CSIRO Marine and Atmospheric Research.)

> Photograph of the benthic invertebrate community on a seamount off the south coast of Tasmania at a depth of about 1000 m. Living and dead stony coral (*Solenosmilia variabilis*) forms the reef matrix, with soft corals, sponges, and other suspension feeders growing over it. (Courtesy of CSIRO Marine and Atmospheric Research.)

An orange roughy swimming over the corals on a seamount off New Zealand. (Courtesy of NIWA New Zealand.)

The dominant seamount fishes from different regions are mostly from different families and orders, indicating the isolation of seamount biogeographic provinces. Different groups colonized this habitat and independently evolved the seamount-associated facies, typically a deep body with strong caudal (tail) fin suited for navigating the strong seamount currents.

> The boarfish (*Capros aper*, Caproidae, Zeiformes) from the North Atlantic. (Courtesy of Robert A. Patzner, University of Salzburg, Austria.)

< Orange roughy (*Hoplostethus atlanticus*, Family Trachichthyidae, Order Beryciformes) of the temperate southern hemisphere and North Atlantic. (Courtesy of CSIRO Marine and Atmospheric Research.)

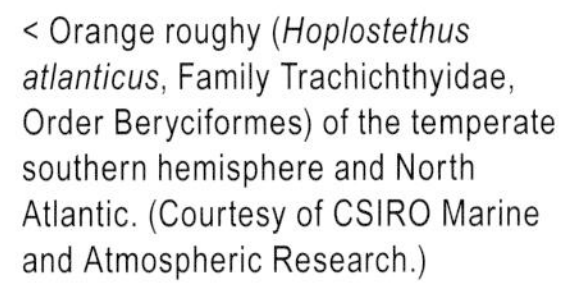

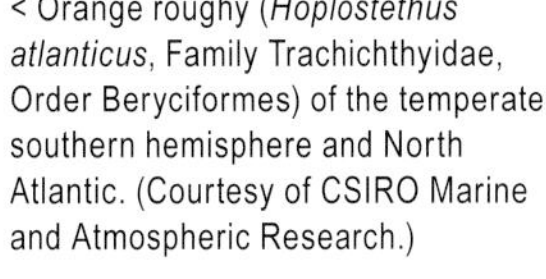

< Alfonsino (*Beryx splendens*, Berycidae, Beryciformes) from the tropics and sub-tropics. (Courtesy of Pablo Reyes, Universidad Austral de Chile.)

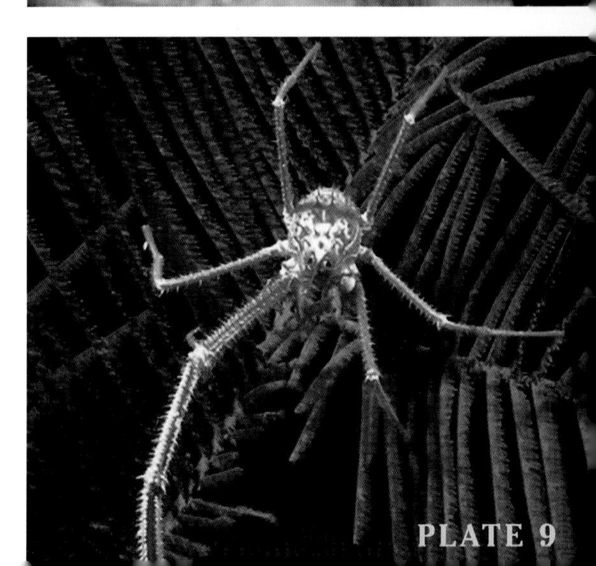

> A galatheid crab on a branch of black coral. Photo from the remotely operated vehicle *Hercules*, operating at 1120 m depth on Corner Seamount in the northwest Atlantic. (Courtesy of Scott Frances, University of Louisiana.)

< The pelagic armorhead (*Pseudopentaceros wheeleri*, Pentacerotidae, Perciformes) of the temperate central North Pacific. (Courtesy of John Randall, Bishop Museum, Honolulu.)

Redfish (*Sebastes* sp.) hovering among a deepwater coral reef (*Lophelia pertusa* and associated species) off the Atlantic coast of Canada. (Courtesy of Department of Fisheries & Oceans, Canada.)

> Galatheid crabs (*Uroptychus cardus*) from seamounts south of Tasmania. (Photo by Karen Gowlett-Holmes, CSIRO Marine and Atmospheric Research.)

> A blobfish (*Psychrolutes microporos*) obtained from a seamount at 1200 m depth on the Norfolk Ridge between Australia and New Zealand. The species is likely an ambush predator. A parasitic copepod hangs from its lip. The function of its large proboscis is not known. (Courtesy of CenSeam and © NORFANZ 2003 Expedition.)

< A conger eel (*Conger oceanicus*), large red crabs (*Eumunida picta*) and sea urchins (*Echinus tylodes*) within a coldwater coral (*Lophelia pertusa*) reef. Photo taken from the Johnson Sea Link submersible off North Carolina at 384 m. (Courtesy of S.W. Ross, K. Sulak, and M. Nizinski, NOAA and US Geological Survey.)

> *Platymaia* sp., a spider crab obtained from a seamount off New Zealand. (Courtesy of CenSeam and Malcolm Clark, NIWA New Zealand.)

> A tusk (*Brosme brosme*), relative to the cod fish, residing within the coldwater coral *Lophelia pertusa* that comprises Sula Reef off Norway. (Courtesy of André Freiwald, University of Erlangen-Nuremberg.)

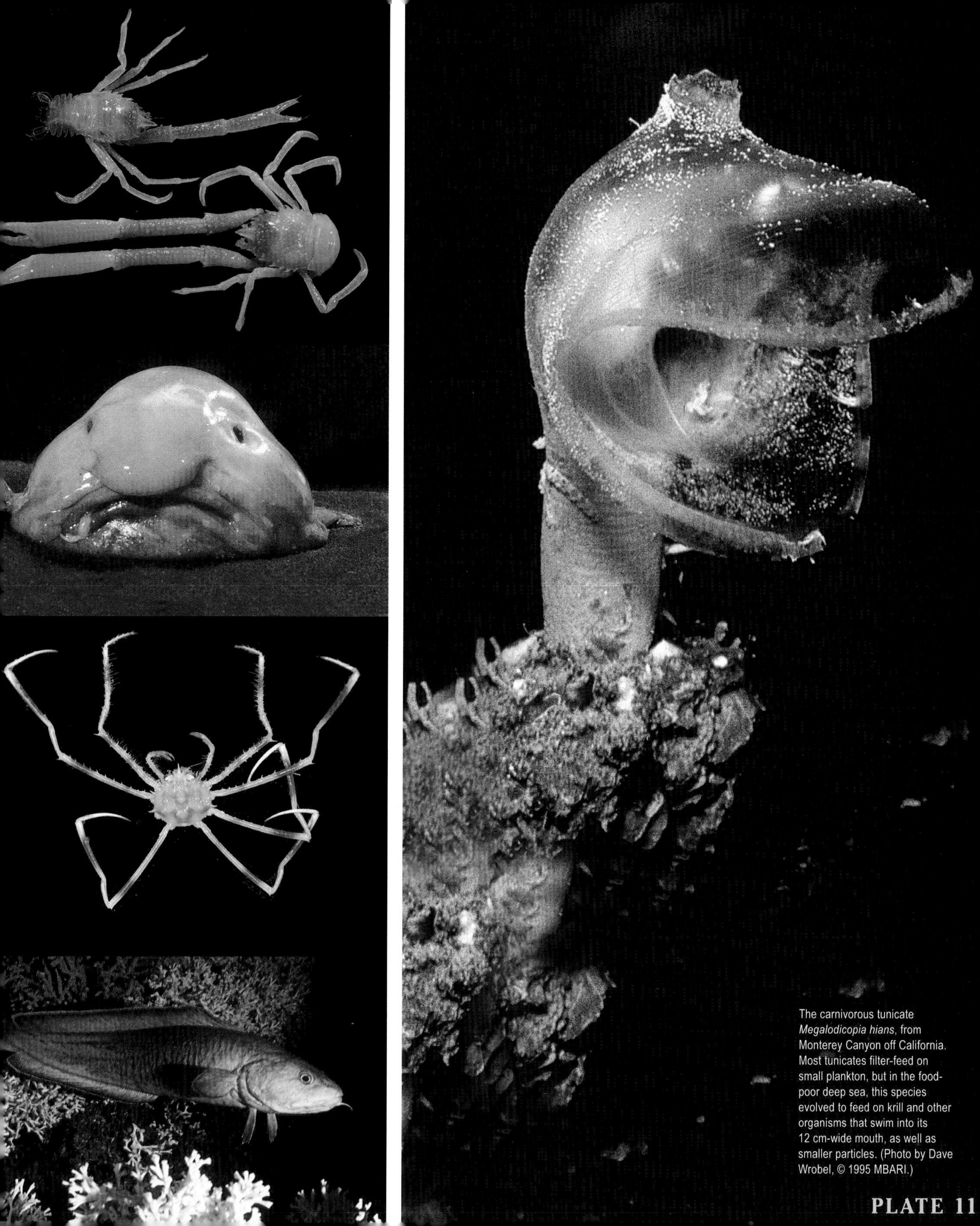

The carnivorous tunicate *Megalodicopia hians*, from Monterey Canyon off California. Most tunicates filter-feed on small plankton, but in the food-poor deep sea, this species evolved to feed on krill and other organisms that swim into its 12 cm-wide mouth, as well as smaller particles. (Photo by Dave Wrobel, © 1995 MBARI.)

PLATE 11

< Black coral with crinoids (*Florometra serratissima*) and orange brisingid starfish that have mounted it, feeding with their arms into the current. Photo taken on Davidson Seamount off California at 1950 m depth. Note the dominance of soft corals in photos from the northeast Pacific on this and the following plate, compared with the hard (scleractinian) corals dominant in the North Atlantic and South Pacific, probably due to the shallower aragonite compensation depth (greater dissolution of calcium carbonate) in the North Pacific. (Courtesy of NOAA/MBARI.)

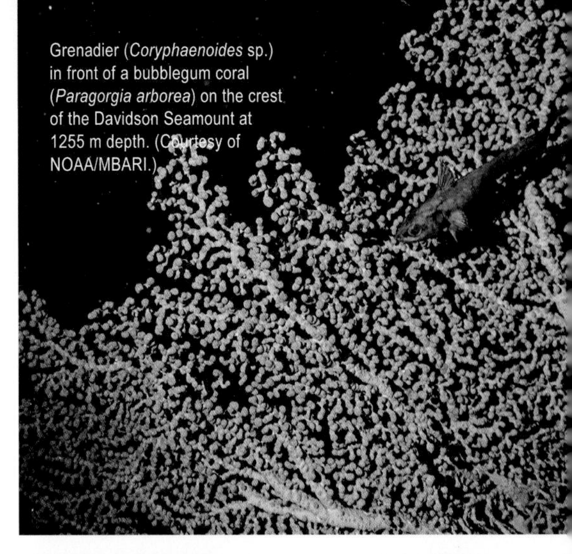

Grenadier (*Coryphaenoides* sp.) in front of a bubblegum coral (*Paragorgia arborea*) on the crest of the Davidson Seamount at 1255 m depth. (Courtesy of NOAA/MBARI.)

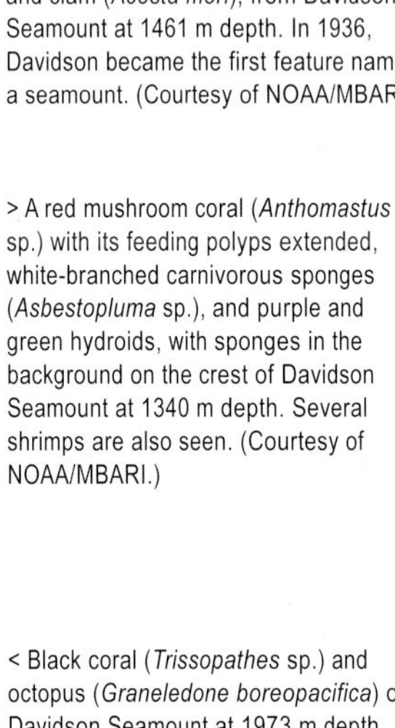

< Benthic octopus (*Benthoctopus* sp.) and clam (*Acesta mori*), from Davidson Seamount at 1461 m depth. In 1936, Davidson became the first feature named a seamount. (Courtesy of NOAA/MBARI.)

> A red mushroom coral (*Anthomastus* sp.) with its feeding polyps extended, white-branched carnivorous sponges (*Asbestopluma* sp.), and purple and green hydroids, with sponges in the background on the crest of Davidson Seamount at 1340 m depth. Several shrimps are also seen. (Courtesy of NOAA/MBARI.)

< Black coral (*Trissopathes* sp.) and octopus (*Graneledone boreopacifica*) on Davidson Seamount at 1973 m depth. (Courtesy of NOAA/MBARI.)

> White-branched sponge (*Asbestopluma* sp.), white ruffle sponge (Farrea sp.), a vermilion crab (*Paralomis verrilli*), and shrimp on Davidson Seamount at 1355 m depth. Sponges virtually all filter-feed on small particles, but *Asbestopluma* evolved carnivory as an adaptation to the food-poor conditions in the deep sea. It entangles small crustaceans in its hairs. (Courtesy of NOAA/MBARI.)

< The flytrap anemone (Family Hormathiidae), so named because of the way it closes on its prey, on the slope of Davidson Seamount at 1874 m depth. (Courtesy of NOAA/MBARI.)

Soft coral and sponge garden and associated fauna at 165 m depth off Adak Island in the central Aleutian Islands, Alaska. (Photograph by Alberto Lindner, NOAA.)

v Bubblegum coral (*Paragorgia arborea*) at the crest of Davidson Seamount (1257 m depth). A basket star and several shrimps are among its branches. *P. arborea* is the largest seafloor organism in the world, with centuries-old specimens to 42 cm in diameter and ~7 m height.[v] On Davidson Seamount, they are observed to 2.5 m (8 feet) in height. (Courtesy of NOAA/MBARI.)

v A close-up of a large bubblegum coral (*Paragorgia arborea*), a gorgonian or sea fan, from the crest of Davidson Seamount at 1254 m depth. A smaller, related species of *Paragorgia* is at its base. Small blue scale worms (Family Polynoidae) can be seen; these are often associated with *P. arborea*. (Courtesy of NOAA/MBARI.)

v Zooanthid soft corals growing on a dead coral with a crinoid (left) and orange brisingid starfish. (Courtesy of NOAA/MBARI.)

^ Coldwater corals, predominantly *Lophelia pertusa*, at 750 m depth on a carbonate mound on the Porcupine Bank in the NE Atlantic, in pristine condition (left), and after trawling (right), showing broken corals and the remnants of a trawl net. (Courtesy of J.M. Roberts, Scottish Association for Marine Science.)

> Model of a deepwater *Lophelia* reef from off Norway, showing *Lophelia pertusa*, sponges, gorgonian corals, fish, and other associated invertebrates. (Courtesy of Jan Helga Fosså, Institute of Marine Research, Norway.)

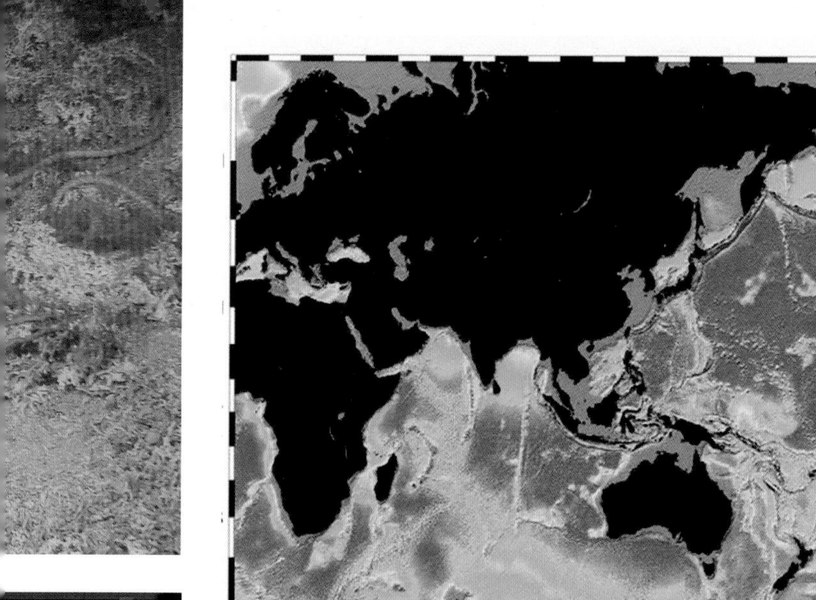

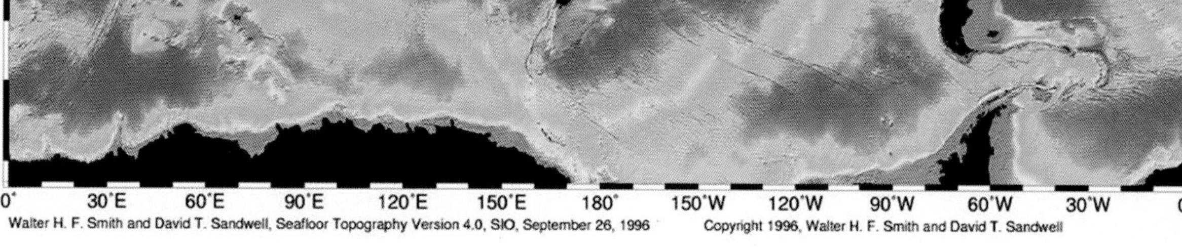

^ The topography of the global seafloor. The Pacific is ringed by trenches, which limit the extent of continental shelf and slope regions, but it has vastly more seamounts. The North Atlantic, on the other hand, has more extensive continental margins around its rim. (Courtesy of W. Smith and D. Sandwell/NOAA.)[vi]

> A close-up of the reef-forming deepwater coral *Lophelia pertusa* with its polyps extended. The specimen was collected from around 300 m depth at the Sula Ridge off Norway. (Photo courtesy of Pål Buhl Mortensen, Institute of Marine Research, Norway.)

^ An estimated 20–30 tonne trawl shot of orange roughy being hauled up the stern ramp of a fishing vessel off New Zealand. (Courtesy of Malcolm Clark, NIWA.)

^ The seafloor on a pristine seamount (above) and a heavily fished seamount (here) south of Tasmania. The pristine reef is dominated by the hard (scleractinian) coral *Solenosmilia variabilis*. A variety of solitary and soft corals, sea urchins, brittle stars, sponges and other organisms are living in association with the reef. Images are from depths of about 1000 m. (Courtesy of CSIRO Marine and Atmospheric Research.)

^Fishermen dumping a large bubblegum coral, caught in their trawl as by-catch, off the stern of the New Zealand trawler *Waipori*. (Courtesy of Greenpeace.)

cells' downward limb. Only the dense ocean crust was recycled within the Earth's mantle; the relatively light continental crust floated on top of the plates and was carried along like a froth, thereby explaining the relatively young age of the seafloor relative to the continents. As Hess noted, the key difference between the new paradigm of plate tectonics or seafloor spreading, and Wegener's hypothesis was that the continents were no longer viewed as 'plowing through oceanic crust impelled by unknown forces'[8] – rather, the seafloor and continental crust rode together on massive crustal plates, impelled by the slow convective movement of the dense, somewhat fluid mantle.

Hot springs in the deep sea

In 1965, a few years after the publication of Hess' landmark paper, J.W. Elder of the University of Manchester predicted that hydrothermal vents – the deep-sea equivalent of hot springs – would be found on the mid-ocean ridges.[9] Hot springs occur on the land where groundwater, heated by underground magma chambers, rises to the surface through faults or other conduits. A similar convective process was envisaged at the mid-ocean ridges. The newly formed seafloor around an active ridge is relatively porous, due to faulting and the cooling and contraction of the young

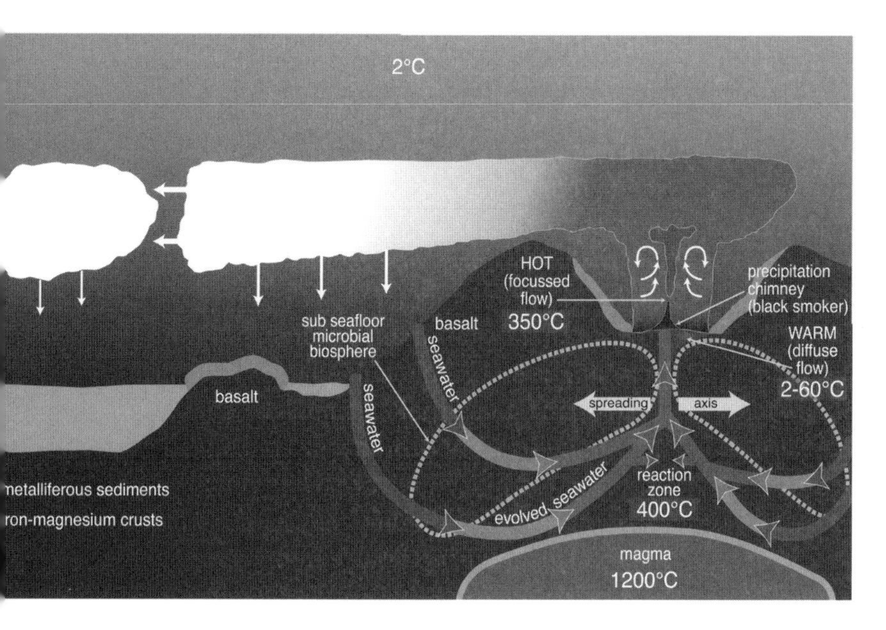

Schematic of a hydrothermal vent convection cell and the chemistry associated with hydrothermal vents. Cool seawater (~2°C) seeps downward through faults and cracks in the seafloor. Magnesium (Mg) and sulfate (SO_4) initially precipitate out. As the seawater approaches the magma chamber, (the high-temperature (HT) reaction zone), oxygen combines to form carbon dioxide (CO_2), sulfate is reduced to sulfide (H_2S), and the water becomes acidic. The hot corrosive vent fluid leaches metals from the basalt (e.g. iron (Fe), manganese (Mn), copper (Cu) and zinc (Zn)). The vent fluid rises to the surface, where sulfides and minerals precipitate as the water cools, forming chimneys (black smokers). The plume rapidly cools, but still carries iron and manganese particles, which sediment onto the surrounding seafloor. (From NOAA Ocean Explorer website, modified from Massoth et al., 1988.)[m]

crust. As seawater percolated through these faults and cracks, it would eventually reach the magma zone, where, once heated, it would be driven back upward through central vents at the spreading center.

Twelve years were to pass before Elder's prediction was verified. During that time, evidence accumulated from a variety of sources, pointing not only to the likely existence of deep hydrothermal vents, but to their having a major role in the chemistry of the oceans, the heat balance of the planet, and even in the affairs of men as the source of major ore deposits. But their most dramatic impact was entirely unsuspected.

The first evidence for seafloor hydrothermal activity extended back almost a century, to the 1880s, when the Russian research vessel *Vityaz* (predecessor to the post–World War II Soviet research vessel) discovered hot, metal-rich brines on the deep seafloor of the Red Sea – a 200-meter-thick pool of 64°C (147°F) water at 2000 meters depth, with a salinity of 320, almost ten times that of normal seawater. At the time, the observations were put down to the curious effects of that region's extreme solar heating and evaporation. But once plate tectonic theory took hold, the Red Sea was recognized to be a young sea, in the process of active seafloor spreading.[10] The remarkable brine formed as hydrothermal fluid rose through evaporate deposits, laid down in the Miocene (8–24 million years ago) during an early stage in the opening of the Red Sea, when circulation was restricted and the proto-Red Sea evaporated.[11]

Metal-rich sediments in the vicinity of the mid-ocean spreading centers provided further evidence of hydrothermal activity. Murray first noted them during the *Challenger* expedition.[12] Then in 1966, the observations were placed in a modern context, when high levels of iron, manganese, copper, nickel and other metals in the sediments on either side of the East Pacific Rise ridge axis were recognized as having been precipitated by 'hydrothermal exhalations.'[13]

Tectonics soon led to the realization that some of the world's most valuable ore deposits were ancient uplifted sections of mid-ocean ridge. One of the best examples is the 5000-year-old mines of Cyprus, chief source of copper to the classic Mediterranean world. (The Latin word for copper, *cuprum*, is derived from the Roman name for the island, *Cyprium*.)[14] The ore deposits, extinct hydrothermal vents subsequently uplifted above sea level,[15] enabled scientists to explore the anatomy of vents and to reconstruct their chemistry in considerable detail, years before they were observed *in situ*.[16]

As seawater descends toward the magma chamber, it is heated to about 350°C (660°F). Depending on the depth – and hence the pressure – seawater does not necessarily boil, even at this temperature, but it reacts with the surrounding basalt, the volcanic rock of the seafloor. The oxygen, sulfur (in the form of sulfate) and magnesium are stripped out of the seawater, rendering it anoxic, rich in hydrogen sulfide, and highly acidic, the pH reduced from the neutral 7.5 of typical seawater to about 3–5.[17] This hot, corrosive hydrothermal fluid continues to react with the basalt, leaching out various metals: iron, manganese, copper, and zinc, among others. As the hydrothermal fluid rises toward the seafloor and cools, these metals precipitate out, generally as sulfides:

for example, iron pyrite or fool's gold (FeS_2), copper sulfide (CuS_2), and other compounds.

Ocean geochemists realized even before hydrothermal vents were discovered that they held the key to longstanding puzzles in ocean chemistry. For example, where does all the manganese in the manganese nodules come from? (Manganese is virtually undetectable in seawater, and little enters through the rivers.) Or the perennial question, why is the sea salt? Assuming that the salt was derived from the rivers, Edmund Halley (of Halley's Comet fame) proposed back in 1715 that the age of the oceans could be estimated based upon the total salt content of the oceans and how much enters each year from the rivers. But neither of these quantities was known at the time, and it was not until 1899 that the Irish geologist, John Joly, took up the suggestion. His estimate – about 100 million years – is far too low.[18] So the real question is why the oceans aren't much saltier, and the answer has to do largely with reactions with the sediments and in the heated reaction chamber beneath the ocean crust, where key ions, such as magnesium and sulfate, are removed.[19] The ocean cycles through the hydrothermal vent system in the order of every 10–100 million years, compared with its cycling through the Earth's hydrological cycle (evaporating and returning to the oceans through the rivers) about every 30 000–40 000 years. Although the flow through the vents is far slower, the chemical transformation within the vent system is dramatic – and for some aspects of the ocean's chemistry, it is comparable to or outweighs the river inputs. The acidity (pH) and salinity of the oceans has been remarkably stable over geological time, a testament to the balance between continental erosion and river runoff at the ocean's margins and the hydrothermal processing beneath the seafloor, as well as processes occurring within the sediments and water column.

One further line of evidence led to the discovery of deep hydrothermal vents. As plate tectonic theory developed, geologists estimated how much heat should be conducted through the newly formed seafloor as it cooled, and then set out to measure this slow form of heat loss. They found, however, that about 25–30 percent less heat was conducted through 'young' crust (less than 65 million years old) than they predicted. The anomaly was greatest in the youngest ocean crust, less than a million years old. The obvious candidate for the 'missing heat' was that it was being rapidly carried to the surface in concentrated hydrothermal convection cells.

The USA has one deep-sea submersible, *Alvin*, commissioned at Woods Hole in 1964 at the dawn of the tectonic revolution. (Its name is derived from Allyn Vine, a key supporter of the submersible project and leading Woods Hole architect of the tectonic revolution.)[20] At the time, the utility of submersibles for doing significant scientific work was still unproven, and the verdict remained unclear for some time. Indeed, after ten years of operation, plagued by problems and dogged by uncertainty, *Alvin* still had not been to 3000 meters, the depth of the mid-Atlantic Ridge. Arguably the most exciting discovery of its first decade of operation was the result of its accidental sinking – the recovery of the unspoiled soup and bologna sandwiches ten months later, which first pointed to the slow rate of microbial metabolism in the deep sea.

All that appeared set to change in 1974 with the politically inspired French-American Mid-Ocean Undersea Study (FAMOUS), the first international collaborative study based around the use of submersibles. The FAMOUS study site on the mid-Atlantic Ridge was 20 miles long and flanked by a volcano at each end, the expectation being that hydrothermal and volcanic activity would be observed live for the first time at a mid-ocean spreading center.[21] However, although the FAMOUS project was judged a success, *Alvin* and the two French submersibles, *Archimède* and *Cyana*, found not a hint of elevated water temperature and no other indication of hydrothermal activity. In the end, this was put down to the mid-Atlantic Ridge being 'slow,' spreading at only about 2 centimeters (an inch) per year. Hydrothermal activity was clearly more difficult to find than originally anticipated – and a submersible, with its limited visibility and a cruising speed of only half a knot (1 kilometer per hour) was probably not the best tool for locating such rare and elusive features.

Three years passed before *Alvin* was funded again to search for hydrothermal vents. This time it was to be in the equatorial Pacific on the Galápagos Spreading Center off the Galápagos Islands, a 'fast' spreading center (10 centimeters per year). In preparation, a cruise in 1976 mapped potential vent sites based on slight temperature anomalies (on the order of 0.1°C) over the ridge, carried out detailed bathymetric surveys, and photographed the regions of interest.[22] Kathy Crane, a graduate geology student at Scripps Institution of Oceanography, who carried out the thermal mapping, noted that large white clamshells appeared in several of the photos just where she found the temperature anomalies. But discarded beer cans also appeared in some of the photos, and the chief scientist, Peter Lonsdale, initially put the clamshells down to refuse from passing shipboard parties. In preparing her map, Crane labeled two of her temperature anomaly sites Clambake I and Clambake II, and during the cruise the following year, she instructed the *Alvin* divers to follow the clamshells, like the yellow brick road in Oz, to the suspected hot springs.[23]

In the interim, it was recognized from the photos that some of the clams were in fact alive, and that they were far larger and more densely aggregated than any previously observed in the deep sea. After initially dismissing them, Lonsdale, a geologist, now rushed to publish the photos while the *Alvin* was at sea diving on the sites the following year – before anyone had an opportunity to recover and examine the clams. Assuming that, like any others, they fed on unicellular organisms and detritus suspended in the water, he concluded that their size and density resulted from unusual feeding conditions: presumably the hydrothermal plumes set up convection cells that strengthened the currents along the bottom, thereby enhancing the flow of food. As Lonsdale acknowledged in a note at the end of his paper, Hessler in reviewing the manuscript cautioned him against concluding too much on the basis of photographs alone.[24] And Hessler was right; the real story was far stranger than anyone at first imagined, and it took scientists several years to unravel.

Life at the hot springs

The clams were the size of dinner plates (up to 26 centimeters long), and proved most unlikely to be served up at a clambake. When opened, like the water sampled from the vent, they reeked of hydrogen sulfide, a highly toxic gas that, like cyanide, blocks the oxygen-binding sites in blood.[25] Yet the gills and meat of the clam were strikingly red: the giant clams had red blood with hemoglobin, and lots of it, unlike any clam the scientists on board had previously seen. (Indeed, the classic distinction between vertebrates and invertebrates, going back to Aristotle, was the presence or absence of red blood.)[26] It was most puzzling: clearly the clams were thriving on a toxic brew of hydrogen sulfide and heavy metals. The density of giant clams, mussels and other organisms in the vicinity of the vents exceeded 10 kilograms per square meter at some sites, utterly unlike anything previously recorded on the abyssal seafloor (Plates 4–7).[27]

And the biology only got stranger. The iconic vent organism was a worm encased in a white tube that grew to 2 meters high, standing straight up from the hard rock seafloor, with blood-red, feather-like plumes protruding from its tube. Remarkably these worms had no mouth, stomach or anus, but proved to be one of the fastest-growing invertebrates on the planet. At a newly vented site revisited periodically in the early 1990s, they reached maturity within two years, growing almost a meter per year![28] Recalling the abyssal clam *Tindaria*, which after 100 years was still less than a centimeter long and required 50–60 years to attain maturity (see p. 84), it is difficult to imagine a creature whose ecology and life history was more atypical of the abyssal seafloor.

Despite the attention that the giant tubeworms received – their photo featured in journals from *National Geographic* to *Nature* – what they were and how they should be classified remained contentious for more than 20 years. Four years passed before Meredith Jones, a specialist in worm taxonomy at the Smithsonian Institute, named the species *Riftia pachyptila*, the thick-plumed rift worm. Jones classified *Riftia* among the vestimentiferans, so-named for their characteristic vestimentum or 'collar,' with which they hold themselves at the top of their tubes, in the hitherto obscure phylum, the Pogonophora (or beard-worms, for their characteristic plume of tentacles).[29]

A minor phylum of tube-dwelling, mostly deep-sea worms, the Pogonophora are remarkable creatures, whose basic biology, taxonomy, and evolutionary history baffled zoologists for virtually the entire 20th century. First sampled during the Dutch *Siboga* expedition of 1899–1900, the specimens were turned over to a prominent French zoologist, Maurice Caullery, who named the species after the expedition vessel and its leader, Max Weber (*Siboglinum weberi*); but after studying them for almost 50 years, he was still unable to decide their taxonomic affinity.[30] Admittedly, they were difficult to work with. Before *Riftia*, most known pogonophorans were thread-like, no more than a few millimeters in diameter – indeed their long fine tubes were mostly discarded as stray bits of trawl twine until the 1950s, when Russian biologists dredged some particularly large specimens from deep

Pacific trenches.[31] Most puzzling, however, was their complete lack of a mouth, gut or anus, a feature not known elsewhere in the animal kingdom. This raised one obvious question – how pogonophorans sustained themselves – but for a time it seemed to resolve another, namely how to classify them. Clearly they stood alone in having a unique body plan, a key defining criterion for a phylum. And so by the middle of the 20th century, the consensus among invertebrate taxonomists was that the Pogonophora represented a new phylum.[32] But this did nothing to resolve the deeper question, the place of the pogonophorans within the evolutionary scheme of the animal kingdom.

Riftia's taxonomy remained a source of confusion for some time to come. Four years after classifying the vestimentiferans as an order of the Pogonophora, Jones decided that they should be considered a phylum in their own right, a proposal that many found exciting – the discovery of a new phylum is a rare and dramatic event. Unfortunately, most of Jones' worm taxonomist colleagues were ultimately unconvinced.[33] But confusion over the place of the vestimentiferans and pogonophorans in the evolutionary scheme extended far deeper.

Animals above the level of the coelenterates (jellyfishes) evolved into two major lineages, known as the protostomes and deuterostomes; the former included the major invertebrate phyla (the annelids, mollusks, and arthropods), and the second led to the chordates, which include the vertebrates. These evolutionary lineages differ in their fundamental body plans and embryological development, differences that are apparent when comparing, say, the vertebrates with the annelids, mollusks and

Riftia pachyptila, the vestimentiferan rift tubeworm discovered at the East Pacific Rise. The tubeworms grow to 2 meters tall. Their characteristic red plume color comes from their unique hemoglobin, which binds and transports both hydrogen sulfide and oxygen as the plume is bathed alternately in water from the vent and the ambient deep sea. (Courtesy of Kim Juniper, Université du Québec à Montréal.)

arthropods, but are less clear for certain groups, such as the pogonophorans. To which group did they belong? One early taxonomist classified them with other plumed polychaete tubeworms in the Annelida.[34] But by the time *Alvin* discovered the vents, the consensus was that the pogonophorans were a curious offshoot of the evolutionary tree of life, more closely related to the vertebrates (including ourselves), along with some other obscure worm-like groups, such as the hemichordates. Pre-eminent invertebrate zoologist Libbie Hyman stated unequivocally in her authoritative 1959 opus *The Invertebrates*: 'It is not open to doubt that the Pogonophora belong to the Deuterostomia.'[35]

Perhaps there would have been a shadow of a doubt if she had seen an entire pogonophoran – the slender worms easily fragmented, and not until 1964 was one collected with the segmented posterior portion of its body intact, complete with setae, reminiscent of a polychaete. But texts into the 1990s continued to place the pogonophorans within the deuterostomes, albeit puzzling over their mixed characters.[36] These issues were not resolved until the late 1990s, when the use of molecular taxonomy (of which more below), combined with further studies of their embryology and morphology, showed that *Riftia* and other vestimentiferans were in fact closely related to the pogonophorans, and both were subsumed with most other marine worms within the polychaetes.[37]

But to return to the 1977 voyage of discovery, vents at that time were primarily of interest to geologists and geochemists, and there was not a single biologist on board. Not until 1979 were biologists able to obtain the use of *Alvin* to explore the vents themselves. But even if the geologists and chemists on board could not identify the giant clams and tubeworms, they recognized the magnitude of what they had discovered. As Jack Corliss, one of the chemists on board, told David Perlman of the *San Francisco Chronicle*, who joined the second half of the cruise, 'I think that what we are finding here will prove to be the greatest discovery in the history of benthic biology since the discovery that life was even possible in the deep sea.'[38]

Chemosynthesis and symbiosis in the deep sea

At the base of all life on Earth are primary producers, capable of 'fixing' organic carbon; that is, able to convert inorganic carbon in the form of carbon dioxide into organic compounds, such as sugars. This process requires energy, and all plants and algae use the energy of sunlight to drive photosynthesis. But there are other sources of energy available for primary production, most notably chemical energy, in the form of reduced compounds, such as hydrogen sulfide (H_2S) or methane (CH_4). That certain microbes utilize the chemical energy released in oxidizing such compounds to fix carbon – a process known as 'chemosynthesis' to distinguish it from photosynthesis – has been known since the 1880s, but until the discovery of the hydrothermal vents, the process was not considered ecologically significant, in that it did not seem to support complex ecosystems containing higher organisms.[39]

Micro-organisms have other rather remarkable capabilities. In 1969 Tom Brock, a microbiologist at Indiana University, discovered a bacterium, *Thermus aquaticus*, thriving in Yellowstone National Park's hot springs at temperatures of 70–75°C (158–167°F) – temperatures previously considered inimical to life.[40] Microbiologists soon discovered microbes – so-called extremophiles – living under a variety of the most exacting conditions on the planet: in salt lakes, in acidic and alkaline environments, kilometers under the surface in oil wells, indeed almost anywhere they chose to look. Several scientists, including John Isaacs, suggested that the geologists and chemists sample for thermophiles at the hydrothermal vents during the 1977 voyage.[41] But no one suspected that invertebrates at the vents would have evolved symbiotic relations with chemosynthetic bacteria, much the way tropical corals have symbiotic algae (zooxanthellae) living within them to enhance their nutrition.

Prior to *Riftia*, pogonophorans were typically recovered from organic-rich, anoxic sediments, so it was speculated that they absorbed organic compounds directly through their tubes and epidermis – a plausible option, perhaps, for a worm no thicker than a pencil lead, but not for the robust, 2-meter-high *Riftia*. How this giant worm grew and sustained itself perplexed deep-sea biologists for several years. In one of those wonderful, if all too rare, moments in science, the answer came in a flash of insight to a first-year graduate student, Colleen Cavanaugh, during the course of a rather dull seminar. Cavanaugh had had research experience working with corals, and in 1980 she was attending a series of informal seminars on the rift system. Holger Jannasch, a Woods Hole microbiologist, spoke about the high densities of bacteria in the vent fluids and suggested that they metabolized hydrogen sulfide. Jones, the Smithsonian curator, then gave a talk about *Riftia*. After puzzling over the question of its nutrition, Jones showed a series of slides of the worm's anatomy. In a cross-section from *Riftia's* trophosome (or feeding body), the central organ in its trunk, Jones pointed out yellowish sulfur crystals. At the time, the trophosome was speculated to be detoxifying the hydrogen sulfide by oxidizing it to sulfur, thereby enabling the worm to survive in the vent environment. Cavanaugh raised her hand and rose to her feet.

'Really, it's clear,' she reportedly said. 'These worms must have bacteria inside them, just as there are bacteria outside them, in the vent fluids.'[42]

Cavanaugh went on to show in her PhD research that the trophosome was devoted to cultivating grape-like clusters of specialized chemosynthetic bacteria.[43] The tubeworm supplied its symbionts with the raw materials for chemosynthesis – hydrogen sulfide, carbon dioxide, oxygen, and nitrogen – and the bacteria nourished the worm with their surplus production. Subsequently, this symbiotic relationship was shown to have evolved independently in several invertebrate groups – indeed in all groups that are dominant at the vents – the vestimentiferan tubeworms, giant clams (*Calyptogena* spp.), vent mussels (*Bathymodiolus* spp.) and a few other bivalve families, the so-called hairy snails (family Provannidae) dominant at vents in back-arc spreading centers in the western Pacific, and the swarming shrimp (*Rimicaris exoculata*) dominant at mid-Atlantic vents (Plates 4–7). The mollusks generally house their symbiotic bacteria in modified gill structures, while the shrimps cultivate bacteria on their carapaces. However, the symbiotic relationship is most highly evolved in the vent tubeworms, which, without a mouth or gut, are entirely unable to feed independently.

The key evolutionary hurdle in the evolution of this symbiotic relationship is the invertebrates' requirement for oxygen, no different from that of other higher life-forms, and the bacterial requirement for the chemosynthetic building blocks, including hydrogen sulfide and carbon dioxide as well as oxygen. These invertebrates must therefore live at the dynamic interface between the anoxic, sulfide-rich vents and the oxygenated ocean waters, whose oxygen is a product of photosynthesis. But how can the organism deliver both oxygen and hydrogen sulfide without poisoning itself? Hydrogen sulfide is highly reactive and normally poisons the oxygen-binding sites in the blood, which is the reason that most organisms in the deep sea cannot tolerate the conditions around hydrothermal vents.

The tubeworms evolved a unique hemoglobin, which in addition to its oxygen-binding site contains a special site for binding sulfide, rendering it inactive until it is delivered to the bacteria within the trophosome.[44] The tubeworms must therefore grow directly around the vent openings, since their hemoglobin-rich plumes must be bathed alternately in water rich in hydrogen sulfide and oxygen. The giant clams, on the other hand, grow within cracks in the basalt around the vents, so their foot reaches down into the sulfide-rich vent water, taking up hydrogen sulfide, while their siphon reaches into the overlying well-oxygenated water column. The clams have not evolved specialized hemoglobin but have a special sulfide-binding factor in their blood. They also have a reduced filter-feeding apparatus and digestive system and little ability to feed independently: when a vent shuts down, they soon die. The vent mussels, on the other hand, are least specialized for this symbiotic relationship, and they filter bacteria and other particles from the water column, as well as receiving nutrition from symbiotic chemosynthetic bacteria living in their gills. As a result, they survive the longest when a vent field shuts down.[45] But the tubeworms compensate for their greater dependence on the vent environment by their ability to achieve rates of productivity 10–100 times higher than the giant clams or vent mussels.[46]

Although the vent habitat is dominated by species that host chemosynthetic bacteria, an entire food web is gathered there, including benthic and pelagic suspension feeders, grazers, predators, and scavengers – but virtually all species are physiologically adapted to vent conditions. Chemosynthetic microbes are at the base of the food chain, including free-living chemosynthetic bacteria, some of which are spewed into the water column, while others form mats on the bottom. Those in the water are fed upon by species of copepods that hover over the vents, and a host of benthic suspension feeders: specialized anemones, a remarkable siphonophore (related to the Portuguese man-of-war) that holds to the bottom with some of its tentacles, enteropneust 'spaghetti' worms, serpulid tubeworms, barnacles, and a primitive filter-feeding limpet, *Neomphalus fretterae*.

Various limpets, gastropods, crustaceans, and polychaetes graze the bacteria on the bottom, while crabs, fish, and polychaetes feed as predators and scavengers.[47] The vents thus stand like small, highly productive oases in the vast expanse of the deep sea, each site typically ranging from a few tens to a few thousand square meters in size – from the size of a living room to a few football fields, as former *Alvin* pilot and vent researcher Cindy Lee Van Dover described them[48] – with perhaps several strung out along a section of ridge to form a field.

Physical conditions at hydrothermal vents vary enormously. The vent sites initially explored in 1977 on the Galápagos Rift, just north of the equator, had maximum temperatures of 10–17°C (50–63°F) – warmer than the ambient temperature at the deep seafloor – 2°C (36°F) – but hardly a 'hot' spring. The following year *Alvin* explored along the East Pacific Rise at 21° N latitude near the Gulf of California and encountered its first 'black smokers,' so-called because the 350°C (660°F) water pouring from them still carried iron and other metal sulfides in solution, which precipitated out as the vent fluid mixed with the frigid seawater, forming a dense black plume. The black smokers develop characteristic chimneys and mounds, built up from anhydrite (calcium sulfate) and the various sulfides precipitated from the plume of hot vent water. The 'Godzilla' smoker chimney on the Juan de Fuca Ridge was 45 meters high before it toppled,[49] and new chimneys can grow 5 meters in a year.[50] Where the vent water temperatures are somewhat cooler (100–300°C), the metals and sulfides will have already been lost from solution, so 'white smokers' form, as whitish minerals (silica, anhydrite, and barite (barium sulfate)) next precipitate.

A sketch of black smokers at the East Pacific Rise, showing clumps of giant tubeworms growing adjacent to the vents and giant clams along fault lines where they can access the vent fluids underneath with their foot, while their gills remain bathed in well-oxygenated bottom water. (Courtesy of Kim Juniper, Université du Québec à Montréal.)

The alvinellid polychaete worms, also known as Pompeii worms, which build their tubes into the sides of black smoker chimneys, are marine biologists' candidate for the 'hottest' invertebrates on the planet. Their open-ended tubes, designed to draw in cooler (20°C) water at one end as heated water exits the other, are routinely bathed by vent waters at 30–70°C and may experience a 60°C range in temperature from one end of their body to the other.[51] One was observed to survive brief immersion in vent water recorded at 105°C (221°F)![52] Alvinellid polychaetes sport a white 'fur coat' of filamentous episymbiotic bacteria on their backs, which are large enough to be seen with the naked eye. However, the worms are also able to operate as deposit feeders, presumably eating bacterial mats, and have a fully functional digestive system. As they live mostly within their tubes, their ecology remains largely a mystery.

Vents not only present a huge range of conditions, they are also unpredictable, highly unstable environments, in marked contrast to the rest of the deep sea. At fast spreading centers, such as the East Pacific Rise, individual vents have lifetimes on the order of only decades, although the overall vent field remains active far longer.[53] Researchers have returned to once-thriving vent sites to find them inactive and the tubeworms and other sessile vent fauna dead and dying; other sites, once re-activated, are rapidly recolonized. Several volcanic eruptions have now been monitored, one 9° north of the equator on the East Pacific Rise and another on the Juan de Fuca Ridge (46° N) off Washington. Unlike after volcanic eruptions on the land or when new volcanic islands are formed, under-sea recolonization is rapid and dynamic. Existing vent communities are obliterated by the lava flows, but spectacular blooms of sub-surface bacteria soon follow. Under these conditions, the vents are commonly described as 'snow blowers,' with 50-meter-high plumes of white, flocculent sulfur-rich filaments, producing a whiteout for the *Alvin* divers. The filaments, produced by a vibrioid bacterium, form a mat around the vents. Within a year, tubeworms and other vent megafauna are observed to be thriving. Tubeworm species appear to be the first to establish themselves at new sites, but whether species succession at the vents is ordered or opportunistic remains unclear.[54] Vent sites on the slowly spreading mid-Atlantic ridge last somewhat longer. One vent field appears to have been there for 125 000 years, but activity is episodic, with each episode of activity lasting in the order of decades to centuries. The communities are also more stable on the mid-Atlantic Ridge, at least on the time-scale that they have been studied to date.[55]

The instability of hydrothermal vents is surely another barrier to colonization of the vents by most deep-sea organisms. To be successful within the vent ecosystem requires the evolution of a life-history strategy directly opposite to the strategy required to survive in the abyssal deep sea: rapid growth, rapid maturation, a high reproductive output, and the ability as larvae to find and colonize new vent sites. Of the more than 400 invertebrate species now recorded from some 30 vent sites, mostly in the Atlantic and Pacific, only 7 percent occur outside this habitat.[56]

Although the species at hydrothermal vents are spectacular, unique, and often found in great abundance, their diversity is notably low. Typically, hydrothermal vent sites are dominated by one or two species, which typically account for 70–90 percent of the organisms. The low diversity at vent sites is reminiscent of other productive but unstable ecosystems, such as estuaries or intertidal habitats, which are found at the interface between two very different environments;

but it's in marked contrast to most deep-sea communities. Samples of deep-sea soft sediments, for example, characteristically contain many species, none of which is notably dominant and virtually all of which are in low abundance (see Chapter 4).

The biogeography of the vent fauna is influenced by the unique distribution of their underlying habitat, as well as by the species' dispersal capabilities and the long-term geological history of the ridge environment. Hydrothermal vents are found along mid-ocean ridges and in so-called back-arc basins behind the ocean trenches, where volcanic activity is associated with seafloor subduction.[57] The mid-ocean ridge system snakes its way for 75 000 kilometers through all the ocean basins – unquestionably one of the planet's most notable geological features, though hidden from our view. Little of it has been explored to date, particularly in the Southern Hemisphere (the ridge systems of the Southern and Indian Oceans and the South Atlantic). Although new species are discovered as each new site is explored, certain underlying patterns are already apparent.

Vent species are mostly sessile as adults, so they depend on the larval stage to disperse. The larvae drift with the currents, which tend to follow the ridge contours near the seafloor, or they may be carried aloft with the hydrothermal plumes. Faunal boundaries are therefore found wherever there are significant breaks in the ridge topography, with the difference between faunas a function of the degree of separation and its history. Where the discontinuity is relatively small, such as between the East Pacific Rise and the ridges of the northeast Pacific (for example, the Juan de Fuca Ridge), the species largely differ, but the genera are mostly held in common. (The East Pacific Rise now terminates within the Gulf of California; it separated from the northerly ridge system about 30 million years ago.)[58] Vents in the back-arc basins of the western Pacific lost their links with the eastern Pacific about 43 million years ago. When first explored, most species there were new to science, but about half belonged to genera known from vents elsewhere. Interestingly, a snail, the hairy gastropod *Alviniconcha hessleri*, replaced the tubeworms, clams, and mussels as the dominant organism containing symbiotic chemosynthetic bacteria. The faunas are almost entirely unrelated, on the other hand, between Atlantic and Pacific vent sites. Tubeworms and vesicomyid clams are notably absent from the mid-Atlantic Ridge, replaced by the swarming shrimp, *Rimicaris exoculata*, which hosts dense populations of chemosynthetic bacteria on its carapace.[59] Only about 40 percent of genera are shared between the Atlantic and Pacific vents.[60]

The archaea and the origin of life

The year of the discovery of deep-sea hydrothermal vents off the Galápagos, 1977, also marked a paradigm shift in taxonomy, a scientific revolution with ramifications that soon extended to our understanding of vent ecosystems and the diversity and evolution of life on Earth.

Since antiquity, in a tradition that goes back to Aristotle, living things have been classified as either plants or animals, a

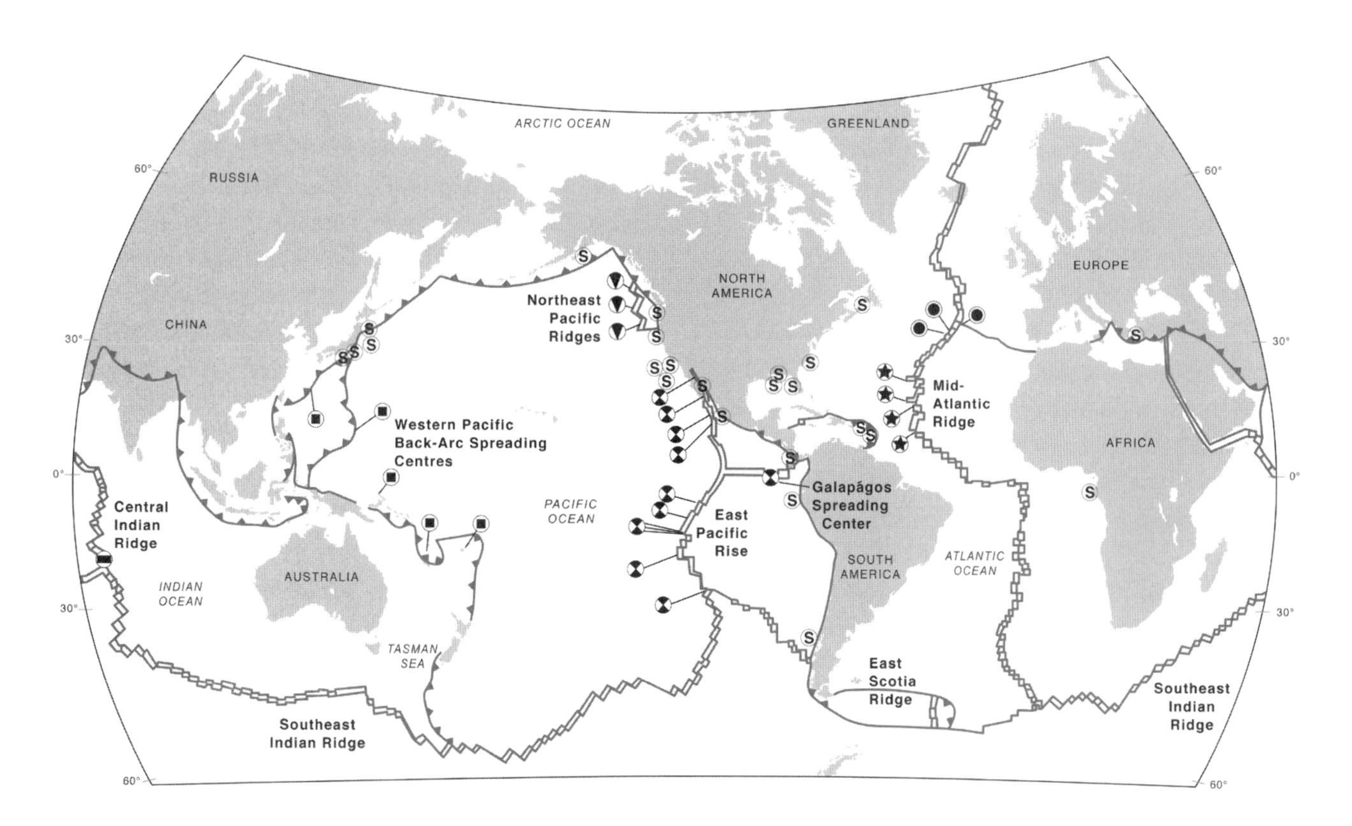

Distribution of known vent and seep sites. The vent sites occur along the ocean's ridge and spreading centers, and different biogeographic provinces are indicated with the different symbols. The seeps (S) occur predominantly along the continental margins. (Modified from van Dover et al., 2002 and Sibuet and Olu, 1998.)[n]

tradition formalized by Linnaeus, the father of taxonomy, in the 18th century.[61] Not until the end of the 1960s was it clearly recognized that the Fungi, Protists (or protozoa), and Monera (or bacteria) represented distinct kingdoms that could not be subsumed under the plants or animals.[62] However, the focus for taxonomy and evolutionary biology in this revised scheme remained on the flowering of diversity in the four kingdoms of higher organisms, the eukaryotes, whose cells all have a distinct nucleus, containing their genetic material. The bacteria or prokaryotes (meaning primitive or pre-nuclear), do not have a distinct nucleus within their cells and were of interest mostly as the basal group, from which the eukaryotes evolved.

The study of taxonomy was predominantly morphological until the 1970s, based primarily on organisms' visible structures. Evolutionary biologists recognized that this approach worked best with the higher organisms, whose structures were large enough to view; bacteria, typically about a micron in size (a thousandth of a millimeter), are too small for us to see much more than their general shape, even under a microscope. The morphological approach suffers as well from the problem of convergent evolution, namely that similar structures – say wings, fins, or a vermiform body plan – may evolve independently and not indicate an evolutionary relationship. Taxonomists, therefore, began to look at the molecular sequences within fundamental intracellular structures that are common to all organisms in order to assess evolutionary relationships.

Not surprisingly, microbiologists were among the first to develop and use these molecular tools, since they had so little else to work with. Their field was also in particular disarray. For such apparently simple organisms, microbes display remarkably diverse metabolic pathways and capabilities, far more diverse than the higher organisms. Plants, animals, and other eukaryotes, for all their morphological diversity, actually have very few ways of supporting themselves. Either they are primary producers, using light energy for photosynthesis, or they consume organic matter (living or dead). (Some eukaryotes live through a combination of the two by forming a symbiotic relationship with algae, as in tropical corals and lichens, or with bacteria, as in the vent fauna.)

The prokaryotes, on the other hand, include photosynthesizers (for example the so-called 'blue-green algae,' or cyanobacteria) and consumers: the familiar bacteria associated with disease and rot. But in addition, prokaryotes also obtain energy from a great variety of inorganic compounds, and they underlie many of the planet's biogeochemical cycles, generating and consuming methane, the various forms of sulfur (sulfides,

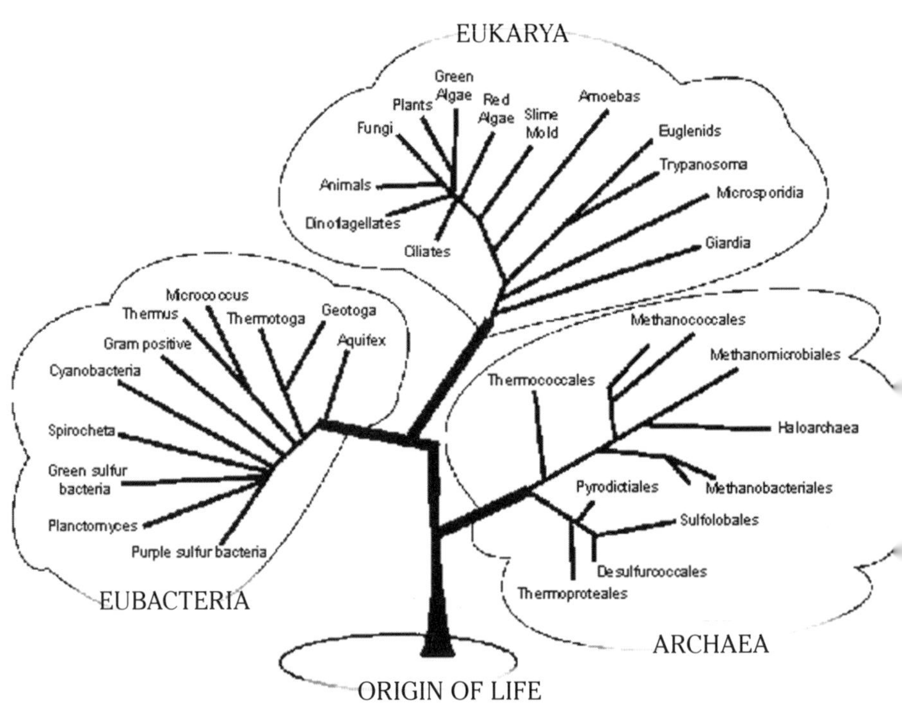

The phylogenetic 'tree of life' showing evolutionary relationships among living organisms based on their rRNA nucleotide sequences.[o] Note the relatively short genetic and evolutionary distance between the traditional plant and animal kingdoms. The far greater genetic diversity found elsewhere in the eukaryote domain and elsewhere among the bacteria and archaea reflects the approximately 3.5-billion-year history of prokaryotic life on earth and the 2-billion-year history of eukaryotes; the plant and animal phyla, however, only appear less than 600 million years ago in the fossil record, at the dawn of the Cambrian Period. (From Paul Koch, University of California, Santa Cruz, modified from Lunine JI, 1999, *Evolution of a Habitable World*, Cambridge University Press.)

elemental sulfur, and sulfate), and nitrogen (ammonia, urea, nitrogen, nitrite, and nitrate), as well as oxidizing and reducing metals (for example, iron and manganese in manganese nodules). Some microbes also flourish under conditions previously considered well beyond the physiological limits of life: thermophiles thriving within hot springs in Yellowstone National Park at temperatures of 70°C (158°F) and above; halophiles that make their home in the Great Salt Lake or the Dead Sea; acidophiles and alkilophiles; micro-organisms recovered from deep-drilled rock strata and oil wells. How are these all related to each other? Simple classifications based on morphology or metabolic potential proved unable to find order in the apparent chaos of microbial diversity.

In 1977, Carl Woese, a microbiologist at the University of Illinois, published two short related papers based on the sequences of nucleotides in a class of RNA (ribonucleic acid) in the ribosome known as 16S rRNA, which is associated with protein synthesis in all living things.[63] Ribosomal RNA is conservative, meaning that it is subject to a low mutation rate, so the similarity in its structure between two species provides a measure of how closely related they are – indeed, with some measure of the mutation rate, one can estimate the length of time since the species diverged. This molecular approach provided biologists with a powerful new tool to assess evolutionary relationships across the full panoply of living organisms, including the universe of micro-organisms.

Woese's results led to a paradigm shift in biology that, though less well known than the Copernican or Darwinian scientific revolutions, may have implications for our everyday anthropocentric worldview that are just as profound. Although Copernicus showed that the Earth was not the center of the universe, and Darwin that mankind was not the product of special creation, many still considered that mankind was center-stage in the evolutionary drama – that evolution was really all about the long slow march from fishes to amphibians to reptiles to mammals that finally culminated with human beings. But Woese's results showed that most evolutionary diversity was not among the plant and animal kingdoms, so obviously arrayed around us in all their glory. Indeed, man and mouse, plants, and fungi (all eukaryotes) proved to be far closer in terms of their evolutionary lineage than were the typical bacteria and a group of prokaryotes, including the methanogens and a number of extremophiles – halophiles and thermophiles – that Woese called the archaebacteria (later, the archaea). Remarkably, their genetic sequences indicated that they were no more closely related to the bacteria than the bacteria were to the eukaryotes. Living organisms were not divided into two fundamental branches, the prokaryotes and eukaryotes; rather, there were three principal *domains* (as Woese termed them): the eukaryotes, the bacteria, and the archaea, which separated early in the evolution of life. Rather than five kingdoms, a current phylogenetic tree of life might indicate dozens, and all but three are unicellular. As Norman Pace of the University of California at Berkeley expressed it, 'animals, plants, and fungi constitute small and peripheral branches of even eukaryotic cellular diversity.'[64]

Hydrothermal vents and the origins of life

One of the most interesting features of the molecular-based phylogenetic tree is that the thermophiles – mostly archaea but there are also some thermophilic bacteria – appear at its base, implying that the earliest prokaryotes were thermophiles. This fact, combined with the discovery of chemosynthetic archaea and thermophilic bacteria flourishing within hydrothermal vents under conditions that have changed little since the early evolution of the planet, naturally raised the question whether life itself evolved under those conditions.

Charles Darwin once spoke of life originating in a 'warm little pond.' In 1953, Stanley Miller, working with Harold Urey at the University of Chicago, modernized Darwin's speculation, showing that amino acids and other organic molecules could form spontaneously under conditions then believed to have prevailed at the Earth's surface during its early history, namely a 'reducing' atmosphere composed largely of methane, ammonia and hydrogen.[65] Miller and Urey hypothesized that the early ocean would have formed an 'organic soup,' which led to the formation of life. The earliest life-forms thus would have subsisted on this 'organic soup.'

The 'organic soup' scenario for the origin of life seems less likely today. The original reducing atmosphere was apparently lost early in the planet's history and replaced by oxidized volcanic gases – nitrogen, carbon dioxide, and water vapor – that would not support the prerequisite organic reactions.[66] Furthermore, the Earth was subject to an intense asteroid bombardment during its early history, sufficient to vaporize much of the ocean and destroy all near-surface life.[67] This bombardment ended about 3.8 billion years ago, but fossilized complex life-forms, apparently cyanobacteria, are found relatively soon thereafter in the oldest sedimentary rocks, which date back to about 3.5 billion years ago.

Soon after returning to Oregon State University from the historic 1977 voyage to the Galápagos vents, Jack Corliss and John Baross (the latter was working with some of the first samples of thermophilic microbes), and a graduate student, Sarah Hoffman, first intuited that life may have originated in this newly discovered habitat. Hydrothermal vents are an ancient habitat, extending back to around 4.2 billion years ago, when the oceans first formed; in fact, hydrothermal activity was probably at least five times more prevalent then. (The radioactivity in the Earth's core, and hence its geothermal energy, has decayed considerably over the past 4 billion years.) Notably, the vents provide a continuous, abundant supply of energy-rich reduced compounds over a wide range of thermal conditions. Was it only coincidence that the archaea, the most primitive organisms on the planet, and thermophilic bacteria, apparently the most ancient bacterial lineage, favor this habitat? Corliss et al. hypothesized that the archaea and thermophilic bacteria evolved within the vents and have lived there ever since. During the Earth's turbulent early history, the vents provided a refuge for early life-forms, from which they emerged and further evolved once conditions were suitable. Based on this view, those seemingly exotic, only recently discovered chemosynthetic metabolic pathways may be the most ancient: life did not originate near the surface of the Earth to consume an organic soup or harvest the sun's energy; rather, it first evolved on the seafloor and subsisted on the geothermally-based chemical energy welling up from deep within the planet itself.[68]

A key stumbling block in early theories of the origin of life has been the transition from organic precursors – an organic soup of amino acids and other organic compounds – to living cells complete with membranes, internal structure, and more complex biochemical machinery. In the late 1980s, Günter Wächtershäuser, a Munich patent attorney with a 1965 PhD in organic chemistry, published a series of papers that posed an interesting potential solution to this puzzle, again pointing to the vents as the cradle of life. Iron pyrite (iron sulfide, FeS_2), commonly known as fool's gold, is deposited in abundance within the hot sulfidic vent environment. Interestingly, it has a slight positive charge and therefore attracts and binds negatively charged ions, such as carbonate, phosphate, and sulfide ions. Held to the pyrite surface by a backbone of these charged ions, longer-chained organic compounds would form naturally. These simple precursors may well have gradually evolved simple autocatalytic metabolism, self-replication, membranes, and so on, all as a surface film on the iron pyrite, before evolving into structures able to exist independently. It is suggestive that RNA and DNA, and many metabolic processes, are based on a backbone of phosphate or sulfide ions, which may once have been held in place on a pyrite surface.[69]

What are the limits to life?

The archaea and certain bacteria are now known to inhabit environments previously believed to be barren of life, particularly the porous hot rocks within hydrothermal vents and deep within the Earth's crust. At this time, so-called hyperthermophiles have been isolated and cultured from vent fluids that grow at temperatures of up to 110°C (230°F). However, high levels of particulate DNA, presumably from microbes, have been obtained within the core of black smoker plumes measured at around 350°C, suggesting that the vents may be 'windows to a subsurface biosphere.'[70] The evidence from vents, combined with the nearly ubiquitous presence of organic molecules indicative of prokaryotes obtained from deep sedimentary rocks, led the maverick astrophysicist Thomas Gold from Cornell University to speculate that there was a deep, hot biosphere that extended 5–10 kilometers into the Earth's crust. Based on a simple back-of-the-envelope calculation, he suggested that the microbial biomass beneath the Earth's surface may be comparable to that of the combined flora and fauna living above it![71]

Regardless of its extent, the clear presence of life within the Earth's crust has added a new dimension to the search for extraterrestrial life. The potential for finding life elsewhere in our solar system was extremely remote when it was restricted to organisms that required adequate sunlight and a relatively narrow range of temperature conditions. However, it is far more likely that we will find chemosynthetic life-forms now that we have so dramatically expanded the horizon of temperatures, pressures, acidity and alkalinity, and chemistries under which we know life can thrive. And we do not have to travel to Mars or the outer planets to find new life-forms – we have almost certainly not yet exhausted the range of possibilities to be found within our own.

Cold seeps

Soon after deep-sea ecologists recognized the chemosynthetic basis for the flourishing oases at hydrothermal vents, they discovered cognate ecosystems elsewhere in the deep sea. A range of such environments are known as 'cold seeps' to distinguish them from the hot vents and to indicate that the reduced compounds – mostly methane and sulfides – are passively seeping from the sediment rather than being actively vented. Unlike the vents, which are found predominantly along mid-ocean ridges, cold seeps occur mostly along the margins of the continents, where there are organic-rich sediments, such that methane is generated and seeps upward to the seafloor. Not surprisingly, seeps are often found in areas rich in natural gas and oil, such as the Gulf of Mexico, where some of the first seep communities were discovered. However, they are also found at so-called 'active' continental margins, such as near trenches, where the sediments are piled up and the pore fluids squeezed out, as well as in a variety of other environments – anywhere, really, where sulfides, methane, or other hydrocarbons come to the surface of the seafloor.

Seep communities were first encountered by chance. Giant vesicomyid clams, bathymodiolid mussels, tubeworms, and polychaete worms were observed from submersibles in dense clumps on otherwise relatively barren seafloor far from mid-ocean ridges. Although for the most part the same species are not found at seeps and vents, the similarity of the faunas was immediately apparent. Also, the density of organisms at seeps ranges between 2 and 20 kilograms per square meter, comparable to the biomass found at vents (Plates 5, 6). Dependence on symbiotic chemosynthetic bacteria was quickly demonstrated: of the 211 species now recorded from some 24 deep seeps in the Atlantic and Pacific, 64 contain symbiotic chemosynthetic bacteria. Most of the seep species are endemic to one or two sites, and only 13 are held in common with the vents.[72]

Although seep and vent ecosystems are both based primarily on chemosynthesis, there are considerable differences between them, beyond the obvious temperature difference, that account for the lack of species overlap. The vent environment is extraordinarily productive but relatively ephemeral, selecting for species that grow rapidly, mature quickly, and reproduce prolifically, in order to colonize new vents. Cold seeps, by contrast, are more stable features, estimated to persist for hundreds to thousands of years – but they are much less productive. Whereas the vent tubeworm *Riftia* holds the record for the fastest-growing marine invertebrate (about a meter in a year), the seep vestimentiferan tubeworm *Lamellibrachia* holds the record for longevity of a non-colonial invertebrate, requiring 170–250 years to grow to 2 meters.[73]

The greater stability of seeps fosters the development of more diverse communities and may have even allowed for the evolution of greater species diversity. Whereas vent sites are characteristically dominated by only one or two chemosynthetic species, as many as 15 chemosynthetic species have been found at a single seep. Although seeps have been less explored than vents, there are more species of giant vesicomyid clams, bathymodiolid mussels and pogonophoran tubeworms known from seeps.[74]

Although methane rather than sulfide is the primary energy source at seeps, surprisingly few species have chemosynthetic symbionts that use methane. The only exceptions are known from the Atlantic: some seep mussels, a pogonophoran tubeworm, and a remarkable carnivorous sponge. Otherwise the seep invertebrates' symbiotic bacteria depend on hydrogen sulfide, much like the vent symbionts. Their sulfide is derived from free-living bacteria in the sediment, which use the methane as their energy source to reduce sulfate to sulfide in the sediment pore water.[75]

Whale falls

In November 1987 Craig Smith was leading an *Alvin* cruise in the Santa Catalina basin off southern California, an area already sampled several times. During the last half hour of the last dive of the cruise, the *Alvin* pilot wandered slightly off course and spotted something unusual. Stretched out on the seafloor, larger than the submersible itself, was the 20-meter-long skeleton of a blue or fin whale, festooned with clams and other fauna. Yet another chemosynthetic habitat was thus discovered![76]

As Smith and his graduate student Amy Baco went on to show, based on studies of whale bones trawled up by deep-sea fishermen, skeletons encountered by chance at the bottom of the sea, and a few dead whales towed out to sea and deliberately sunk, the lifespan of communities living on a large whale carcass on the seafloor exceeds 50 years, comparable to the lifespan of the living whale. Over this period, it hosts a succession of faunas, some remarkably diverse – 415 species have now been recorded from a handful of whale specimens – including a core group of some 29 species known from no other environment (Plates 6, 7).

When a dead whale falls to the seafloor, it first falls prey to a host of generalized deep-sea scavengers: sharks, hagfishes, macrourid (grenadier) fishes, crabs, amphipods, and the like. Depending on the size of the whale, the process of stripping the soft tissue from the bones can take between several months and 1.5–2 years, with some 40–60 kilograms of whale meat and blubber consumed per day. The organic-rich sediments and bones are then colonized for a further few years by dense aggregations of polychaete worms, some similar to those found at sewer outfalls and other sites of organic enrichment. Densities of 20 000–45 000 per square meter have been observed – the highest macrobenthos densities ever recorded below 1000 meters depth.

Thus, at the end of 3–4 years, all that remains on the seafloor are the bones. But whalebones are exceedingly rich: the vertebrae of large baleen whales are potentially more than 60 percent lipid by weight. Microbial decay of these energy-rich hydrocarbons depletes the oxygen within the bones, so anaerobic microbial decomposition takes over. Much as at the seeps, the lipid fuels the reduction of sulfate to sulfide, which is slowly released from the bones, leading to colonization by mats of chemoautotrophic bacteria and their grazers, as well as vesicomyid clams and high densities of mussels with endosymbiotic chemosynthesizing bacteria. The bones remain productive for decades (around 50 years), comparable to the longevity of hydrothermal vents. During this stage, some 185 species on average were recorded on each whale skeleton, far higher than the number observed at seep and vent habitats, making the whalebone macrofauna one of the most diverse in the deep sea. (Only about 230 species have been recorded from seeps worldwide.) To date, whale skeletons have only been intensively studied off California, so the number of species known from this habitat will rise substantially when whale falls are studied more widely. Although many of these species are known from other environments, there is a core group, including at least 29 known only from whale skeletons, that appear to have evolved specifically to live on this unique habitat.[77]

Osedax (literally 'bone devouring') is the most curious creature discovered on whalebones to date, a thread-like tubeworm (its body only 0.2–0.5 millimeters thick) with the characteristic red crown or plume protruding from its tiny tube and without a gut – but also without the trophosome where pogonophorans characteristically house their endosymbiotic bacteria. Instead, *Osedax* hosts its bacterial symbionts within a root system, which grows into the bone from the posterior end of the worm. However, its symbiotic bacteria do not utilize sulfide or methane. They are, unlike the symbiotic bacteria of all other known pogonophorans, not even chemosynthetic. Rather they belong to a different bacterial order (the Oceanospirillales) that specializes in breaking down complex organic compounds. Thus they directly consume the whalebone lipids to nourish their host. In a further unique twist, the tubeworms in the two known species of *Osedax* all proved to be female. The males were minute dwarves (in *O. frankpressi*, only 0.15–0.25 millimeters long), much resembling typical annelid larvae except that they were filled with sperm. As many as 111 males were discovered in a single female's tube. Analysis of *Osedax*'s mitochondrial nucleotides suggests that it diverged from the vestimentiferans and other pogonophorans in the Eocene, about 42 million years ago, around the time that whales evolved. These minute worms and their unique bacterial symbiotic relationship thus likely co-evolved alongside the great whales.

To understand how an entire fauna can have evolved that specializes to live on decaying whalebones, Smith and his colleagues examined the prevalence of whale falls, past and present. Off California, home to the gray whale, where they have done most of their work, they estimated that the average distance between gray whale skeletons was only about 5 kilometers, well within the dispersal capability of many larvae. Prior to commercial whaling, the density of whale falls may have been around six times greater worldwide, leading some to speculate that commercial whaling may have already led to the extinction of some species that specialized on whale falls.[78]

Large cetaceans have been extant for more than 40 million years, and during the age of dinosaurs there were ichthyosaurs and other large marine reptiles whose skeletons may have supported a similar chemosynthetic community. Fossilized mollusks similar to modern chemosynthetic species have been found associated with whale fossils extending back 30–35 million years, indicating a long co-evolution of these communities with their whale hosts.[79]

As marine ecologists became attuned to the potential for chemosynthetic symbioses, they found other habitats that support sulfide-based chemosynthetic faunas. For example, the pogonophoran species originally found in low-oxygen, organic rich sediments also host chemosynthetic symbionts. Their ecological strategy is similar to that of the vent and seep clams. Their long slender tube is buried deep in the anoxic, sulfide-rich sub-surface sediments, where they take up hydrogen sulfide, while their plume, acting as a gill, extends into the oxygenated water above.[80] Sunken wood, which can take more than 10 years to decompose, is colonized by several small (4 millimeter) mussel species that house chemosynthetic symbionts in their gills. One mussel species (*Idas washingtonia*) also densely populates whalebone. Although these diminutive mussels are in different genera from the giant mussels endemic to vents and seeps, their molecular phylogeny, derived from their ribosomal RNA, indicates that they are part of the same evolutionary lineage, distinct from non-chemosynthetic mussels. In fact, species associated with sunken wood may have been the initial stepping stone in the evolution of chemosynthetic mussels. The close similarity of rRNA sequences across all the chemosynthetic mussels also indicates that their invasion of vent and seep habitats may have been relatively recent.[81]

How ancient, in fact, are the vent, seep and other chemosynthetic faunas, and what is their evolutionary history? One of the most striking features of these chemosynthetic-based environments is the extraordinarily high level of endemism, particularly at higher taxonomic levels. Not only are more than 85 percent of vent species found only in chemosynthetic habitats, but a large proportion are not closely related to any other species in the deep sea: half the genera and some 24 families (20 percent) are known only from these environments. These endemic higher taxa range across the vestimentiferans, polychaetes, gastropod and bivalve mollusks, and crustaceans.[82] The evolution of a distinct chemosynthetic-based fauna is also characterized by the virtual absence of other groups, such as holothurians (sea cucumbers), starfish, and coelenterates (except for anemones).[83] In part this is due to the physiological stresses associated with these environments: high and fluctuating temperatures, low oxygen, and high hydrogen sulfide and heavy metal levels. Adaptation has also often involved the evolution of complex symbiotic relationships.

The antiquity of this community, however, remains an open question. Unfortunately, the fossil record is quite sketchy, with only 19 fossil hydrothermal vent sites known – mid-ocean seafloor is relatively seldom raised above sea level. The oldest known vent site, in the southern Urals of Russia, is from the Silurian Period in the Paleozoic (408–438 million years ago) and has yielded fossilized brachiopods and primitive mollusks known as monoplacophorans, groups that largely went extinct at the end of the Paleozoic, as well as tubes that resemble those of vestimentiferans.[84] Brachiopods and monoplacophorans have seldom been noted at contemporary vent sites, so clearly the entire vent community does not extend back to the Paleozoic.

Even the antiquity of the vestimentiferan tubeworms is in doubt. They are almost certainly the most ancient chemosynthetic lineage, being most fully evolved, morphologically and physiologically, in their symbiotic relationship with the chemosynthetic bacteria. When they were believed to represent a distinct phylum, there was little question that they extended back into the Paleozoic along with the other phyla. Now, however, when the pogonophorans are generally viewed as only a family of polychaetes,[85] their origin is no longer clear. The molecular genetic 'clock' based on genetic sequence data indicates that vent tubeworms probably evolved from a seep ancestor much less than 100 million years ago.[86] Are the Silurian fossils from another type of tubeworm – there are many – or is the molecular 'clock' running slow? At this time, it is difficult to say. Of course, 100 million years is still quite long ago – it takes us back to the Cretaceous within the Mesozoic Age of Dinosaurs – but it is not as ancient as some had believed.

A key issue is whether the vents served as a refuge during the great waves of extinction that periodically swept over the planet. William Newman, a Scripps taxonomist and evolutionary ecologist, first proposed this in 1985 based on the number of relict groups – living fossils – he and others discovered at the vents, some dating back to the Mesozoic or possibly the Paleozoic: several of the most primitive barnacle species (and genera) known, as well as the primitive filter-feeding limpet, *Neomphalus*, mentioned earlier.[87] Because the vent community does not depend on photosynthesis for its sustenance, it was likely spared the consequences of the asteroid impact and ensuing 'nuclear winter' that led to the extinction of the dinosaurs and many other groups at the end of the Cretaceous (Mesozoic). But the mass extinction at the end of the Paleozoic was far more severe, leading to the extinction of an estimated 95 percent of marine species. This extinction event had far greater impact on the deep sea, in particular: the oceans became strongly stratified and the deep sea became anoxic for about 20 million years.[88] The vent fauna today has a high tolerance of low oxygen, high hydrogen sulfide, and high carbon dioxide conditions, so it may have been better able to survive than most, but whether it too was ultimately overwhelmed remains unknown.

Fortunately, the molecular and fossil data are more consistent for the vesicomyid clams and bathymodiolid mussels, which appear to have evolved during the recent Cenozoic era, extending back only about 50 and 22 million years, respectively.[89] Their evolutionary pathway, like that of the tubeworms, seems to have proceeded from seep to vent environments. On the other hand, the swarming bresiliid shrimps, which seem to have adapted to the vent environment less than 20 million years ago, probably first evolved there.[90]

Less than 30 years after the discovery of the remarkable deep oases of life at hydrothermal vents, we now realize that chemosynthetic ecosystems are relatively diverse and widespread in the deep sea: on spreading centers and back-arc ridges, along continental margins, within trenches, and wherever great whales or large amounts of organic debris happen to fall. Whereas the well-known symbiosis between marine algae and metazoans is today found in only one group – the warm-water corals – symbioses between chemosynthetic bacteria and higher invertebrates have independently evolved in more than half a dozen different groups: the pogonophoran and alvinellid polychaetes, the vesicomyid clams, the bathymodiolid mussels and several other bivalve families, the bresiliid shrimps, and the hairy (provannid) snails. Regardless of the antiquity of the higher organisms presently living in these environments, the archaea and thermophilic bacteria at the base of these ecosystems almost certainly extend back billions of years, perhaps to the very origins of life on Earth. With so little of the deep ocean and virtually none of the sub-seafloor habitat explored, there is little doubt that many outstanding pieces of the puzzle of life remain to be discovered in these environments.

CHAPTER 6

Seamounts and Deepwater Coral Reefs

Deepwater corals – an introduction

For most people, 'coral' brings to mind the tropical reefs of the Indo-Pacific, Caribbean and Red Seas. And no wonder – these reefs are the most wondrous biological-*cum*-geological structures on the planet. With our snorkeling and scuba gear, we are fortunate in being able to appreciate these organisms and their associated ecosystem: like us, they are restricted to relatively shallow waters; they live in a symbiotic relationship with algae known as zooxanthellae, which require sunlight to photosynthesize. The more than 700 species of the so-called zooxanthellate stony corals have become an icon of the rich biological potential of life on our planet. But in fact, there are approximately as many species of stony corals *without* zooxanthellae: corals not restricted to shallow sunlit waters, which live mostly in the cold and dark of the deep sea. And if one considers all coral groups – the black corals, soft corals, hydrocorals, and so on, as well as the stony corals – two-thirds are deep-water species.[1]

Even marine biologists have only recently come to appreciate the incredible diversity of deepwater corals – and of the entire ecosystem associated with them, the deepwater analogue of tropical reefs. Indeed, of the 706 known species of azooxanthellate stony corals, more than 200 – almost a third of those known – have only been described since 1977.[2]

The earliest known reference to deepwater corals is from *The Natural History of Norway*, published in 1755 by the Right Reverend

Erich Pontoppidan, Bishop of Bergen. The corals, or 'sea trees' as he called them, were collected by fishermen and used both as food and 'powerful medicaments,' being 'absorbent, refrigirative, emollient, astringent and strengthening.' Coral beads worn as a necklace served as 'a preservative against the apoplexy, the plague and other contagions.'[3] The Right Reverend was almost certainly referring to *Lophelia pertusa*, the dominant reef-forming deepwater coral of the North Atlantic, which Linnaeus himself described in 1758 in the 10th edition of his *Systema Naturae*. Norwegian fishermen have long known about these reefs, where long-line catch rates of deepwater fishes, such as redfish, ling, and tusk were several-fold higher than elsewhere.[4] Still, the extent of Norway's deepwater reefs was only realized in the last 25 years, following extensive acoustic and video surveys of the seafloor by Statoil (the Norwegian state oil company) in preparation for laying pipelines to its offshore oil fields. The Norwegian reefs, found mostly between depths of 200 and 400 meters, are on average 200–500 meters wide, reach up to 35–60 meters in height, and cover about 1500 square kilometers of the continental slope.[5] A veritable Great Barrier Reef of the deep!

Deepwater corals – along with the sponges, anemones, sea pens, hydroids, bryozoans, barnacles, tunicates, crinoids (or sea lilies), and many of the brittle stars and brisingid starfish living in association with them – are suspension feeders, filtering or capturing plankton and detritus 'suspended' in the water. This is a third major deep seafloor mode of consumption, in addition to the deposit feeding of most of the fauna living within the sediments and the chemosynthetic symbiosis present at vents and seeps. Suspension feeding is best suited to relatively rich or energetic waters, where prey is concentrated or swept past the organisms in sufficient quantity. But where are these conditions to be found in the deep sea?

Seamounts

Until very recently – the last 15 years or so – ecologists generally considered that, except for the weird and wonderful world of vents and seeps, the deep sea consisted almost entirely of vast sedimented slopes and plains, populated by a sparse megafauna and a diverse but Lilliputian macrofauna that fed mostly on the sediments. Although additional environments were recognized – deep canyons that cut across the continental slopes, and a veritable maze of ridges, banks, and seamounts that variously bisected and dotted the deep-sea landscape – these were extremely difficult to sample. Such rugged topography was a veritable graveyard for the trawls and dredges that biologists dragged blindly across the seafloor. 'Untrawlable' was the catchword for such ground. These environments had been sampled from time to time since the *Challenger* expedition, and they had yielded some striking material, including corals the size of small trees. But the risks associated with sampling these habitats were sufficiently extreme that biologists remained content for the most part to pass them by, with the tacit assumption that such environments were of marginal significance to the ecology of the deep sea.

When I first arrived in Hobart in July 1989 to work on the still-developing Australian orange roughy fishery, the CSIRO (Commonwealth Scientific and Industrial Research Organization) deepwater fishery group had just completed a two-year trawl survey of the continental slope around Tasmania. Large sections of the slope were dotted with seamounts; deemed untrawlable, they were passed over during the survey. Unfortunately for the group's credibility, just as it duly reported its estimate of the region's orange roughy abundance – a modest 15 000 tonnes or so – based on the trawl catch rates, fishing boats began returning to Hobart so low in the water that their decks were almost awash, their holds filled with orange roughy. The fishermen had found orange roughy spawning on one of the local seamounts and, given their value – a boatload of about 100 tonnes was worth around a million dollars – they had learned to trawl the untrawlable. The biomass of orange roughy on this single seamount was clearly far greater than the CSIRO estimate for the entire region; soon it would be apparent that some 90 percent of the fish were living on and around the region's seamounts. No longer could biologists assume these environments to be insignificant; no longer could they simply pass them by (Plates 8–16).

Seamounts are submerged volcanoes rising some hundreds to thousands of meters above the seafloor, mostly extinct but a few still active, such as Loihi Seamount off Hawaii. When a submarine volcano rises above the sea surface, it forms an island; those that never reach sea level or that sink back beneath the waves are seamounts. Reflecting their volcanic origin, seamounts are typically steep-sided, with downward slopes, or declivities, of up to 25–50 degrees, and many are conical, although others are elongate.[6]

Seamounts were first sampled during the *Challenger* expedition, but those early naturalists, with only a lead-line to laboriously obtain single-point soundings in the open ocean, had no way of knowing the shape of these relatively shallow features – and no way of knowing what they were. Their rare chance encounters with seamounts, however, yielded an extraordinarily rich fauna; so distinctive, in fact, that these were the only stations the naturalist Moseley singled out to describe in his memoir of the voyage:

> We obtained the same animals from the depths in the most widely separated places over and over again, with tedious reiteration. There were, however, one or two localities which we hit upon which are worth referring to, because they are especially rich in deep-sea forms, and because these occur there in comparatively shallow water.[7]

The discovery of isolated submerged mountains rising some thousands of meters from the deep seafloor had to await the invention of the echo sounder by the US Navy in 1919. By 1925, the US Coast and Geodetic Survey had adopted the instrument and was soon regularly discovering, naming and charting seamounts along its survey lines, particularly in the waters of the northeast Pacific.[8] However, recognition of them as major seafloor features only followed the wide-spread use of echo sounders across the Pacific during World War II. Two outstanding marine geologists of the 20th century, Harry Hess and Henry Menard, best known for the part they played in developing the revolutionary theory of seafloor spreading and plate tectonics, earlier played key roles in bringing seamounts to the attention of the marine scientific community.

As noted in the previous chapter, Hess, a Princeton geologist who served as a naval commander in the Pacific during World War II, was notorious for his seemingly obsessive collection of seafloor data, well beyond the call of his naval duties. Shortly after the war, he published a seminal paper, 'Drowned ancient islands of the Pacific basin,' in which he advanced a remarkably prescient theory to explain the curious flat-topped seamounts that he encountered, as he put it, during his 'random traverses incidental to war-time cruising.' Although the theory of plate tectonics, with its comprehensive understanding of continuous seafloor creation, spreading, and subsidence was still more than a decade away, Hess recognized that these seafloor features, though typically 1000–2000 meters beneath the sea surface, must have originated as volcanic islands, whose peaks were subsequently eroded by wave action before the islands subsided. He named them 'guyots,' in honor of the first Princeton geology professor, Arnold Guyot.[9] We know today that seamounts are often formed where a mid-ocean 'hotspot' pushes up a series of islands and seamounts as the ocean plate rides over it, much the way a machine might extrude soft-whipped ice-cream into cones on a moving conveyor belt. As the plate moves along and settles, the volcanic islands often subside beneath the sea surface, leaving a chain of seamounts

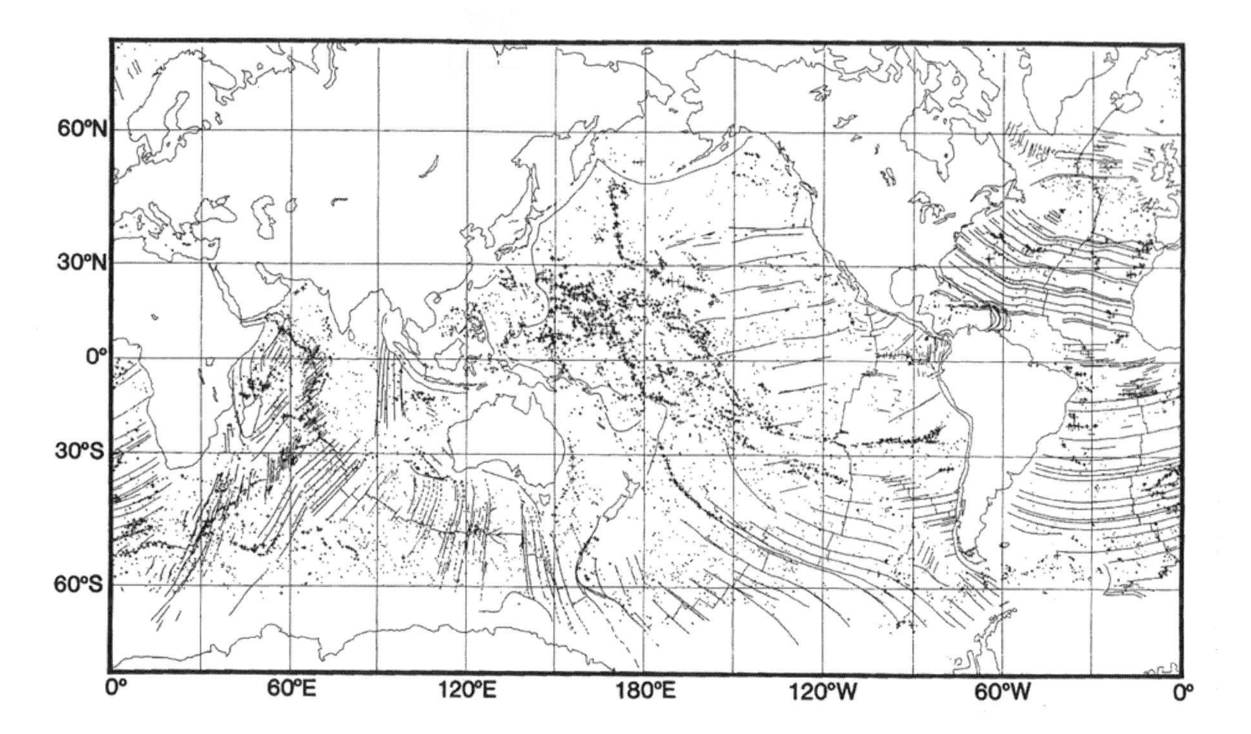

The global distribution of seamounts, based on ships' echo-sounder observations and satellite observations of gravity anomalies for large seamounts. (From Smith, 1991.)[P]

and guyots trailing behind the currently active volcanic island. The Hawaiian Islands and their associated seamount chain are a classic example. Hess's guyots supplied some of the first tantalizing evidence of large-scale seafloor movement.

But how important are seamounts to the geology and ecology of the oceans? Hess had concerned himself with guyots, a particular subset of seamounts; he had encountered only 20 and found 140 more on the available charts. How common were seamounts generally?

Henry Menard of the Scripps Institution of Oceanography took it upon himself to determine this, taking the straightforward but not necessarily obvious route of asking those who knew: the US Navy and local fishermen. Surprisingly, his persistence paid off, and he was able to learn from these most secretive groups the positions and descriptions of seamounts in their primary areas of operation in the northeast Pacific. Extrapolating from the number of seamounts along a mere 7400 kilometers of sounding lines between Mexico and Hawaii, Menard estimated that there were about 10 000 seamounts greater than 1000 meters in height in the whole of the Pacific Ocean.[10]

Today we realize that Menard's estimate, which might at first seem rather high, was in fact far too low. The region for which Menard could obtain data didn't turn out to be representative of the ocean as a whole; in fact, seamounts are most highly concentrated around the island chains of the western and South Pacific. Present estimates are closer to some 30 000–50 000 or more seamounts greater than 1000 meters high in the Pacific Ocean. And because there is an inverse exponential relationship between the number of seamounts and their size – meaning that large seamounts are relatively rare and that there are vastly more small seamounts than large ones – the number of seamounts of all sizes in the Pacific is in the order of 600 000–1.5 million.[11]

These estimates are only approximate, because the ocean seafloor is still poorly mapped. Indeed, better maps are available on the scale of seamounts for Mars or the backside of the moon than for the ocean basins of our own planet. Current data are derived from a combination of acoustic traces along ship's tracks and gravity anomaly measurements from satellites. (Where there are large seamounts beneath the sea surface, there is greater mass and hence greater gravitational attraction, so precise measurements of gravity over the earth's surface provide an approximate map of the distribution of seamounts.)

Nothing so dramatically illustrates the geological differences between the oceans as their seamount statistics. The Pacific, ringed by its so-called 'rim of fire,' where oceanic plates are subducted in a network of trenches beneath the continental plates, is home to much of the planet's volcanic activity and not surprisingly contains the vast majority of its seamounts. Deep ocean trenches and subduction processes are virtually absent, however, in the North Atlantic. In comparison with the approximately 1 million seamounts to be found in the Pacific Ocean, there are only 810 in the North Atlantic greater than only 100 meters in height. There are still no published estimates of the numbers of seamounts in the South Atlantic, Indian, or Southern oceans.

Ocean circulation around seamounts

It is fundamental to oceanography that the ecology and evolution of marine communities are deeply entwined with their physical environment: to understand ecology, it is necessary to start with the physics of the water, which in turn is inextricably linked with the underlying geology and seafloor topography. Nowhere is this better illustrated than with seamounts.

Seamounts influence their ambient physical environment – the flow of currents in their vicinity – as profoundly as mountains create their own wind-climate on the land, replete with the oceanic equivalent of updrafts and downdrafts, enhanced gustiness, and so on. Virtually every form of water movement in the ocean – currents, waves, and tides – interacts with the topography of the seafloor. And because seamounts present the sharpest relief across the floor of the ocean basins, they have the most dramatic influence. This special physical environment in turn influences every aspect of the ecology and evolution of its associated community: the body plan of its typical species, their physiology and metabolism, modes of feeding, the dispersal of their larvae, and reproductive strategies, leading over geological time to the evolution of a unique and diverse fauna.

Except in areas influenced by major deep currents, such as the Gulf Stream, currents in the deep sea are typically quite weak, on the order of a few centimeters per second. (A current of 1 knot, or 1 nautical mile per hour, is equal to 50 centimeters per second.) When such a current encounters a seamount, the water rises over the slope, flows over the top and descends on the leeward side. The current accelerates as the flow field squeezes over and around the seamount, setting up counter-rotating vortices on its upstream and downstream flanks. If the ambient currents are somewhat stronger, a gyral circulation may form directly over the seamount; this is known as a Taylor column. These eddies generate upwelling and downwelling, with obvious implications for the productivity of the overlying waters. Such effects can influence the water column far above the seamount summit, leading to the apparently mysterious aggregation of tunas, seabirds, sea turtles, and other near-surface marine predators over seamounts whose summits may lie hundreds of meters beneath the waves.[12]

The seamount topography can also interact with tidal currents, which due to their oscillations can set up resonance, leading to the formation of eddy features known as *seamount-trapped waves*, with up- and downwelling cells created over the seamount slopes. *Internal waves*, which ride along density gradients between water masses in the ocean's interior, can also interact in a nonlinear way with the seamount topography, leading to turbulent water movements along its flanks.

The strength of these interactions of ocean currents, tides, and waves with seamounts varies depending on the height, size, and shape

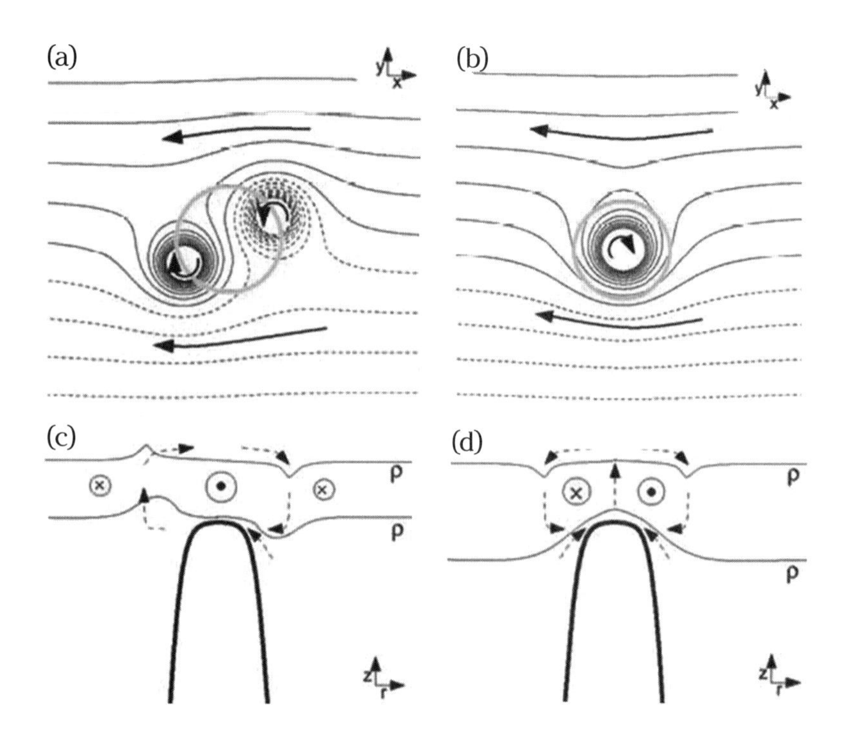

Currents flowing past a seamount are intensified and influenced by the seamount topography. A weak ambient current forms (a) counter-rotating vortices, sometimes referred to as 'butterfly wings' as it flows over a seamount, whereas a strong current forms (b) a Taylor column or gyre directly over the seamount. The vertical circulation is also affected, such that upwelling and downwelling cells form over a seamount, related to (c) weak or (d) strong ambient currents. Upwelling may bring nutrient-rich deep water into near-surface waters, enhancing primary productivity. (From Beckmann and Mohn, 2002.)[q]

of the seamount, its latitude, the stratification of the water column, and the strength and periodicity of the water movements. Not surprisingly, these motions vary considerably both in their amplitude and in the types of eddies created. However, it is not unusual to find currents in the immediate vicinity of seamounts in the order of several tens of centimeters per second, which is about ten times faster than the ambient deep-ocean currents.

The strength of these currents – and the difficulty of fishing on a seamount – was impressed upon me by watching commercial fishermen attempt to land their trawls on a particular patch of a seamount where they saw a body of fish on their echo-sounder. A commercial trawl net has a mouth about 20 meters wide and 5 meters high. This is held open as the vessel steams ahead by the pressure of the water against the trawl doors, massive metal vanes weighing several tonnes each. Fishing at a depth of 650–1000 meters with a trawl warp to depth ratio of 2:1 means that the trawl is dragged a kilometer or so behind the vessel. As the trawl is 'flown' down onto the seamount, it is not unusual to see the warps (the massive cables attached to the doors and net) suddenly swing off to one side of the boat or the other as the net encounters these currents and, like a huge kite, is carried away by them.[13]

Seamount ecology

Although seamounts can influence the water column for hundreds of meters above their summits, their greatest impact is on the seafloor itself and on the flow of water immediately above. The currents on seamounts (and also in submarine canyons and certain regions of the continental slope) winnow away the falling sediment, leaving bare rocky substrate unsuitable for the deposit feeders of the open seafloor but ideal for colonization by deep-sea suspension feeders. In addition to the few species of stony corals (only about six) that form true deepwater reefs, such as *Lophelia pertusa*, there are hundreds of species of deepwater corals in a dazzling array of groups that grow on the hard substrate directly or colonize the reef-forming corals themselves, growing on and over them to reach up into the current to snare prey drifting past. There are solitary stony corals; black corals; various groups of soft corals, such as the bamboo corals, gold, pink, and red corals; gorgonians or sea fans; and hydrozoans or lace corals. Other suspension-feeding groups grow on or climb to the top of the reef, including sponges, anemones, bryozoans, and a variety of echinoderm groups: the primitive crinoids or sea-lilies, and certain brittle stars, basket stars, and brisingid starfishes whose 'legs' have evolved to extend into the currents to feed. These communities vary considerably depending whether they are on seamounts or the margins of a continent, their depth, the surrounding topography, ambient currents and the productivity of the overlying water. Today, exploration of these environments is one of the most active areas of deep-sea ecology. Remarkably, however, several decades passed between the first awakening of scientific interest in these environments and their full appreciation as veritable oases of the deep.

Carl Hubbs, an intrepid ichthyologist and a colleague of Menard's at the Scripps Institution of Oceanography, was the first marine ecologist to become interested in the newly discovered seamount habitat in the years following World War II. Hubbs was stimulated by the impressive catches Japanese and west-coast American tuna fishers were taking over seamounts, as well as by Menard's data showing the vast number of seamounts spread across the floor of the deep sea. Hubbs made the first collections of benthic fishes from seamounts along the coast of North and South America and in 1959 posed the questions that set the agenda for seamount ecological research to the present day:[14]

- What species live on banks and seamounts?
- How do they become established there?
- Do seamounts serve as biogeographic stepping stones?
- Does the isolation of seamounts lead to the evolution of new species?
- What is the productivity of seamounts? Are the fishes there sufficiently abundant to support fisheries?
- What are the physical and ecological factors responsible for the abundance of life in these environments?

In brief, what role do seamounts play in ocean ecology, biogeography, and evolution?

John Isaacs, also from Scripps, was a second key early figure in seamount research, and the first to point to their role in fisheries. Once a fisherman himself, Isaacs wrote in his landmark paper published in *Science* in 1965:

> Most predators, including fishermen, depend on the schooling behavior of their prey. Moreover, the frequency of these small-scale aggregations varies sharply between adjacent areas tens to hundreds of kilometers apart, even where little difference in the physical and chemical nature of the water is discernible; mechanical interactions, such as those between circulation of waters, topography of the sea floor, and behavior of the organisms, must be implicated.[15]

During six cruises off Cape Colnett, Baja California, Isaacs had observed 'sonic spires' of rockfish (*Sebastes* spp.) over Banco San Isidro, which rises steeply from 730 to about 100 meters depth. What sustained these aggregations? Why were they found only over the bank?

In this region, as elsewhere, Isaacs observed on his ship's echo sounder the daily vertical migration of the krill *Euphausia pacifica*, one of the dominant phytoplankton grazers in the California Current. The krill came to the surface to feed at night, and swam down in the day to between 150 and 300 meters depth, where they formed a distinct deep scattering layer. Isaacs realized that the bank was a natural migrator trap. Each night krill drifted over the bank while feeding, only to be intercepted by the relatively shallow seabed when they swam back down at dawn; unable to obtain refuge in the darkness of deeper waters, the krill fell easy prey to the rockfish gathered there awaiting them. In the evening, when from all around the bank the krill rose again toward the surface, from the bank itself, to paraphrase Lewis Carroll, arose there none, for they had been eaten, every one.

Perhaps not every one, but Isaacs estimated that the consumption rate of the krill trapped over Banco San Isidro was 40 times higher than the average in the California Current.

John Isaacs was a polymath engineer who never paused long enough in his career to obtain a PhD, yet published papers on a vast range of topics from cosmology to the structure of marine food webs and novel means to extract energy from the ocean. With such disparate interests, and ideas generally outside the mainstream, much that he wrote failed to fall on fertile ground, or took considerable time to bear fruit. Seamounts proved exceedingly difficult to sample, and Isaacs' observations seemed too anomalous. Seamount research fell into a hiatus for several decades, until technological developments and a dramatic resurgence in seamount fisheries re-awakened scientific interest.

In 1987, near the end of this lull in seamount research, two Scripps researchers reviewed everything known about the global seamount fauna and its diversity, bringing together every published record of species sampled on seamounts to date. In fact, the record was quite modest. Over the century since the *Challenger* expedition, a total of 596 species of invertebrates and 449 species of fishes were recorded from approximately 100 seamounts. However, most seamounts were sampled cursorily, with often only a few samples and perhaps a single group, such as the fishes, recorded.[16] In fact, just five seamounts accounted for 72 percent of the species in the record. Vast regions were virtually unsampled – for example, only 27 species had been recorded from the entire southwestern Pacific with its maze of islands and seamount chains, and only 13 species from two seamounts over the whole Indian Ocean.

A little more than a decade later, in 2000, I published a paper in *Nature* in collaboration with Bertrand Richer de Forges of the Centre IRD (L'Institut de recherche pour le développement) in French New Caledonia and Gary Poore of Museum Victoria, based on sampling ten seamounts found on two ridges (the Norfolk ridge and the Lord Howe rise) south of New Caledonia and 14 seamounts in a cluster off the south coast of Tasmania. From this small corner of the southwest Pacific, we recorded more than 850 invertebrate species, 42 percent more than were previously known from all seamounts globally. Some 29–34 percent of the species were new to science and therefore potentially endemic (that is, restricted) to the region and its seamount environment. Moreover, the number of species recorded from each seamount was proportional to the number of samples obtained from it (p. 125). With no leveling off in this relationship, it was apparent that many more species remained still undiscovered, even from this small group of seamounts. Seamounts were a previously unrecognized 'hotspot' of biodiversity in the oceans.

The most interesting aspect of this study, however, was the apparently limited distribution of many of the seamount species. As discussed in Chapter 4, species in the deep sea characteristically have extensive distributional ranges because environmental conditions are more uniform and there are relatively few topographic barriers. Deep water masses, such as North Atlantic Deep Water and Antarctic Bottom Water, extend over the seafloor throughout the deep ocean; and at intermediate depths (approximately 1000 meters) masses such as Antarctic Intermediate Water may also extend over vast distances. Planktonic organisms, and creatures that spend a significant portion of their lives among the plankton – fishes that spawn in open water and crustaceans, such as lobsters, crabs and prawns – use stable current flows to maintain the integrity of their populations.[17] Biogeographic patterns – the distribution of species and of entire communities or assemblages of species – across the vast pelagic ecosystems of the world's oceans largely follow the distribution of the ocean's water masses, which define pelagic habitats by their distinct temperature and salinity characteristics; the water masses also enable species to integrate their life histories within stable circulation features. Thus fish and invertebrate megafauna species living over the seafloor plains at abyssal and even bathyal depths may be distributed across entire ocean basins, and their families and many genera, such as the grenadier *Coryphaenoides*, have a circum-global distribution.[18]

Some of the more notable biogeographic patterns in the deep sea can only be understood within the context of deepwater circulation patterns. For example, Antarctic Intermediate Water originates at about 50°S, has a circum-global distribution in the Southern Hemisphere, and can be traced into the North Atlantic but not the North Pacific. The orange roughy is the dominant fish living within the core of the Antarctic Intermediate Water (which roughly spans 750–1200 meters depth) at temperate latitudes right around the Southern Ocean, from New Zealand and southeastern Australia in the southwest Pacific, across the Indian Ocean to Namibia in the southeast Atlantic and Chile in the southeast Pacific. It also extends, along with 24 percent of the common species found at these depths off Tasmania, into the northeast Atlantic; yet none of these species extends to an appreciable extent into in the North Pacific.[19]

In striking contrast to the broad distributions generally observed in the deep-sea fauna, in our seamount study there was reasonable overlap (21 percent) in species between seamounts along a ridge, but on average only 4 percent of species were shared between seamounts on parallel ridges about 1000 kilometers apart, and there was no overlap of species between the seamounts south of Caledonia and those off Tasmania.[20] For comparison, 60 percent of benthic decapod crustaceans (the group including shrimps, crabs, and lobsters) on the continental slope off southeastern Australia were also found in the tropical Indo-Pacific.

Such restricted distributions imply extremely limited dispersal. Not only do these species remain at a particular site as adults, but many have a very limited larval period when they can disperse more widely, or none at all.[21] Furthermore, the circulation around seamounts promotes their insularity. The eddies and currents tend to follow the contours of the seamounts and ridges – physical oceanographers refer to this as 'topographic rectification' of the currents – so exchange between disjunct seamount chains is extremely limited.

Seamounts, unlike hydrothermal vents (which we discussed in Chapter 5), are stable environments, mostly many millions of years old. For many species there is little evolutionary incentive to broadcast progeny over vast oceanic distances – there is only a remote chance they will encounter another seamount, and a suitable environment is already close at hand. Isolated islands, such as the Galápagos, have long been famed as sites for localized species evolution. Seamounts are in fact submerged islands, and many function as the Galápagos of the deep, achieving genetic isolation through a combination of topographic effects on the local currents and evolutionary pressure to reduce larval dispersal – another example of the causal chain linking topography, physical oceanography, ecology, and evolution in the deep sea.

A similar view of seamounts as hotspots of biodiversity and evolution in the deep ocean emerged from the southeastern Pacific, based upon studies carried out by the former Soviet Union that were only recently published in the Western scientific literature. In the 1970s and 1980s, Soviet scientists studied 22 seamounts along some 2500 kilometers of the Nazca and Sala y Gómez ridges, which extend from Chile to Easter Island in the southeastern Pacific.[22] Despite their proximity to South America, however, the seamount fauna was derived from the Indo-Pacific region, having spread across the Pacific basin apparently by using the many intervening seamounts as stepping stones, as Hubbs hypothesized. But the isolation of these seamounts and the considerable distances involved led to the evolution of a distinct fauna. Approximately half of the invertebrates and fishes were new to science and endemic to the region's seamounts: 51 percent of the 192 invertebrate species and 44 percent of the 171 fish species identified.

But do seamounts in fact serve as islands for the evolution of new species in the deep sea or are they only oases of higher productivity and diversity? It is characteristic of the exceptional diversity of seamount faunas that many of the species are exceedingly rare and may be encountered only once or twice in the samples from an entire cruise or even from a number of cruises. So long as seamounts and the deep sea remain poorly sampled, it is difficult to ascertain whether such species are in fact quite localized or are rare, but more widely distributed. However over the past 20 years, French scientists have conducted some 28 cruises, obtaining about 1500 samples from seamounts and the continental slope of New Caledonia. The best-studied seamounts are along the Norfolk Ridge, which extends southward from New Caledonia.

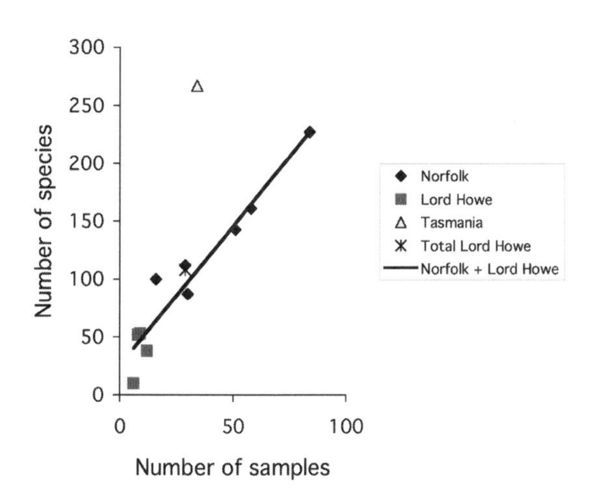

The relationship between the number of samples and the number of species recorded from seamounts in the Coral and Tasman Seas in the southwest Pacific. Although more species were obtained here than were known from all seamounts previously sampled worldwide, the lack of any leveling off in the relationship indicates that many species remain to be discovered. The Tasmanian seamount data appear as an outlier, probably because this point is a composite from sampling 14 seamounts, each with no more than three samples. (From Richer de Forges et al. 2000.)[r]

Focusing on a single, highly diverse family, the Galatheidae (or squat lobsters: see Plates 9, 11), a recent study observed that seamount samples contain on average about three times more galatheid species than samples from the slope, but that all 62 galatheid species known from the Norfolk Ridge seamounts are also found on the continental slope of New Caledonia.[23] But this is perhaps not a strong test of endemism, since the Norfolk Ridge is contiguous with the New Caledonian slope; also, the galatheids are characterized by a reasonably long dispersal phase in the plankton. Further such studies, combined with studies of the population genetics of species between different seamounts and seamount chains, should soon resolve the extent to which seamounts serve as centers of speciation, as well as of diversity in the deep sea.

The issue of seamount endemism has profound conservation implications. If a significant proportion of species has only localized distributions on seamounts, this raises for the first time the specter of substantial extinctions in the deep sea. Large-scale disturbance to a chain or cluster of seamounts, such as from deepwater trawling, could then lead to the loss of entire suites of species. The hydrographic isolation of seamounts and the limited dispersal capability of much of the fauna also indicate that recovery from disturbance is likely to be extremely slow. The design of Marine Protection Areas to protect seamount faunas requires understanding how widely these species are distributed and what proportion of a regional fauna is likely represented on each seamount. The impact of trawl fisheries on seamount and deepwater coral faunas and the associated conservation and governance issues are the subject of Chapters 10 and 11.

Coldwater coral reefs along continental margins

The strong currents that prevent sediment accumulation on seamounts and nourish their corals are also encountered on certain oceanic banks and continental slopes. Examples include the coast of Norway, mentioned previously, which is bathed by the North Atlantic Current, and the Florida Straits and Florida-Hatteras slope where the Gulf Stream sweeps past, maintaining currents often greater than a knot (50 centimeters per second) even at depths of 500–850 meters. Only about half a dozen species of stony corals form true deepwater reefs, which can grow to tens of meters high and extend for kilometers. Based on the slow growth rate of deepwater corals (estimates range from 0.5–2.5 centimeters per year, about a tenth the rate of growth of shallow-water tropical corals), the larger deepwater reefs are estimated to be thousands of years old.[24]

Because of the continuity of continental margins and of the currents sweeping past them, the species living on these deepwater reefs are widely distributed. However, these reefs provide habitat structure and refuge from predators much like shallow-water reefs, and they may support comparable species diversity. An ongoing study of the species living in association with *Lophelia pertusa* reefs in the northeast Atlantic has thus far encountered more than 1300 species.[25] Twenty-five small blocks of living and dead *Lophelia* weighing between 0.2 and 2 kg collected off the Faroe Islands yielded some 298 species.[26] A similar study that sampled *Oculina*, a deep reef-building coral found off Florida, collected 20 000 invertebrates from among the branches of 42 small colonies; some 348 species of molluscs, echinoderms, and decapod and amphipod crustaceans were identified.[27] And in some regions no less diverse, for example the brilliant coral and sponge 'gardens' recently discovered along the Aleutian chain in the North Pacific or in canyons off the Atlantic coast of Canada, reef-forming corals have not taken hold but the current-swept hard substrate is dominated by a dazzling variety of soft corals, sponges and allied groups.

Why are extensive deepwater stony coral reefs formed by *Lophelia* and *Oculina* found around the rim of the North Atlantic but not the North Pacific, where soft corals dominate instead? (Compare Plates 8–11, 14–15 with 12–15.) Deep-sea scleractinian (stony) corals virtually all occur above the aragonite saturation horizon – that is, the depth at which this form of calcium carbonate ($CaCO_3$) begins to dissolve.[28] Scleractinian corals must precipitate $CaCO_3$ from seawater to build their skeletons. The aragonite saturation horizon is between about 1500 and 2600 meters depth in the North Atlantic but only between 120 and 580 meters in the North Pacific: the deep North Pacific, without its own sources of deepwater, is more stagnant, leading to oxygen depletion, the buildup of CO_2 and greater acidity.[29] The potential impact of increased CO_2 levels in the atmosphere and its uptake by the oceans on the future viability and distribution of scleractinian corals in the deep ocean may be profound and will be examined in Chapter 9.

A striking feature of deepwater coral reefs is the large quantity of dead coral present. In part this is caused by the corals themselves. Those that create these reefs are all densely branching (Plates 8–10). The polyps emerge from the branches or stolons, which grow continuously upward into the nourishing current. As a natural consequence, the polyps underneath eventually starve. There are also a number of natural predators on the corals (starfish, boring sponges, and polychaete worms), as well as a variety of parasites (a foraminiferan, specialized parasitic copepods and other crustaceans).[30] In some areas where only dead coral is found – for example my group found only dead coral below about 1400 meters on the seamounts south of Tasmania – one wonders whether changes in climate and hydrography are responsible.

Yet another type of structure that supports deepwater reefs, which has only been discovered this past decade, is deepwater mounds. In 1998 the UK carried out a series of surveys to the west and north of Scotland, baseline monitoring in preparation for expansion of the offshore oil and gas industry into the deep slope waters of the region. The surveys recorded dense growth of sponges, some tens of centimeters in diameter, practically carpeting the seafloor at about 500 meters depth. These sponge grounds, known to the fishermen as *ostebund*, seem to form in areas of enhanced turbulence, such as the shelf break, where internal tidal waves are propagated. Most notable, however, were some hundreds of mounds about 5 meters in height and 100 meters in diameter, discovered at about 1000 meters depth over a 20-square-kilometer area to the northwest of Scotland, almost halfway to the Faroe Islands. Named the Darwin Mounds after the research vessel *Charles Darwin*, they were covered with the deepwater coral *Lophelia* and possessed long downcurrent 'tails' densely populated with giant protozoans known as xenophyophores. These single-celled organisms, up to several centimeters across, protect themselves by cementing 'tests' or shells around themselves from the sand, mud and gravel.[31]

The discovery of the Darwin Mounds on the south flank of the Wyville-Thomson Ridge, first encountered by Thomson, fairly reeked of the history of oceanography. The prevalent species of xenophyophore, *Syringammina fragilissima*, was the first of its group to be described and was first encountered in that very area in 1882 by Gwyn Jeffreys, a colleague of Forbes and Thomson and one of the last of that merry band of amateur British naturalists who pursued dredging in the latter half of the 19th century.

Unfortunately, video surveys carried out by one of the 1998 expeditions found widespread evidence of trawling down to depths of over 1000 meters. In the face of vociferous opposition from fishing groups, the Darwin Mounds region was finally closed to bottom trawling in 2003, but only after a subsequent survey showed further trawl damage to the reefs.[32]

Lophelia-covered mounds are also known from the Porcupine Seabight southwest of Ireland and at depths of 400–500 meters in the Gulf of Mexico. The mounds all appear to be formed by fluid seeps. In the case of the Darwin Mounds, the seeps bring sand to the surface, and the enhanced flow over the mounds appears sufficient to lead to their being colonized by *Lophelia*. The mounds in the other regions are larger: those in the Porcupine Seabight and the Gulf of Mexico range from 10 to 300 meters in height, almost as high as the small seamounts fished off New Zealand and Tasmania. These appear to be formed by the build-up of biogenic carbonate rock, as hydrocarbons seeping upward are metabolized by microbes in the sediments. The carbonate, in turn, provides hard substrate for *Lophelia* to colonize.[33]

Fishes of seamounts and deepwater reefs

Fishes are also associated with deepwater reefs, coral gardens, and seamounts. Some appear to use the coral structures for refuge, such as juvenile and gravid rockfishes (species of *Sebastes*) found hovering within thickets of *Lophelia* off Norway. Ling and tusk, which are commercially fished in the North Atlantic, also occur in greater abundance around *Lophelia* reefs, while in the North Pacific, several other rockfish species and Atka mackerel are noted for their association with the coral and sponge gardens along the Aleutians and in the Gulf of Alaska (Plates 10, 11, 13).[34]

Massive aggregations of benthopelagic fishes such as orange roughy, oreos, alfonsino, rockfishes, pelagic armorhead and other species hover over seamounts, sometimes extending more than 100 meters into the overlying water column (Plate 8). For the most part, these fishes appear to be associated with the seamount rather than the corals *per se*, typically feeding on midwater fishes, squids, crustaceans, and other prey carried along by the currents, as well as on vertically migrating prey intercepted and concentrated by the shoal topography.[35]

Seamount-associated fishes are strikingly different from the weak-bodied fishes typical of other deep-sea habitats: the viper and angler fishes typical of the midwater, or the grenadiers and deepwater cods living over the flat seafloor. Seamount fishes are typically deep-bodied, with broad caudal (or tail) fins: these are adaptations for strong-burst swimming, enabling them to maneuver in the swift currents associated with seamounts (Plates 8–10). And these differences are more than skin deep. The typical bathypelagic fish or grenadier is indeed weak-bodied: their flesh is literally watery, with a water content greater than 80 percent, and has little protein (the primary component of muscle) or lipid. These flesh characteristics render them almost neutrally buoyant, as well as relatively unpalatable, at least from the human point of view. Their metabolism (the rate at which they use energy) is generally only between 1 and 10 percent that of shallow-water fishes.[36]

Seamount-associated fishes, on the other hand, have flesh characteristics and metabolic rates closer to those of fishes that live on the continental shelf or that actively migrate into near-surface waters to feed[37] – characteristics that, once discovered, made these fishes prime targets for commercial fishing. But there was an additional bonus: some seamount species, orange roughy in particular, carried their lipid content predominantly in the form of wax esters, rather than the traditional fish oils. Although these oils, with their omega-3 fatty acids, provide one of the main health benefits of eating fish ('good for the heart'), they are also responsible for the 'fishy' taste and smell of most fish. Not having these oils, orange roughy could be frozen, thawed, filleted and refrozen, all without notable loss of quality. A fish that didn't smell or taste like a fish was perfect for those markets, such as the USA, where most people don't really like fish but believe it's good for them. Of course the folks in marketing didn't point out that the bland flavor was due to the lack of healthy fish oils. Nor was it publicized that these long-lived predators, near the top of the deepwater food web, had mercury levels just marginally under the maximum permissible level of 0.5 milligrams per kilo.[38]

The grenadier (*Coryphaenoides*) and orange roughy are typical fishes of their respective environments, the relatively flat seafloor and seamounts. Both feed largely on invertebrates and small midwater fishes found in the water column, and both are commercially fished.[39] In other words, they are dominant fishes at a comparable trophic level in the deep-sea food web in their respective environments. But if we look beyond these superficial similarities, their ecological and physiological strategies could not be more different. Adapted to life in the slow lane of a food-poor environment, the grenadier is built for slow, sinuous cruising, and its metabolic rate is only about 5 percent per unit weight that of the orange roughy. As a consequence, although the daily food intake of a grenadier is remarkably low – only about 0.05–0.1 percent of its body weight, compared with about 1 percent of body weight for the orange roughy – its growth efficiency may be as high as 50 percent (i.e. 50 percent of the food it consumes is converted to biomass), compared with only 5 percent for the orange roughy.[40] Although the orange roughy is living in a richer environment, the cost of living there is higher as well. At the end of the year, its energetic balance sheet is barely positive: growth is virtually nil and, prior to being thinned out by fishing, the Australian population had sufficient energy reserves to spawn only every other year.[41]

The precarious nature of life on seamounts is best illustrated by the pelagic armorhead (*Pseudopentaceros wheeleri*), which prior to fishing was the dominant fish in the relatively food-poor region of the Emperor Seamounts to the northwest of Hawaii. The armorhead grows to maturity in the open waters of the North Pacific and then settles onto the seamounts, where it spawns over the next several years of its life. But the pelagic armorhead actually lives in energetic deficit, progressively growing thinner, such that fishery scientists initially considered the fat new recruits to the seamount and the wasted older fish to be different species.[42]

Evolution works on the life history of species as well as their form and physiology; that is, it works on the patterns of growth, maturation, and reproduction, and even on the longevity of species. For reasons that are still not entirely clear, the seamount environment has led to the evolution of a life history unique to seamount-aggregating fishes. Among mammals, there is a good general relationship between size, metabolic rate, and longevity. Across the size range from mice to elephants, the smaller creatures generally live faster – having a higher metabolism but a shorter life span, leading to the popular perception that a species is allotted only so many heartbeats. With deepwater fishes, however, this is not the case. The meso- and bathypelagic fishes, despite their greatly reduced metabolism, generally live less than ten years, much the same as an active pelagic species of similar size; for example an anchovy or pilchard.[43] Benthopelagic fishes living over the continental slope, such as the macrourids (*Coryphaenoides* spp.) or sablefish, have a greatly reduced metabolism but also generally live for around 50–75 years – which is much longer than comparable species on the continental shelf, such as the Atlantic cod, which lives to about 25 years.[44] At the far end of this continuum, several of the key species living on seamounts, such as the orange roughy and oreos, though growing to only about 50 centimeters, can live for an extraordinary 100–150 years.[45]

So what accounts for the longevity of these fishes? Obviously mortality is very low, around 4 percent per year, which can be related in part to a relative lack of predators. Several species of deepwater sharks caught in association with the orange roughy likely prey on them, and there is speculation that the giant squid, *Architeuthis*, also preys on orange roughy, several having been caught by orange roughy trawlers over the years. Sperm whales are also observed diving on seamounts with orange roughy; they feed largely on squid, which may also be associated with seamounts (including the giant squid), but they also eat fishes. Could they be one of the dominant predators of orange roughy? Despite the large number of sperm whale stomachs examined during the days of commercial whaling, there are no records of orange roughy being found. I was about to consign the idea to the great dustbin of interesting but unproven speculations when I came upon a large jar of oreos, a fish commonly aggregated on seamounts in association with orange roughy, on a shelf in the Moscow University fish collection; the label stated that it had been obtained from the stomach of a sperm whale.

There is strong circumstantial evidence that orange roughy have co-evolved with large predators and have developed striking, albeit still poorly understood sensory systems and behavioral adaptations to avoid them. Some of the earliest observations from submersibles of the behavior of midwater fishes noted that they often seemed to hang there in a kind of torpor.[46] But not the orange roughy. During my group's acoustic surveys, we attempted to photograph the aggregations to confirm that they were in fact composed of orange roughy. As we lowered our camera, we could observe it on our ship's echo sounder. It passed through successive deep scattering layers without eliciting any noticeable reaction, but when the camera was more than 100 meters above the orange roughy, which were in plumes rising 50–100 meters above the seafloor, the entire aggregation vanished from view, either descending directly onto the seafloor, a typical flight response of fish over a reef, or swimming outward 30–40 meters or more, beyond the range of our acoustic beam.[47]

Clearly orange roughy have evolved a marked avoidance reaction to predators descending toward them from above. But how had they sensed us, and how was the aggregation able to respond in such a coordinated manner, much like a near-surface fish school but without the visual contact that enables fish schools to respond to each other? The common name for the orange roughy family of fishes is the 'slime heads.' Their heads contain an extensive system of internal canals filled with wax esters, which turn into an oily liquid at surface temperatures and pressure. These canals seem to comprise a sensory system, most likely for detecting movement around them. Orange roughy also have a characteristically deep lateral line, a series of pits running along the side of their body that allows them to detect the movements of fishes around them, enabling them to communicate and coordinate the aggregation's flight response in the absence of light. Orange roughy may be subject to predation by a variety of large midwater animals, but they have adapted to minimize this to a degree that stands out among deep-sea fishes.

Unfortunately for the orange roughy, their flight response, similar to that of shallow-water reef fishes that dive for cover from predators within the shelter of the reef, proved utterly maladaptive in the face of the modern deepwater trawl. The descending trawl pursues fish to the bottom, directly into the corals if need be, where it proceeds to scoop them up with remarkable efficiency. Hauls of 10–50 tonnes of orange roughy were often achieved in the early days of orange roughy fisheries, with the trawl actually on the bottom for only a few minutes (see Plate 16).

Ecological theory suggests two further hypotheses to explain the evolution of such remarkable longevity. The first postulates a continuum in species' life histories between 'weedy' and 'competitive' species. Weedy species are good colonizers, adapted to unstable environments. They characteristically mature early and have brief adult lives, during which they direct their surplus production into high reproductive output. At the other end of the spectrum are species adapted to highly stable environments, with little opportunity for colonization. Here it is advantageous to maximize one's ability to compete over an extended adult life. In the plant kingdom, this theory is exemplified by the weedy roadside annual at one end of the continuum and climax forest tree species at the other. In the world of fishes, the relatively unstable pelagic realm is generally populated by short-lived species with high reproductive output, such as the silvery schooling species (anchovies, pilchards, and herrings), mackerels, and salmons. The sedimented plains and seamounts of the deep sea are highly stable environments, characterized by long-lived species, whether invertebrates or fishes. On the other hand, the relatively ephemeral deep-sea hydrothermal vent environment is characterized by species with dramatic growth rates that rapidly mature, as we saw in the previous chapter. Seamounts are clearly an intensely competitive environment, with dense aggregations of invertebrate suspension feeders growing and climbing over each other to extend their feeding appendages into the fresh current; this is a clear analogue to the climax forest environment, dominated by tall, ancient trees that vie for sunlight in the upper canopy. Competition among the massive aggregations of seamount fishes is no less severe: as in any schooling situation, feeding is best at the front. Opportunities for young fishes to successfully recruit into this environment may be as limited as chances for young trees to grow up beneath a mature forest canopy. Little wonder, then, that orange roughy do not recruit to the seamount environment until they are adults, about 30 centimeters in length and 25–30 years of age, and that their reproductive life then extends for the next 100 years or so. As noted earlier, competition is so severe among the adults that they may build up the requisite energy reserves to spawn only every other year.

Another variant of life-history theory hypothesizes that great longevity results not simply from environmental stability but the combination of low adult mortality and highly variable mortality during early life. To return to the example of the forest, despite the immense seed production of mature trees, it may be many years before an opening in the forest canopy allows new young trees to grow to maturity. There are few studies of juvenile mortality among deep-sea fishes, but some of the scant evidence available suggests that seamount fishes also experience highly episodic bursts of recruitment to their populations. When orange roughy were first fished off Tasmania, the most common age (the mode) in the population was between 40 and 60 years, although young fish recruit to the population when about 25–30 years old. In virtually all fish populations that receive regular inputs of young fish, the youngest fish are most abundant, due to the cumulative effects of mortality as the fish grow older. The implication for the Tasmanian orange roughy, however, was that there had been significant input of young fish to the population 10–30 years previously and very little since. Similarly, the world's largest orange roughy fishery, off the Chatham Rise of New Zealand, has seen virtually no change in the size structure of the population since the mid-1980s. Most intensively fished populations experience a rapid reduction in mean size, as the large older fish are removed and the population comes to be dominated increasingly by small, young fish. The lack of such change in the Chatham Rise population indicates that it has been fished down without significant input of new recruits to the population over the past 20 years. This pattern has fairly obvious implications for fisheries management, which we will explore in Chapter 10.

Earlier we spoke of seamount clusters and chains as the 'Galápagos of the deep': how topographically and hydrographically-induced isolation led to species evolution within restricted regions, previously unknown for the deep sea. However, this is found only among those benthic invertebrates that have evolved to limit their dispersal by having brief larval periods in the plankton or none at all. The dominant seamount-associated fishes, on the other hand, are open-water spawners, whose eggs and larvae, though little studied, are presumably within the plankton for weeks or months.[48] As we saw, the distribution of these species is often on the scale of the intermediate water masses, sometimes spanning the temperate zones of the Southern Hemisphere and North Atlantic, as in the case of the orange roughy and associated species. However, the relative isolation of these intermediate water masses has led to one of the few instances of convergent evolution in the ocean. The oreos and orange roughy of the temperate Southern Hemisphere, the pelagic armorhead of North Pacific seamounts, the alfonsinos that dominate tropical seamounts, and the rockfishes (*Sebastes* spp.) around the rim of the northeast Pacific and northwest Atlantic all have a similar body plan – deep-bodied and well-suited to strong bursts of swimming – but they evolved from entirely different families and even orders of fishes (Plate 9).[49] Whereas the sedimented plains are colonized around the world (except near Antarctica) by a single benthopelagic fish fauna dominated by gadiform (or cod-like) fishes (the grenadiers, morid cods and their relatives), the seamounts at intermediate depths (approximately 500–1500 meters) were colonized by distinct groups that evolved independently in these hydrographically isolated deepwater regions. Once considered rather obscure taxa, atypical of the deep sea, these fishes in fact represent a distinct ecological and evolutionary stream, adapted to the high-energy deep-sea environments found on seamounts, in canyons and along certain current-swept continental margins.

Our understanding of this environment is still rudimentary. Virtually nothing is known of the reproduction and early life history of these species, of the genetic relationships between populations living on neighboring seamounts or seamount chains, of the diversity of these communities (except at a very few locations), or of the ecological interactions between the benthic, benthopelagic, and overlying pelagic environments. Many of my conclusions are necessarily still tentative and likely to evolve as our understanding improves. However, there has been a veritable explosion of interest in this environment, largely because of the grave threats facing it, mostly from deepwater fisheries (see Chapter 10), which has lent a deep sense of urgency to ongoing efforts to study and conserve it. Much of this magnificent environment will have vanished before we even know what was there.

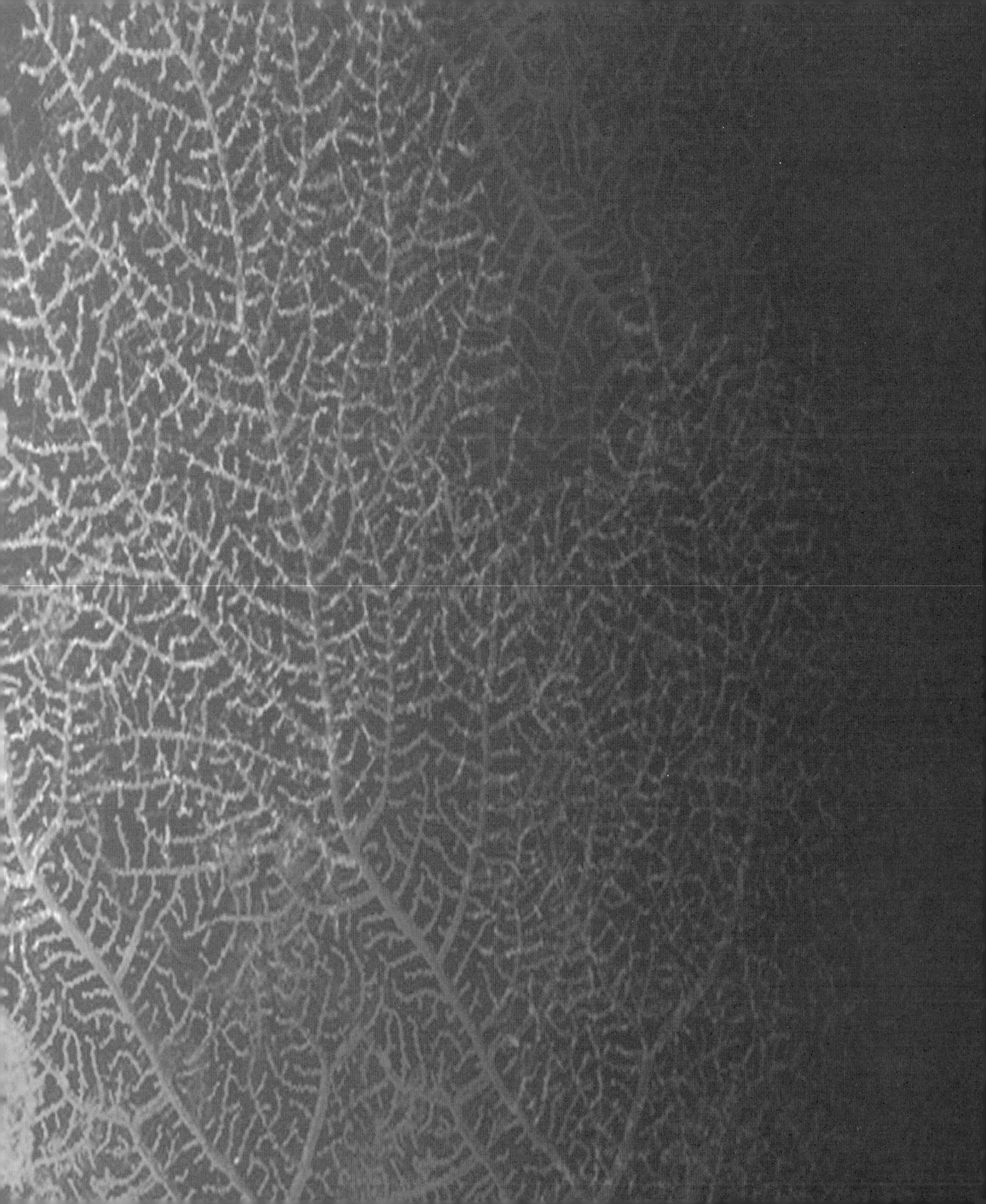

Part III

The Human Footprint across the Deep Sea

Fifty years ago, the deep sea seemed almost as remote and far removed from man's activities as the far side of the moon, and such views linger today. Other than as a subject of scientific curiosity, what is in the deep sea of value and human interest? As the matron in Charles Saxon's *New Yorker* cartoon wonders, why should we care about the deep sea?

For a long time, the remoteness of the deep sea seemed its most valuable asset: it provided the ultimate dump site for humankind's most deadly wastes. Disposal at sea was cheap, easy, out of sight, and unlikely ever to return to haunt us: the turnover time of water in the deep sea – its exchange rate with surface waters – is on the order of 1000 years. No one considered that human pollutants might ever significantly affect the deep sea – its diluting capacity appeared virtually infinite. Rachel Carson never mentioned the oceans, much less the deep sea, in *Silent Spring*, published in 1962.

Today such views have been stood on their head. Notwithstanding the hype of the aerospace industry, the deep sea now stands as mankind's next – and most likely its last – great frontier. For half a century global fishing fleets have, without fanfare, scoured the deep sea in search of the world's last virgin stocks. Greater reserves of hydrocarbons are now known to be stored within deep ocean sediments than are to be found anywhere on the land. Vast tracts of the abyssal seafloor are virtually paved with metal nuggets, and certain hydrothermal vents contain some of the world's richest gold deposits.

Vast tracts of the deep sea remain the most pristine and undisturbed habitat on the planet. But the baseline is shifting; today, virtually no part of the deep-sea fauna remains unaffected by man's activities, and these activities are set to expand dramatically in the 21st century. Human impacts on this most extensive of earth's environments arise from pollution and dumping, fisheries, and mineral and energy resource extraction. Potentially massive impacts of human-induced climate change are not fanciful on time-scales of a century and more, the scales of consequence in this debate.

In sum, the broad and often bewildering array of human activities affecting the deep sea is similar to those affecting more familiar environments. But we know less about the deep sea and its ecology and have often paid scant attention to our potential impacts there. As stewards responsible for our planetary environment, we must begin by understanding where we stand and what lies ahead. That is the purpose of the following chapters: to review what we know – and don't know – about the impacts of human activities on the deep sea.

CHAPTER 7

Dumping and Pollution

The deep sea as the ultimate dump site

All of us are familiar with the problems of waste disposal on land: land-fill sites filled to capacity and with nowhere to go; toxic wastes leaching into the soil, groundwater and waterways; old and unwanted cars, white goods, furniture and so on dumped in ditches and ravines at the end of dirt roads; a landscape littered with empty bottles, drink cans, plastics, and fast-food packaging. It's generally accepted today that such practices cannot continue if the landscape aesthetic is to be reasonably maintained. For a time, however, the deep sea seemed the ultimate dump site for individuals and governments alike: virtually limitless and out of sight.

My first summer job after graduating from the University of Washington in 1973 was to assist with a trawl survey of the southeastern Bering Sea, conducted by the National Marine Fisheries Service out of their field station in Kodiak, Alaska. Along with the tonnes of pollock, halibut and other flat fishes, king and tanner crabs, shrimp, starfishes, and whelks that came up in our nets were the occasional boot, can or bottle, and bits of fishing gear discarded by the various fleets. One glossy black bottle (Japanese sake?) was sufficiently exotic and attractive that I took it home and converted it to a lamp base. Less attractive, however, was a beach on one of the remote and uninhabited Shumagin Islands; we found it littered with bits of torn netting, nylon twine, and plastic detergent and motor-oil containers. This synthetic refuse, highly resistant to breakdown, was eventually recognized as not only unaesthetic but highly deleterious to marine life: pieces of styrofoam, plastic bags, cigarette filters, condoms, and other debris choke sea life and block intestinal tracts; netting and other plastics entangle or 'necklace' seals and sea lions, turtles and sea birds. In 1988, what is known as Annex V of MARPOL (Marine Pollution), the International Convention for the Prevention of Pollution from Ships of the International Maritime Organization (or IMO) banned the disposal of plastics at sea.[1]

The loss of a few ships, such as the *Titanic*, are memorialized in our history and popular culture, but in fact some 10 000 merchant ships, the equivalent of more than 40 million tonnes, were lost due to war or accidental sinking between 1914 and 1990.[2] Due to increased global shipping, approximately a third of this total sank in peacetime between 1971 and 1990, according to Lloyd's register – on average, remarkably, a ship was lost every two days. These ships carry down with them their fuel, and often poisonous cargoes. Most ships are lost in coastal waters, and there is widespread awareness about how they may foul our shoreline, but their impact on the deep sea has never been studied.

Munitions dumping at sea

The bottom of the sea, for a long time, served as the ultimate dump site for a range of military and industrial waste. Records of such activities are often poorly maintained: the types of munitions, the quantities, and where they were actually dumped. However, since the end of World War II, well over a million tonnes of defective, obsolete, and surplus conventional and chemical munitions have been dumped into the waters off Europe. Although many such exercises remain secret, Britain has reported dumping 137 000 tonnes of chemical weapons at several sites between 45 and 400 nautical miles off its shores at depths of 500 to 4500 meters. Some 38 ships containing more than 150 000 tonnes of chemical munitions were scuttled off the coast of Norway.[3]

Much of the material consisted of German chemical munitions uncovered after World War II, including tear gas, nose and throat irritants, lung irritants (phosgene), mustard gas, and the nerve gas tabun. Many of these chemicals break down relatively quickly when exposed to seawater, so the nerve gas and tear gas do not pose a long-term environmental risk. However, the nose and throat irritants contain a high proportion of arsenic, which disperses into the marine environment. Mustard gas is highly insoluble and forms lumps and crusts on the seabed. Its impact on the fauna is not known, but trawl fishers have reported a number of incidents. In the Baltic between 1957 and 1991, German trawl fishermen reported ten incidents of burns from mustard gas, and since 1976 more than 400 contaminated Danish trawl catches have had to be destroyed.[4] A survey carried out in 1996 of a major munitions dumping ground off the UK, the Beaufort Dyke, revealed no significant contamination of the sediment or fish fauna. However, 25 explosions have been detected in this area since 1982. When a 60-centimeter-deep trench for a gas pipeline was dug in 1995 between Scotland and Northern Ireland, well to the north of the Beaufort Dyke, some 4500 incendiary devices were dislodged from the seabed. These floated to the surface and were blown into the Firth of Clyde; after drifting onshore and drying, the phosphorus spontaneously ignited. The British Royal Naval Explosive Ordnance Teams commonly defuse munitions washed or trawled up: they worked on some 10 301 articles between 1973 and 1977 alone.[5]

The Contracting Parties to the London Convention, as it is known (more fully, the Convention on the Prevention of Marine Pollution by Dumping of Wastes and other Matter) agreed in 1972 to refrain from munitions dumping at sea, but 'sovereign immunity' provisions provided an escape clause. Britain ended the practice in 1992, some 20 years later.

Nuclear waste disposal at sea

In 1946, immediately following the end of World War II, the United States began to dispose at sea some of the nuclear waste it had accumulated from the development of atomic weapons. Its first ocean dump site was off the Farallones, small rocky islands home to thousands of seals, sea lions, and sea birds, just 50 kilometers off the coast of California, due west of San Francisco. The apparently pristine character of the islands and their surrounding waters was belatedly recognized by the US Federal Government, which in 1981 declared it a marine sanctuary. But by that time the continental slope around the islands harbored, in addition to a riot of mostly poorly known species, some 47 000 fifty-five-gallon steel drums of nuclear waste, holding an estimated 15 000 curies[6] at the time of disposal, lying at depths of approximately 800–1600 meters and spread out over many hundreds of square kilometers.[7]

The United States ended its ocean dumping of nuclear waste in 1970, by which time it had so disposed of some 75 000 drums containing 100 000 curies of radiation over 29 major dump sites and many smaller dump sites in the North Atlantic and Pacific.[8] But by this time most of the major industrial nations – the UK, France, Germany, Japan, Italy, the Netherlands, and Sweden, along with New Zealand in the Southern Hemisphere and even land-locked Switzerland – were dumping their nuclear waste into the ocean as well.

Some 25 years after the US initiated ocean dumping of nuclear waste, concern over this unregulated and proliferating practice led the London Convention in 1972 to ban the ocean dumping of high-level radioactive waste, to develop guidelines for the containers to be used and to stipulate that the dumping should occur only at special sites, at depths greater than 4000 meters. In fact no country admitted to the dumping of high-level radioactive waste – that is, waste that is physically hot to handle. In 1983 the London Convention agreed to a voluntary moratorium on all dumping of radioactive waste at sea, which the Convention made binding in 1993. An inventory of the dumping to date showed that approximately 1.24 million curies had been disposed of in the oceans by this time, spread over some 73 sites, predominantly in the North Atlantic and North Pacific. The European nations had dumped some 220 000 drums, a total of 142 000 tonnes of nuclear waste material, in the waters of the northeast Atlantic.[9]

Remarkably little is known about the potential ecological impact of nuclear waste disposal on the seafloor. No proper monitoring was ever conducted at any of the dump sites. Some of the early dumping used containers deliberately designed to implode, releasing and presumably dispersing the waste material. When some early containers failed to sink, they were shot full of holes, which quickly brought the waste into contact with the environment. Subsequent surveys of some of the dump sites have shown that the drums are in varying states of decomposition and often leaking, and that they provide hard substrate which has been colonized by sessile invertebrates. Elevated levels of radioactivity have been found in a range of organisms at US deep dumpsites: sessile suspension feeders such as anemones, deposit feeders (starfish and sea cucumbers), and mobile feeders associated with the sediments, such as grenadier fishes and decapod crustaceans.[10]

This sad chapter in man's legacy to the marine environment came to a dramatic head with the shock revelation in 1993 that the Russians, who until then maintained that they had never dumped nuclear waste at sea, in fact had violated their commitments under the London Convention in the most gross and careless manner. In a report prepared for President Yeltsin by Alexey Yablokov, his top environmental advisor, and made public almost immediately – a true testament to the spirit of *glasnost* – it was revealed that some 17 nuclear reactors had been dumped into the Arctic waters around the island of Novaya Zemlya. Six were from nuclear submarines, dumped with their spent nuclear fuel still in place: removal was judged 'impossible due to the damaged condition of their cores. For the same reason, 125 irradiated fuel assemblies could not be removed from the core plate of the OK-150 reactor unit on the nuclear icebreaker *Lenin*.'[11] As if these revelations of the dumping of high-level waste were not shocking enough, the report revealed that only one of these reactors had been dumped in water of any depth, at 300 meters; the others lay in a mere 20–50 meters of water within various inlets around this remote Arctic island. An estimated 2.4 million curies of radioactive waste were so disposed, almost twice the total dumped by all other nations combined. The fact that Soviet nuclear naval vessels and the nuclear icebreaker had also been dumping large quantities – more than 300 million liters – of contaminated cooling water from their reactors directly into the sea, again in contravention to the London Convention, might seem to pale into insignificance, except that the Russian navy brazenly continued this practice. In October 1993, following the release of the Yablokov report, a naval vessel dumped hundreds of tonnes of radioactive water into the Sea of Japan under the gaze of Greenpeace activists and the Japanese media. These revelations had one positive outcome, however. In November 1993 the London Convention passed a formal total ban on any further dumping of radioactive waste at sea.[12]

Is this issue then finally closed? Perhaps – but perhaps not. The problem is that, despite decades of effort and billions of dollars spent, there is not yet a recognized solution to nuclear waste disposal. The amount of waste continues to accumulate from nuclear power generation: in the United States alone there are now well over 40 000 tonnes of spent fuel stored at nuclear power plants, and the amount grows by over 2000 tonnes per year.[13] As nuclear weapon stockpiles are reduced, the need to find safe storage for nuclear material increases dramatically. The most toxic waste – plutonium-239, used in reactor fuel and atomic weapons – has a half-life of 24 360 years – for all practical purposes, forever – and it's exceedingly difficult to guarantee that these wastes will remain isolated on land; that earthquakes, seepage into groundwater or human trespass will never occur. Although efforts have focussed on land-based solutions for some time, none has yet been found to everyone's satisfaction; not surprisingly, no one wants such wastes stored in their backyard.

This apparent stalemate has led some to propose a re-examination of seabed disposal. Options include burial within holes drilled (or punched by bullet-shaped canisters) into stable mid-ocean sediments,[14] or even some form of deep-ocean storage that could be recontainerized if problems emerge,[15] although it's difficult to imagine repacking nuclear waste from crumbling containers at some thousands of meters depth. Proponents of seabed dumping or storage point to the sad history of land contamination by nuclear waste and the lack of evidence for widespread contamination around present seabed dump sites. But the fact remains that the impacts of radioactive contamination in the sea have barely been studied. As we will soon see, some marine organisms have a remarkable but poorly understood ability to concentrate a range of substances, including natural radio-isotopes, at some hundreds, thousands, even millions of times their background level in the environment.

Industrial and sewage waste dumping

The London Convention in 1972 also called upon signatories to end the dumping at sea of land-produced industrial waste. An estimated 17 million tonnes of industrial wastes were still dumped at sea in 1979, which was reduced to 6 million tonnes in 1987.[16]

The dumping of sewage sludge offshore is another form of dumping with the potential to have significant impacts on the deep seafloor biota. Some 17 million tonnes of sewage sludge were dumped at sea in 1980, decreasing to 14 million tonnes in 1985.[17] Between 1986 and 1992, 8–9 million tonnes of sewage sludge were dumped each year at a site 185 kilometers off the coast of New Jersey in water depths of about 2500 meters. Early physical oceanographic models indicated that the sewage would not affect the bottom because of dilution and dispersal of the material in the water column, so initial monitoring only examined impacts within the water column. But Cindy Lee Van Dover, Fred Grassle and colleagues at Woods Hole Oceanographic Institution (WHOI) published a study in *Nature* showing, based on the unique stable isotopic composition of the sludge compared with plankton-derived organic matter, that the sludge had entered the deep-sea food chain and was being actively consumed by the dominant surface deposit-feeders, a sea urchin and a sea cucumber. The sewage was likely affecting an area up to 350 kilometers downstream and was effectively doubling the organic inputs to the sediments 50 kilometers downstream of the dumpsite. The dumping ceased in 1992.[18]

TBT and sex change in mollusks

In 1975 the US National Research Council estimated that ships routinely dumped 6.3 million tonnes of garbage into the oceans each year.[19] Much of the impact of this trash is aesthetic; more insidious, however, may be the impact of certain debris falling away from these passing ships that no one knowingly casts overboard.

During a month-long survey of the deep benthic fauna of the eastern Mediterranean in 1993, B.S. Galil and A. Golik of the Israeli National Institute of Oceanography and M. Türkay of the German Forschungsinstitut Senckenberg took the opportunity to also examine their beam trawl samples for evidence of litter.[20] In all they obtained 17 trawl samples between Israel, Greece, and Italy, of which 70 percent contained litter. The most common item was not plastic, although this was common as well, but paint chips, flaked off the hulls of the passing ship traffic. The authors made no more of this curious, seemingly innocuous finding.

Ed Goldberg, the Scripps marine chemist who pioneered the MusselWatch program to monitor coastal pollution, described tributyl tin (TBT) in 1994 as 'the most toxic substance ever deliberately introduced into natural water.'[21] An anti-fouling agent in marine paints, widely applied to the hulls of vessels and aquaculture nets since the early 1970s, TBT disrupts reproduction in certain molluscs, notably whelks, at concentrations as low as a few nanograms per liter of seawater (parts per trillion) and is toxic to the embryonic and early life stages of many organisms at levels of 10–100 nanograms per liter. In whelks and other snails it causes a curious condition known as imposex or pseudo-hermaphroditism, which consists of the development in females of male primary sexual characteristics, such as a penis, and also leads to infertility.[22] TBT pollution and the incidence of imposex in molluscs was widely reported from harbors and other nearshore environments in the 1980s, leading to most developed countries banning its use on vessels less than 25 meters in length. In 1990, the Marine Environmental Protection Committee (MEPC) of the IMO similarly banned the use of TBT-based anti-fouling paints on vessels smaller than 25 meters. Larger vessels were exempted because of the lower levels of TBT in offshore waters.[23]

TBT pollution and its potential impacts are virtually unstudied in the deep-sea environment. However, despite bans on its use on smaller vessels, recent reports indicate that TBT may be continuing to accumulate in offshore sediments[24] and entering deep-sea food chains. A close correlation between shipping intensity and the incidence of imposex has been reported in whelks from the open waters of the North Sea.[25] Deep-sea fish, crustaceans, cephalopods, echinoderms, and gastropods collected from 980 meters depth in Suruga Bay, Japan contained butyl tin concentrations of 980, 460, 460, 130, and 21 nanograms/gram of wet weight, respectively, levels similar to those observed in shallow-water areas with heavy shipping where chronic toxicity effects of TBT have been observed.[26] And significant butyl tin levels (approximately 2 parts per million in dry weight) were observed in Ballenas Basin, a deep sedimentary basin (377 meters depth) 25 kilometers offshore of Vancouver, Canada. Interestingly, there was little diminution in TBT concentrations to a depth of 30 centimeters in the sediment core, corresponding to deposition in the 1970s, which may indicate very slow degradation in the sediments: a half-life on the order of decades or longer.[27] Clearly the potential impacts of shipping on the deepwater biota need to be examined in areas such as the eastern Mediterranean, where shipping traffic is concentrated and the bottom is littered with paint chips slowly raining down from ships' hulls.

Chlorinated hydrocarbons and radioactivity in the deep sea

Whereas TBT was deliberately introduced into the oceans, other key pollutants that affect the global environment – heavy metals, such as mercury, and the chlorinated hydrocarbons, such as the PCBs (polychlorinated biphenyls), DDT (dichlorodiphenyltrichloroethane), and the 'drins' (aldrin, dieldrin, and so forth) – were generally used on the land, and that was where ecological concerns were initially raised. Rachel Carson based her book *Silent Spring* on widespread reports of freshwater fish- and bird-kills in the 1950s. Similar incidents were not observed in the ocean until somewhat later, and the likelihood that these pollutants might affect the deep sea seemed particularly remote.

The chlorinated hydrocarbons of concern are relatively high molecular weight compounds that are highly toxic and persist in the environment for decades. Although virtually insoluble in water, they are highly soluble in lipid and so are stored within organisms' lipids and accumulate through food chains. DDT and the 'drins' are pesticides, whereas the PCBs were widely used in the manufacture of paints and plastics and as dielectrics (insulators) in transformers and capacitors. The use of DDT was curtailed in the developed world in the early 1970s but it continues in the developing world. American production of DDT through 1974 was about 1.4 million tonnes; global production up to then was probably about twice that.[28] Over a million tonnes of PCBs were produced and, ultimately, largely dispersed into the environment, before production was cut back in the 1970s in the USA and in the mid-1980s in Europe.

Initial assessments concluded that the deep ocean was effectively shielded from these pollutants by a series of formidable physical and chemical barriers. First, it was assumed that these pollutants mostly entered the sea through runoff, so any impacts would be limited to coastal waters. Second, because of their relative insolubility in water (their concentration at saturation is about 1 part per billion), it was believed that their impact on life in the sea would be negligible. Further, it was assumed that even if these pollutants made their way into the open ocean, there was extremely limited exchange between the surface and the depths of the ocean: deep convective mixing between surface and deep waters occurs only in winter in the Greenland and Labrador seas and off the Antarctic continent. The residence time of water in the deep ocean – the average time that a water parcel remains in the deep sea – is on the order of a thousand years. Presumably even such persistent chemicals as DDT and the others would have largely degraded before they penetrated to the deep ocean, and once there, they would be diluted to infinitesimal concentrations: with more than 90 percent of the ocean's volume, the deep has near-infinite diluting capacity.

Unfortunately, these apparent barriers all proved to be theoretical constructs, resting on oversimplifications and incorrect assumptions. These pollutants not only rapidly penetrated to the deep interior of the ocean, they ended up concentrated in the biota there, such that the deep sea may be more accurately seen as the ultimate sink for these pollutants, rather than as a pristine sanctuary.

One of the first of the key assumptions to be revised dramatically concerned the route by which these pollutants make their way into the ocean. The most dramatic instances of marine pollution have been caused by deliberate (and sometimes officially sanctioned) discharges. Minamata disease, a severe form of mercury poisoning characterized by a range of horrific consequences, including mental retardation, severe birth defects, deafness, blindness, paralysis, and death, arose in the 1950s in the eponymous coastal town in Japan, where the fish and shellfish, heavily consumed by many of the locals, were contaminated by mercury discharges from chemical plants.

Another severe instance of coastal pollution was uncovered in the late 1960s off southern California. Until 1970 Monsanto, the world's largest producer of DDT, had a permit to dump its wastes directly into the Los Angeles-area sewage system; from there, much of it was then pumped into the ocean. The resulting ocean discharge of chlorinated hydrocarbons was estimated to be 91 tonnes per year, about ten times the amount of pesticides carried into the Gulf of Mexico by the Mississippi River.[29] By the time the practice was stopped, the pesticide residues had permeated the region's marine food chain. Canned jack mackerel (*Trachurus symmetricus*) and Pacific bonito (*Sarda chiliensis*) from California were condemned by the US Food and Drug Administration for having DDT concentrations several times higher than the maximum tolerance of 5 parts per million (ppm). The pelican population in the region appeared to be headed for extinction, with almost total reproductive failure reported in 1969 in the colony nesting on Anacapa Island 35 miles west of Santa Monica Bay, due to thin-shelled eggs that broke beneath the brooding birds. The eggs contained an astounding 1818 ppm of DDT in their lipids.[30] (A key impact of DDT and related compounds is on calcium metabolism.) In 1972, pelican colonies along the California coast enjoyed their first successful breeding in about ten years, as DDT discharges ended and its agricultural use was curtailed. Reports in the late 1960s of sea bird reproductive failure and of sea lion deaths were also associated with runoff from regions of high agricultural use of DDT: in the Laguna Madre off the coast of Texas and in Monterey Bay, California, which receives runoff from the Salinas Valley.[31]

Unfortunate as such ecological disasters are, they are highly visible and arouse communities to take action that ends the most egregious practices. For the most part, however, these pollutants enter the oceans not through outfall pipes or the rivers, whose effluent can be monitored, regulated and eventually turned off, but invisibly through the atmosphere. Both mercury and the relatively high molecular weight chlorinated hydrocarbons are semi-volatile and so are naturally released into the atmosphere. DDT and the 'drins' were widely sprayed onto crops, forests, and waterways, from which they then largely evaporated. The PCBs were also released into the environment through their use in paints and plastics and as lubricants, as well as through the improper disposal of transformers and large capacitors in which they were used. Although these compounds were not deliberately dumped at sea, a National Academy of Sciences report published in 1971, 'Chlorinated Hydrocarbons in the Marine Environment', predicted that 25 percent would enter the oceans within the year of their production. Because of their virtual insolubility, a negligible proportion of these compounds enter the sea through the rivers: perhaps 20 percent of the PCBs and only an estimated 2 percent of the DDT compounds. The remainder arrive via the atmosphere. Once released, these compounds are soon carried by winds around the planet, where most eventually wash out with the rain into the ocean. As a result, significant concentrations of DDT and PCBs are reported from ocean waters in the Arctic, off Antarctica, and in remote regions of the Southern, Indian, and South Atlantic oceans, as well as in the North Atlantic and Pacific closer to the major sources of these pollutants.

The impact of chlorinated hydrocarbons on global marine ecosystems will probably never be known. The contamination was ubiquitous and marine ecosystems are too poorly monitored and understood to allow us to separate natural from anthropogenic variability. Most of the functioning of marine ecosystems occurs at the base of the food chain among the microbes, phytoplankton, and microzooplankton, which are all extremely small – on the order of microns to tens of microns – unicellular, and with high surface to volume ratios. This renders them highly exposed to trace elements, some of which they in fact depend upon – taking them up in extreme trace amounts for their essential nutrition.

The sensitivity of the organisms at the base of marine food chains to DDT, PCBs, and related compounds has turned out to be nothing short of remarkable. There are reports, for example, that PCB concentrations of only 1 part per billion depress the productivity of laboratory phytoplankton cultures by 50 percent.[32] The sensitivity of marine microbes, phytoplankton and microzooplankton to trace pollutants was probably most clearly brought home to marine scientists when it was realized that the metabolism and productivity of these key groups were affected when oceanographers routinely collected them in ostensibly clean water 'bottles': simple open tubes with stoppers at either end. Known as Niskin bottles, they are lowered through the water in the open position to a particular depth, where the stoppers are released and the bottle snapped shut with a piece of rubber or latex tubing that runs through the tube between the stoppers and serves as an elastic band. The latex – seemingly simple and innocuous – released minute amounts of PCBs into the 10-liter water samples, sufficient to depress standard laboratory measurements of microbial, phytoplankton, and microzooplankton productivity by up to 50 percent![33]

In assessing the impact of PCBs on near-surface marine ecosystems at the time of their peak use, a team of chemists and biologists from the Woods Hole Oceanographic Institution concluded:

> the concentrations of PCB that we have found to prevail in Atlantic Ocean organisms during 1970–1972 are within the range that could produce effects ... Plankton samples show the highest of marine concentrations, with PCB levels in their body lipids ranging up to hundreds of parts per million (thousands, in samples rich in phytoplankton). We believe that such levels must be affecting the marine biota. Our belief is supported by recent laboratory studies that showed that some PCB concentrations we have measured in Atlantic water (100 ng/l) caused changes in the species composition of mixed algal populations and possible interference with algal uptake of various nutrients.[34]

But how do these substances penetrate the deep sea? The early models started from a first-order physical and chemical understanding and ignored potential biological interactions, which can be hugely complicated and difficult to deal with. The first indication that the biology, however messy, cannot be ignored came from the global geochemical tracer experiments carried out in the early 1960s by the USA and the USSR, when they exploded within the Earth's atmosphere a series of ever more powerful thermonuclear bombs.

Radioactive particles produced by the bombs were largely injected high into the stratosphere and transported around the world at mid-latitudes within about 15–25 days.[35] Relatively short-lived radio-isotopes in the fallout, such as zirconium-95 and niobium-95, with a half-life of 1–2 months, were therefore used to examine short-term mixing events. Atmospheric fallout from nuclear testing is found mostly in the form of very small particles, less than 4.5 micrometers (millionths of a meter). If inert, these particles would require many years to sink several thousand meters to the seafloor in the open ocean. However, the fallout disappeared rapidly from the ocean's surface waters, leading ocean chemists to think that the particles may be adsorbed onto larger phytoplankton particles, which sink at the rate of about 100 meters per month. Based on this assumption, however, the fallout would still require several years to sink to 3000 meters. Imagine, then, oceanographers' surprise when these fallout products turned up in almost as high levels (relative to long-lived radio-isotopes) in bottom-feeding sea cucumbers at 2800 meters depth off the coast of Oregon as in sea cucumbers living on the continental shelf at 200 meters. The fallout had entered the deepwater food web not within years, not months, but within a week or so.[36] What the oceanographers' early models had failed to take into account was the over-riding importance of the 'bio' component in the biogeochemistry of the oceans.

In 1961, within two months of the resumption of Soviet atmospheric testing, samples of the euphausiid (or krill) *Euphausia pacifica* from 72 kilometers (45 miles) off the coast of Oregon showed a dramatic spike in radioactivity from several of the radionuclides that initially dominate nuclear fallout: cerium-141, zirconium-95 and its decay product niobium-95, all with half-lives of between 32 and 64 days.[37] Euphausiids such as *E. pacifica* are key phytoplankton grazers in much of the world's oceans. But unlike the copepods, which are generally in the order of a millimeter or so, euphausiids are typically a centimeter or more in length. At night they feed in near-surface waters, and here they ingested the radionuclides adsorbed to the surface of phytoplankton and detritus particles. With the coming of dawn, they descend with full guts to several hundred meters depth to avoid the gamut of epipelagic planktivores, from mackerel to salmon, for whom they are preferred prey. And while at depth, they digest, defecate, molt, and are preyed upon by a variety of midwater planktivores, thereby rapidly transporting the fallout particles into the twilight depths of the ocean.

In fact, the daily vertical migration of the krill is not actually critical to the rapid biological transport of fallout from near-surface to deep waters, as demonstrated in the aftermath of the nuclear explosion at the nuclear power station in Chernobyl, Ukraine. On April 25–26, 1986, this tragic accident released some 50–185 million curies of radiation into the atmosphere – several times more than was generated by the bombs dropped over Hiroshima and Nagasaki. As it turned out, less than a fortnight before, researchers at the International Laboratory of Marine Radioactivity associated with the International Atomic Energy Agency in Monaco and the French Centre des Faibles Radioactivités had set out an automated time-series particle trap off the coast of Corsica in the Mediterranean. The trap, set at 200 meters, contained six collector cups, programmed to sample consecutively for 6.25 days each. Air sampling over Monaco showed that the radioactivity arrived within days; rain on 4–5 May then deposited the radioactivity in a single pulse. A week later, a surge of fallout radionuclides 10–100 times higher than background levels was collected in the particle trap. Copepods collected in near-surface plankton samples were incubated; their fecal pellets proved highly enriched in many of the dominant fallout radionuclides.[38] The fecal pellets of copepods only a millimeter or so in length are still sufficiently large (on the order of a few hundred microns) to sink rapidly through the water column – approximately 20–60 meters a day – while the sinking rate of the larger euphausiid fecal pellets is about ten times greater.

But rapid removal from near-surface waters is only one part of the impact of marine food webs on the biogeochemistry of radioactive fallout. The other part, of even greater importance, concerns the dramatic enrichment or concentrating effect of food web processes. Researchers from the University of Cape Town in South Africa and the International Laboratory of Marine Radioactivity at the Musée Océanographique in Monaco initially observed that a naturally-occurring radionuclide, polonium-210, was over 100 times more concentrated in the fecal pellets of *Meganyctiphanes norvegica* (one of the dominant euphausiids off the coast of Europe) than in the whole organism; more than ten times higher than in their potential food; and about 200 000 times higher than in the surface seawater! The euphausiids excreted 99 percent of their plutonium uptake in their fecal pellets. Although plutonium is present at extremely low levels in surface seawater, its residence time in the surface layer is estimated to be on the order of only a year due to the process of being scavenged by small particles, concentrated and repackaged within the marine food chain, and sent by rapid transit (in the form of fecal pellets) into the deep interior of the ocean and to the seafloor.[39]

Given this account, it no longer comes as a surprise that generally higher levels of radioactivity are found in midwater fishes, squids, and crustaceans than in species living near the surface.[40] This has been shown most clearly for polonium-210, a natural nuclear decay by-product of radon, which emanates from the earth's crust. Polonium-210 supplies more than 90 percent of the natural background radioactivity in most marine organisms. Whereas a typical concentration of polonium-210 for fish from surface waters is 1.5 pico-curies per gram (pCi/g), the values for some of the most common midwater fishes, such as lantern fish and hatchet fish, ranged from 24–39 pCi/g. (A pico-curie is a trillionth (10^{-12}) of a curie.) The natural radioactivity levels from polonium-210 for midwater shrimps were mostly in the same range, but one species, *Gennadas valens*, displayed 117 pCi/g, the highest natural radioactivity level ever recorded for an organism. Significantly, the polonium was not evenly distributed throughout the organisms; rather, it was concentrated in the liver of the fish and in the hepatopancreas of crustaceans and squids (the gland in invertebrates that carries out liver functions). The polonium concentration in the hepatopancreas of *G. valens* was a remarkable 856 pCi/g. In general there was a concentration factor of about 10 000 in going from the polonium in seawater to its concentration in phytoplankton, and in the order of a million for these midwater crustaceans. To explain these higher levels in the midwater biota – and in certain organisms in particular – the researchers pointed to evidence that some midwater crustaceans apparently feed particularly heavily on detritus and may even selectively consume the hepatopancreas of other shrimps.[41] *G. valens*' overall background radiation dose is about 310 rem per year, and the dose received in its hepatopancreas is over 2600 rem/year, almost 1000 times higher than the natural radiation received by humans.[42]

A number of basic questions remain unanswered in these studies. Why is polonium concentrated in the hepatopancreas and liver of these midwater organisms?

Does it serve some unknown purpose, or is it a chemical artifact, along the lines of strontium-90, which is taken up in bone tissue as a chemical analogue of calcium? What are the genetic and other implications of this high level of radiation? How close are these creatures to the margin of significant radiation-induced mutation and mortality? As yet there are no answers to these questions.

Clearly the environmental implications of any further radiation burden need to be considered in light of both the extremely high natural levels already present in some organisms and their remarkable ability to concentrate radioisotopes in deepwater food webs. Studies to date have focussed on the rapid scavenging of radionuclides, such as polonium and plutonium, by small particles, mostly phytoplankton, in near-surface waters. These particles are then grazed, with the radionuclides becoming concentrated in fecal pellets, and subsequently further concentrated within deepwater food webs. If plutonium is released from wastes on the deep seafloor, will it be scavenged by the thin rain of detritus that drifts downward through the deep water column – the marine snow, flocs of aggregated phytoplankton cells, fecal pellets, crustacean molts and other material – as well as the sediments and detritus that near-bottom currents periodically raise off the seafloor? Such questions are difficult to resolve, but these are the uncertainties that fuel public anxiety over deep-sea dumping. The dumping of low- and intermediate-level nuclear waste was carried out so haphazardly, with so little monitoring or understanding of its potential implications, and with so little public scrutiny, it is no wonder that public trust has been eroded, and that the public is unwilling to believe that governments and the nuclear industry will in future properly dispose of the huge accumulated waste stockpiles from the nuclear power cycle and nuclear weapons excess to present requirements. The host of surprises that overturned initial scientific expectations, and the scant understanding that we still have of the biogeochemistry of radionuclides, also contribute to present concerns over deepwater disposal. If, like the evil genie, these wastes escape their deepwater containment, they will never be coaxed back in.

Chlorinated hydrocarbons

Once in the ocean, DDT, PCBs, and related compounds have a high affinity for phytoplankton and other living organisms. Unlike the radionuclides, they are highly soluble in the lipids found in living organisms, from phytoplankton and zooplankton to fish, birds, and mammals, including ourselves. As a result, these compounds are sequestered in organisms' lipids and concentrated as they pass up the food chain. They are also transported quickly into deepwater ecosystems.

Once taken up by phytoplankton, PCBs and DDT become highly concentrated in the fecal pellets of herbivores and rapidly removed from the euphotic zone. The concentration of PCBs in the feces of the euphausiid *M. norvegica* was 1.5 million times higher than their concentration in the surrounding seawater and 3.5–21 times higher than in their prey.[43] The entire process of transport into the deepwater realm was far more rapid than scientists initially predicted. George Woodwell and his group from the Brookhaven National Laboratory in New York initially estimated, based on studies of carbon dioxide transport, that DDT would remain in the upper mixed layer for about four years.[44] Subsequent measurements by a Japanese team revealed that the mean residence time of these compounds was less than a year in open oceanic waters, and as little as a few weeks in productive coastal waters.[45]

As a result of their rapid removal from near-surface waters, DDT and PCBs may have as great or even greater impact on deepwater as on near-surface ecosystems and as much impact on open ocean as on coastal systems. Thus although these pollutants are deposited predominantly over coastal regions adjacent to their sources,[46] the higher particulate load of coastal waters – on the order of 100 times higher – means that the pollutants are rapidly scavenged from the water column, deposited on the seafloor and largely buried.[47] Sedimentation rates on the floor of the open ocean are negligible – as low as 1 millimeter per thousand years, compared with several centimeters per year inshore.[48] As a result these pollutants, once scavenged from the water by the plankton, consumed and transported below the euphotic zone, tend to remain within deepwater food webs.

Over the period 1970–1974, as the USA and European countries halted the use of PCBs in products such as plastics, inks and paints, which rapidly leached into the environment, PCB concentrations in near-surface waters dropped dramatically – about 40-fold.[49] The bad news, however, was that the accumulated PCB production of the previous 10–15 years was now found throughout the deep ocean, such that its concentration was now virtually uniform through the water column.[50]

Several recent studies have examined levels of PCBs and DDT in midwater fishes, both in areas reasonably close to sources of pollution, such as off Japan and southern California and in the Gulf of Mexico, and also in remote, ostensibly pristine regions of the mid-North and South Atlantic.

Not surprisingly, high levels of DDT were found in midwater fish, such as the lantern fish *Stenobrachius leucopsaurus*, sampled 40 kilometers off Los Angeles in the late 1960s and early 1970s: about 4 ppm DDT for the whole fish (dry weight), which is equivalent to about 7.3 ppm on a lipid basis. (The US Food and Drug Administration set a maximum tolerance of 5 ppm for DDT in fish to be used for human consumption.) Comparable levels of DDT (2.7 ppm) were found in the liver of Pacific hake (*Merluccius productus*) caught by a Soviet trawler off northern California and Oregon, an indication of how far the pollutant had moved through the region, once within mobile components of the marine community.[51]

A team of researchers from the Universities of Florida and South Florida investigated PCB levels across a range of mesopelagic species – lantern fish, hatchet fish, and other plankton feeders, both vertical migrators and non-migrators – in the Gulf of Mexico. The midwater fish from an offshore region of the Gulf of Mexico, largely influenced by the Gulf Stream Loop Current, generally had levels of chlorinated hydrocarbons 10 times higher than the concentration in near-surface plankton. PCB levels ranged from 3.2–42.9 ppm (lipid basis). Significant concentrations of DDT and dieldrin were also found, but their concentrations were generally less than the PCB concentrations by a factor of ten or more. The study concluded that 'concentrations of pesticides and PCBs from midwater fishes were well within levels shown to be physiologically significant or even toxic to other fish species: the literature is extensive as to the secondary effects and potential genetic and population consequences of DDT, dieldrin and PCB residues.'[52]

A Woods Hole research group found comparable levels of PCB and DDT within a range of mesopelagic fishes across the North Atlantic: PCB levels generally greater than 1 ppm on a lipid basis, with a range between 0.14 and 38 ppm, indicating widespread contamination of the open ocean deepwater environment. DDT concentrations were consistently lower, generally by a factor of about five.[53]

These pollutants soon find their way to the seafloor and into benthic as well as deep pelagic food webs. A large fraction arrives via fecal pellets and other particles not intercepted by the deep pelagic fauna. Once on the seafloor, these organic-rich particles are rapidly ingested – they are the primary food source for much of the benthic community. Many benthopelagic predators migrate upward into the water column to feed on midwater prey as well.[54] Meso- and bathypelagic food webs and those at the deep seafloor are all closely linked, and so is their cycling of pollutants.

As a result, despite the seeming remoteness of the deep sea, pollutant levels in demersal (bottom-dwelling) species, whether inhabiting the deep-sea or the continental shelf, are remarkably similar. Thus in the early 1970s, just prior to and after the US discontinued use of DDT, studies carried out on both the eastern and western sides of the North Atlantic found DDT and PCB levels in the liver of the morid cod, *Antimora rostrata*, living at depths of 2000 meters and greater were similar to the levels reported in the liver of Atlantic cod (*Gadus morhua*) living over the continental shelf off the coast of Canada. Total DDT levels (that is, levels of DDT and its various metabolic breakdown products) in the morid cod were 7.1 ppm in the liver (on a wet weight basis), and PCB levels about a third of that.[55]

Several key points emerged from these studies. First is the rapidity with which terrestrial pollutants found their way into the ocean, mediated by physical and chemical processes. Second, despite the example of ocean dumping of DDT off southern California, the process was predominantly diffuse, virtually invisible and without dramatic incident. Finally, once in the ocean, these compounds were rapidly incorporated into marine food webs and transported into the deep interior of the ocean, where they largely persist. Today we can take some comfort from the fact that PCBs are no longer manufactured; that DDT is no longer used in developed countries, although its use continues in the developing world; and that sewage and industrial wastes are no longer disposed of in the deep sea. But it is deeply disturbing that there is no reasonable answer to the big question: did these pollutants in fact significantly affect biological and ecological processes in the deep sea? Lack of evidence

– the fact that we did not look, or did not look hard enough to rigorously test for potential impacts – should not be taken as evidence of a lack of effect. The recent development of long-term studies of deep-sea communities in the North Atlantic and North Pacific, designed to examine potential impacts of future climate change, are signs of progress (see Chapters 4 and 9).

Trace metals

Trace metals are a third class of pollutants of concern in the deep ocean. However, unlike the chlorinated hydrocarbons, whose biogeochemistry and impacts are roughly comparable and can be treated as a class, the trace elements differ considerably from each other, so discussing them is more complicated. Several are required in trace amounts by various organisms, although they become toxic at higher concentrations. Iron, for example, is an essential and often limiting micro-nutrient for phytoplankton production, as well as being a key element of the respiratory pigment hemoglobin. Copper serves a similar function in hemocyanin, the respiratory pigment of many molluscs and higher crustaceans, and vanadium is found in the respiratory pigment of tunicates. Zinc and cobalt are also required in trace amounts.

Mercury

Mercury – and to a lesser extent cadmium – are the trace metals of greatest concern in deepwater and other marine ecosystems: both are highly toxic and have no known biological function. Unlike other trace metals, mercury is lipid-soluble, magnified through the food chain, and builds up over time in long-lived organisms. Mercury is also volatile; like the chlorinated hydrocarbons, most mercury in the oceans – an estimated 90 percent – enters through the atmosphere.[56] As with other pollutants, there is concern for both its ecological and human impacts, as we increasingly assume the role of top predator in deepwater ecosystems. Human intake of methyl mercury, a particularly toxic form of mercury, occurs predominantly through consumption of seafood.

Mercury is used widely in a variety of industries: as a catalyst in the chlor-alkali industry and in paints, batteries, and other electrical apparatus. It is also emitted as a trace contaminant in the burning of fossil fuels, particularly coal, which accounts for about 65 percent of anthropogenic mercury emissions; waste incineration accounts for another 25 percent.[57] Cadmium has also had widespread industrial use, as well as being emitted as a by-product of lead and zinc mining and refining, of the iron and steel industries, and as an impurity associated with zinc in that metal's many uses. Mercury production was reduced substantially as its environmental impacts became better known; world production fell to a little over 3000 tonnes in 1992 from a peak of 10 600 tonnes in 1971. On the other hand, cadmium production has remained relatively constant at about 19 500 tonnes per year; its increased use in nickel-cadmium batteries has offset its reduced use in other areas.[58]

These trace metals occur naturally, unlike the chlorinated hydrocarbons. Tracing their biogeochemical cycles depends on global models, which are based on relatively few measurements and many assumptions and extrapolations, with key processes still poorly understood.

The natural input of mercury to the atmosphere, mainly its degassing from the earth's crust, was once believed to dwarf human inputs, but data from lake sediments, peat bog cores and atmospheric measurements spanning a quarter century now indicate that the situation is the reverse: anthropogenic emissions are 2.5–4 times greater than natural emissions. Half of anthropogenic emissions are deposited locally; the remainder is taken up into the atmosphere and redistributed around the world. As a result, mercury concentrations have increased by a factor of two to three over the last century in the surface ocean and atmosphere. The residence time of mercury in the atmosphere is only about a year, so this increased atmospheric burden of mercury is deposited each year on the land and sea; the annual deposition is estimated to have increased since the beginning of the 19th century by a factor ranging from 0.5–5.1.[59] Because of the slow and extensive recycling of mercury between ocean, atmosphere, and land, it is estimated that 'elimination of the anthropogenic [human-generated] load in the ocean and atmosphere would take fifteen to twenty years after termination of all anthropogenic emissions.'[60] Anthropogenic emissions of cadmium to the environment are also believed to be at least as great as natural inputs.[61]

The long-term outlook may be more hopeful. Whereas measurements at sea undertaken between 1977 and 1990 had shown atmospheric mercury concentrations increasing at about 1.5 percent annually,[62] this trend now appears to have been reversed. Between 1990 and 1994, atmospheric mercury concentrations suddenly decreased by 22 percent over Europe and the Atlantic, and measurements taken up to 2001 indicate atmospheric mercury levels are now comparable to those measured in 1977–80.[63] This is attributed to reduced mercury production and use, particularly in the developed world, leading to reduced emissions from waste incineration, along with increased use of desulfurization in coal-fired power plants, which reduces emissions of mercury as well as of sulfur dioxide.

The biogeochemical cycling of mercury in the ocean is particularly complex, because of its many forms: elemental mercury, inorganic ionically bound mercury, colloidally bound mercury and organic methylated (dimethyl and monomethyl) mercury. Not only do their chemical dynamics differ in surface and deep waters, but their uptake within the food chain, their toxicity, and the degree to which they are excreted or stored and accumulated within organisms all differ as well.

In the atmosphere and ocean surface waters, mercury is found primarily in its volatile elemental state. Much like the chlorinated hydrocarbons, elemental mercury is relatively insoluble in water but is adsorbed on the surface of phytoplankton, bacteria and other small particles, and is readily dissolved in lipids. However, elemental mercury seems to be bound mostly to the outer membrane of phytoplankton, little of which is assimilated when consumed by zooplankton, so an estimated 85 percent of the elemental mercury consumed by zooplankton is transported to deeper water in their fecal pellets.[64] As a result, mercury concentrations are substantially greater below the thermocline than in the upper mixed layer.[65]

Once in deeper water, whether transported biologically or through physical mixing, microbes (and possibly further processes still poorly understood) transform the mercury into its organic form, methyl mercury, which is the dominant form in deep water.[66] Methyl mercury, unfortunately, is particularly toxic; and when mixed or upwelled into surface waters, it is taken up within the cytoplasm of phytoplankton, and, when consumed, over 60 percent is assimilated by zooplankton. Methyl mercury continues to be preferentially assimilated in further trophic transfers up the food chain to fish, so although the concentration of methyl mercury in surface waters is only about 5 percent that of inorganic mercury, it is about 16 times more prevalent in plankton-feeding fish.[67]

Mercury concentrations in fish generally increase with depth, trophic level, size and age. Thus mercury concentrations in mesopelagic plankton-feeders, such as lantern fishes, are several-fold higher than in pelagic fish at the same trophic level, such as mackerel. Heightened mercury levels in deep-sea organisms, relative to those from coastal environments, are now reported from the North Atlantic, North Pacific and Mediterranean[68] across a wide range of groups: crustaceans, such as shrimps and krill, demersal fishes (morid cods and grenadiers compared with cod), and benthic invertebrates, such as sea stars and sea cucumbers.[69]

As fisheries extend beyond the continental shelf, there have been increasing concerns about the elevated mercury levels in deepwater species. The USA and EU countries have set 0.5 ppm as the maximum permissible level for mercury in food. The mean levels in long-lived deepwater species near the top of the food chain, such as rough-head grenadier (*Macrourus berglax*), abyssal grenadier (*Coryphaenoides armatus*), Pacific hake (*Merluccius productus*) and orange roughy (*Hoplostethus atlanticus*) approach this level (0.3–0.4 ppm), and larger, older individuals often attain or exceed it.[70] Mercury levels appear particularly problematic in areas such as the eastern Mediterranean, where the concentration in larger individuals of five species of deepwater sharks were about 5 ppm, which is ten times higher than the current accepted levels.[71] In the Tyrrhenian Sea off Tuscany, where there is input from mercury-bearing ores, large hake have mercury levels up to 3.2 ppm.[72] Assuming a concentration of methyl mercury in surface waters of 0.1 parts per trillion, these values represent a concentration factor of 5–50 million.

To place these values in an ecological context, a simple early laboratory experiment showed that mosquito-fish exposed to sub-lethal amounts of mercury were less able to escape predatory bass and thus suffered higher mortality. These effects were observed for fish with mercury levels in their flesh as low as 0.67 ppm, levels comparable to those observed in some hake and orange roughy today, and considerably less than the levels in deepwater sharks and hake in parts of the Mediterranean.[73] Will there be significant impacts on ecosystem functioning? This key issue has not yet begun to be addressed.

Present models indicate that mercury

levels in fish have likely tripled, along with the approximate tripling of mercury concentrations in the ocean in the modern relative to the pre-industrial era.[74] In addition to questions of impacts on ecosystem functioning, this has led to concerns about human health, since the source of most methyl mercury in humans is from eating seafood. A recent study published in the *Journal of the American Medical Association* indicated that although average mercury levels in adult women were not a cause for concern, mercury levels were four times higher in fish consumers (defined as women who had eaten three or more servings of fish in the past 30 days) compared with those who had not eaten any, and approximately 8 percent of women had concentrations higher than the US Environmental Protection Agency's recommended reference dose (5.8 micrograms per liter), below which exposures are considered to be without adverse effects. The report recommended that pregnant women or those intending to become pregnant should follow federal and state advisories on fish consumption.[75]

Cadmium

Cadmium is sometimes found at quite elevated levels in particular regions or species, for reasons that remain obscure. A study of this element in the plankton of the North Pacific found generally low levels (less than 5 ppm, dry weight) but levels in the plankton off Baja California and extending hundreds of kilometers out to sea were consistently higher than 10 ppm.[76] Equally mysterious, the mesopelagic shrimp *Systellaspid debilis*, in the Northeast Atlantic Ocean, has been consistently found to have far higher cadmium levels than other midwater crustaceans from the region, between about 11 and 32 ppm.[77] Typical values for pelagic shrimps in coastal regions are on the order of 1–2 ppm; such high cadmium levels are only recorded for highly contaminated sites, such as the Bristol Channel.

Cadmium, unlike mercury, does not appear to be concentrated up the food chain. Levels of cadmium in the muscle of dominant midwater commercial fishes in the Northeast Atlantic, such as hake, blue whiting, and black scabbard fish, were 0.004–0.034 ppm (wet weight basis), a small fraction of the values recorded for midwater crustaceans. However, pollutants are often taken up by the liver, and median cadmium levels in the liver of these fish ranged from 0.22 ppm for the hake to 6.98 ppm for the black scabbard fish, which considerably exceeded the European limit for human consumption of 2 ppm. Although fish livers are not commonly consumed, black scabbard livers are considered a delicacy on Madeira and in Portugal.[78]

Summary – pollutant impacts

Neither distance nor depth shields the deep sea from pollution. Transport of chlorinated hydrocarbons, mercury, and many radionuclides is predominantly through the atmosphere – so they rapidly achieve a global distribution and rain down all over the planet. These substances are generally relatively insoluble in water but are adsorbed onto particles and dissolve in lipid, such that they are readily taken up by the phytoplankton at the base of the food chain, grazed, and rapidly transported into deep water via the grazers' fecal pellets or their vertical migrations. Sequestered within the marine biota and concentrated many thousand times relative to their concentration in the water column, the large volume of the deep ocean no longer provides significant diluting capacity. Indeed, the deep-sea biota becomes one of the ultimate sinks for these pollutants. As some marine chemists and ecologists have noted, 'the deep-sea fauna may be the first global biotic environment that faces a long term danger from contamination; molecules have plenty of time.'[79]

The biogeochemical cycling of these pollutants and their interactions with marine organisms are still not well understood. Are there synergistic interactions among the various pollutants? At what levels do physiological or behavioral effects occur, leading to the loss of more vulnerable species or those with the greatest burden of pollutants? If such impacts were to occur, leading to changes in the species composition of deepwater ecosystems – much as phytoplankton communities exposed to chlorinated hydrocarbons shift in laboratory experiments[80] – would we recognize that it was happening? Even if the deep sea were adequately monitored, could we distinguish the effects of, say, climate change from those induced by pollutants? Certainly not at this time.

What is the long-term outlook for pollutant impacts on the deep sea? The chlorinated hydrocarbons such as PCBs, DDT and related pesticides are exceedingly long-lived, but fortunately their use has been sharply curtailed since their environmental impacts were recognized. Over time their concentrations in the deep sea should diminish, as they did within the ocean's surface waters in the 1970s and 1980s. Dumping of industrial and radioactive waste into the ocean is also largely a matter of the past. However, mercury pollution appears less tractable, since it is released largely as a by-product of coal burning, waste incineration and smelting ores. Although mercury use has decreased, the anthropogenic contribution to the global environment is still about twice its natural inputs.[81]

CO_2 sequestration in the ocean

Carbon dioxide (CO_2), the by-product of burning fossil fuels, is an emission for which there seems no reduction in sight. About 5.4 billion tonnes (or gigatonnes) of carbon are now released into the atmosphere each year as CO_2 from the burning of fossil fuels. About a third is absorbed by the ocean, and some may be taken up by the growth of terrestrial plants – although over the past two centuries, deforestation and land-clearing have contributed significantly to the flux of carbon dioxide into the atmosphere. In any case, most of these emissions remain in the atmosphere, leading to an increase from 280 ppm in the pre-industrial atmosphere (circa 1800) to about 360 ppm today, with a projected increase to some 850 ppm by 2100 under the current 'business as usual' scenario: a trebling of pre-industrial levels in the atmosphere (see Chapter 9).[82]

There have been many proposals about how to ameliorate the build-up of carbon dioxide in the atmosphere, several of which have involved sequestration in the deep ocean. Superficially at least, this appears attractive. CO_2 differs from other gases in the atmosphere in that virtually all the CO_2 in the combined ocean-atmosphere system (98.5 percent) is already in the ocean. By comparison, only 0.6 percent of the oxygen is within the ocean. The main reasons for this are that CO_2 is very soluble (about 30 times more soluble than oxygen), and that it forms a weak acid called carbonic acid in seawater, which readily reacts to form bicarbonate (HCO_3^-) and carbonate (CO_3^{2-}) ions, leaving only about 1 percent of the dissolved inorganic carbon in the form of non-ionic CO_2. So the ocean readily takes up excess atmospheric CO_2.[83]

The ocean also contains vast quantities of calcium carbonate, the building block of coral reefs and the shells of molluscs and other organisms, whose remains (primarily the shells of tiny phytoplankton known as coccolithophores) comprise much of the sediments over a large portion of the deep ocean floor: there is an estimated 30 000 times more carbon locked up within the ocean sediments than is present as CO_2 within the atmosphere (see Chapter 9). The ocean thus has an immense buffering capacity: the deep ocean is able to absorb on the order of 1000 to 10 000 gigatonnes of carbon, a very large amount relative to the 6 gigatonnes of carbon being released annually into the atmosphere. On a time-scale of centuries, about 80 percent of the excess carbon dioxide in the atmosphere will be taken up by the oceans.[84] So, one may ask, why not speed up this process?

Two very different approaches have been proposed to sequester humanity's excess CO_2 production in the ocean. One approach, direct injection of CO_2 into the deep ocean, was first proposed in 1977.[85] The second involves massively enhancing phytoplankton production, leading to a drawdown of CO_2 from the atmosphere into the ocean. This latter proposal would not seem relevant to a book on the deep sea. But while phytoplankton account for about half of the earth's photosynthesis, they comprise less than 1 percent of the photosynthetic biomass; unlike the forests, they hold out little prospect for carbon storage. The enhanced surface production must therefore be removed to the deep ocean through enhanced sedimentation, for example of fecal pellets or of the phytoplankton itself, as occurs under bloom conditions. Unless the carbon is removed to the deep ocean it will simply be metabolized and return to the atmosphere. Another unique feature of CO_2 is that its solubility *increases* in colder water, as well as with increased pressure, so the deep ocean is undersaturated with CO_2 and has considerably more capacity to take up CO_2 than does the surface water.

But how could we significantly enhance ocean productivity? The key lies in a new understanding of the nutrients critically limiting phytoplankton growth. Oceanographers once considered that nitrogen was the key nutrient limiting phytoplankton production in the ocean.

However, some 30 percent of the world's oceans – much of the Southern Ocean and both the subarctic and equatorial Pacific – have consistently low phytoplankton concentrations despite the relatively high levels of nitrate in their surface waters. Only in the 1990s has it become apparent that iron, which is required in trace quantities to synthesize chlorophyll and for several other key biochemical processes, appears to be deficient in these waters, limiting phytoplankton growth. A series of spectacular mesoscale experiments have now iron-fertilized patches of ocean comprising tens to hundreds of square kilometers in the open Southern Ocean and the subarctic and equatorial Pacific. In each case, phytoplankton productivity and concentrations increased several-fold, and nutrients such as nitrate, phosphate, and silicate were concomitantly reduced, along with concentrations of CO_2 in the overlying atmosphere and dissolved inorganic carbon in near-surface waters.[86] In regions with considerable excess silica in the water, large diatoms mostly dominated these blooms, which generally aggregate and sink into deeper waters once nutrient depletion sets in, although the experiments typically did not last long enough to observe this.

John Martin of Moss Landing Marine Laboratory in California, who pioneered the iron-limitation hypothesis, also hypothesized that iron may ultimately regulate atmospheric CO_2 levels and hence global climate, through its impact on ocean productivity. Air bubbles from cores of the Antarctic ice sheet sample the atmosphere from the time the ice was laid down and show, not surprisingly, that global CO_2 levels were much reduced (about 200 ppm) during glacial periods compared with interglacial periods (about 280 ppm prior to our own industrial period). Interestingly, higher levels of iron-containing terrestrial dust are also present during glacial periods, because of a five-fold expansion of arid tropical regions and enhanced global winds.[87] The dust blown off the land may provide sufficient iron to stimulate ocean productivity such that the excess nutrient, today found over almost a third of the ocean, is taken up by the phytoplankton. If this added productivity ends up in the deep ocean, atmospheric CO_2 will be drawn down, leading to a new equilibrium between CO_2 levels in the atmosphere and ocean.

The Kyoto Protocol, by which most nations have agreed to limit greenhouse gas emissions, assigns each country an emission limit which it can exceed if it takes measures to absorb atmospheric carbon. Some countries have also established carbon emission taxes, by which businesses are taxed according to the amount of carbon they emit into the atmosphere. A number of embryonic industries have emerged, proposing different ways to reduce atmospheric CO_2. Certain geo-engineering projections indicate that iron fertilization of the ocean is the least expensive solution to carbon sequestration, costing only about US$1–2 per tonne of carbon – less by a factor of 10–100 than the next cheapest option, forestation.[88]

These projections are based on laboratory estimates that for every tonne of iron added to the ocean, some 30 000–100 000 tonnes of additional carbon will be photosynthesized and stored in the deep ocean. However, this assumes both that all the iron will be taken up by phytoplankton and that all additional carbon photosynthesized sinks into deep water ungrazed, rather than being recycled in near-surface waters. In a recent, fairly typical Southern Ocean experiment, only an additional 1600 tonnes of carbon per tonne of iron – just a few percent of the theoretical expectation – sedimented even to the relatively shallow depth of 100 meters.[89]

The biological mill in the ocean mostly grinds exceedingly fine. Under natural conditions, only a few percent of the organic carbon fixed by phytoplankton ends up in the deep ocean. The rest is ingested and re-ingested in the upper layers of the ocean, where at each step in the food web, most of the carbon is metabolized and respired; that is, released as CO_2, which may then return to the atmosphere. Massive sedimentation of uningested phytoplankton into the deep ocean depends upon a special set of conditions leading to a bloom of large diatoms at sufficiently high concentrations that the cells form flocs that sink rapidly out of the surface layers. (Diatom cell walls are composed of silica, which is denser than seawater.) However, a diatom bloom requires excess silica, as well as nitrogen and phosphorus, in the water, and the most accessible iron-deficient ocean regions contain only moderate excess silica, limiting the potential for continued diatom blooms.

The key concern with iron fertilization of the oceans, if undertaken on a massive scale, is that it may lead to massive changes in the ecology of the oceans: changes in the plankton communities in near-surface waters and oxygen depletion in sectors of the deep ocean, if carbon sedimentation greatly increases.[90] This is because sedimented organic carbon is eventually consumed and metabolized by marine microbes and other organisms in the deep ocean, using up the limited oxygen available there. Furthermore, several ocean models now indicate that iron fertilization, even carried out on a massive scale, will have only a modest impact on atmospheric carbon dioxide levels (17 percent or less) and that the carbon dioxide would return to the atmosphere within decades, once fertilization ceased.[91] Unfortunately, the lure of carbon credits has led to commercial proposals for large-scale iron fertilization of the oceans, although it's uncertain whether its environmental impacts, or even its efficacy, will be assessed or monitored adequately.

Others have proposed injecting CO_2 directly into the ocean. In order that winter mixing does not return it to near-surface waters and thence back into the atmosphere, most proposals favor deep disposal.[92] The potential risks to deep ocean ecosystems are difficult to evaluate: several disposal methods have been proposed, whose underlying physical dynamics – the basic interaction of the CO_2 and seawater – are very different and still poorly understood. The potential ecological impacts present further issues that have only begun to be explored.

Carbon dioxide undergoes phase changes to its liquid and solid forms at the temperatures and pressures found in the deep sea. At depths between 300 meters, where CO_2 first turns liquid, and about 3000 meters, liquid CO_2 is buoyant, so it returns to the surface, some of it dissolving along the way. However, at depths below 3000 meters, it becomes denser than seawater. As noted above, the deep ocean is highly undersaturated with CO_2 because its solubility increases at lower temperatures and higher pressures. As a result, CO_2 disposed of at depths below 3000 meters would form a carbon dioxide 'lake' on the seafloor that would gradually ice over and eventually dissolve.[93]

The presence on the seafloor of large deposits of CO_2 hydrate (the solid form of carbon dioxide, which in the sea forms a water–CO_2 complex), releasing CO_2 into the seawater, poses two primary ecological hazards. First, the additional CO_2 will lower the pH of the seawater – make it more acidic – despite the considerable buffering capacity of seawater. Some injection methods would reduce the pH by more than one pH unit, thereby increasing acidity by more than a factor of ten.

Deep-sea organisms have lived in a highly stable physical environment and appear to be sensitive to small changes in acidity. Due to their reduced metabolic rates, deepwater fishes and squids seem to have only 1 percent of the intracellular buffering capacity of shallow-living species. Decreased pH severely affects the ability of respiratory pigments, such as hemoglobin, to take up oxygen, as well as affecting other cellular functions. A drop of 0.2 pH units in the arterial blood of several deepwater crustaceans has been found to lead to a 25–50 percent drop in their ability to bind oxygen, and fish hemoglobin appears to be even more sensitive. A drop in seawater pH of 0.5 was found to affect oxygen uptake in a midwater shrimp (*Gnathophausia ingens*), indicating a very limited ability to regulate its internal pH in the face of changes in its environment. Other studies have observed that small increases in CO_2 and resulting decreases in pH lead to metabolic suppression in a number of deep-sea organisms, a mechanism whereby organisms shut down protein synthesis (and hence, growth and reproduction) when physiologically stressed. It is estimated that sequestration of sufficient CO_2 in the deep ocean to stabilize atmospheric CO_2 at twice the pre-industrial levels (550 ppm) would lower the pH of the entire ocean by about 0.1 pH units by 2100.[94]

Second, a recent field experiment showed that, attracted by bait, deep-sea fish scavengers entering an area of CO_2 hydrate are rendered unconscious by the enhanced CO_2 concentration. This experiment raises the further specter of a deep-sea death trap, the equivalent of the La Brea tar pits: a miasma that organisms wander into and die, only to attract further victims.[95] However, observations around Loihi seamount near Hawaii, which actively vents large amounts of CO_2, indicated that some scavengers, such as lysianassid amphipods and basket eels, detected and avoided the CO_2 plume, even when bait was placed within it.[96]

What are the prospects for deep-sea disposal of CO_2? The economic driver for deep-ocean and other forms of CO_2 sequestration is the carbon tax now being levied on industry for CO_2 emissions. For example, the Norwegian government charges NOK300 (about US$45) per tonne of CO_2 emitted. The cost of deep-ocean sequestration appears to be approaching this.[97] However, as indicated by the recent cancellation of a Norwegian experiment to examine the impact of releasing five tonnes of CO_2 at various depths off its shores, there are considerable legal hurdles – does ocean CO_2 sequestration contravene international agreements on dumping at sea? – as well as environmental questions that need to be addressed before such schemes are accepted by the global community. The continued build-up of CO_2 in the atmosphere poses one of the gravest global environmental threats, and it will loom ever larger the longer it continues unabated. There will therefore likely be increasing pressure to sequester CO_2 in the deep ocean or other geological deposits, such as saline aquifers or old oil and gas fields.[98]

Conclusion

Whether wittingly or in ignorance, the deep ocean often serves as a sink for our society's most noxious by-products, from the deliberate dumping of chemical weapons and nuclear wastes to the inadvertent build-up within the deepwater biota of synthetic organic chemicals and mercury. To a large extent our concern for these by-products ended as they were removed to the deep – offshore and out of sight. While many – but not all – of these practices have been stopped or reduced, it remains our responsibility to monitor and understand the ultimate fate and impacts of pollutants. As coastal fisheries continue to be depleted and fishers venture into deeper waters, it is increasingly likely that these unwanted by-products will return to haunt us, set before us on our dinner plate. As our uses of the deep sea continue to grow, so will our need to understand our impacts upon it.

CHAPTER 8

Mining the Deep Sea

Manganese nodules and deep seabed mining

Among the more curious items collected by the *Challenger* in its dredges were large numbers of metallic nuggets, ranging in size from small pebbles to cobbles. Composed predominantly of manganese and iron (two parts manganese dioxide to one part iron oxide), but also containing small amounts (1–3 percent each) of cobalt, copper, and nickel, they have come to be commonly known as manganese or polymetallic nodules. Widely distributed at abyssal depths of 4000–6000 meters, they are found over some 70 percent of the deep seafloor.[1] As might be expected, their size, shape, abundance, and mineral composition vary considerably from one region to another. Where they are most abundant, the seafloor is virtually paved with nodules: up to 90 percent of the seafloor is covered, with densities of 10–60 kilograms per square meter.

The great chemical oceanographer Wally Broecker, of Columbia University and the Lamont-Doherty Geological Observatory, referred to 'the great manganese nodule mystery.'[2] The various metal constituents of the nodules are found in such minute quantities in seawater that until the relatively recent development of 'ultra-clean' analytical techniques, they could not be reliably measured, due to trace contamination from the glassware, the wire used to lower the sample bottles, the ship environment, and so on. And yet there they are, the most abundant material on the seafloor after the 'clay' and the sedimented calcite and opal (silica) remains of planktonic organisms. What is the origin of the nodules? What part might they play in the ecology of the deep ocean? Lying there apparently for the taking, what might be their value, and the implications of mining them?

John Murray noted some larger organisms on nodules recovered during the *Challenger* expedition.[3] But samples obtained from nodule-rich regions in the early period of deep-sea exploration were generally recovered in poor condition, the nodules grinding and crushing much of the fauna within the cod end during the long process of trawling or dredging them from the seafloor. So it was not until the mid-1980s, when modern box cores were used to retrieve virtually undisturbed samples of the seafloor, that Lauren Mullineaux, a deep-sea ecologist then working as a post-doctoral fellow with Bob Hessler at Scripps, discovered their fragile fauna – a unique ecological community contained within the universe of the nodules. Most of the fauna inhabiting deep-sea sediments are tiny but mobile sediment feeders. The nodules, however, provide the only hard substrate available over vast reaches of sediment-covered abyssal plain, and the nodule fauna is predominantly sessile, using the nodules as a base from which to extract food from the currents wafting past. A very small proportion of the organisms are multi-cellular invertebrates such as Murray had noted: sponges, hydrozoans, bryozoans, mollusks, ascidians, and others. Some 98–99 percent of the organisms consisted of foraminifera, single-celled amoeboid protists that secrete a calcium carbonate shell around themselves. Some forams (as they are commonly known) live within the water column, others on the seafloor. Generally, though, they are well known to sedimentologists, who use various species to date sedimentary deposits, such as during exploration for oil; and major chalk deposits, such as the white cliffs of Dover, are predominantly composed of forams. But many of the forams that formed mats, tunnels, and tubes over the surface of the nodules were from groups – entire families – never seen before.[4] Like the higher invertebrates, many of the forams also appeared to be suspension-feeders, extending their pseudopodia up into the currents to extract particles. Hjalmar Thiel (a deep-sea benthic ecologist from the University of Hamburg) and colleagues subsequently uncovered yet another group of organisms, mostly nematode worms and minute harpacticoid copepods, living in the sediment within cracks and crevices in nodule surfaces.[5] The nematodes appeared to subsist on bacteria that also colonize the nodules. Worlds within worlds!

How these nodules form, and why they are found in such abundance in some regions, remain some of the most enduring mysteries of the deep sea. The nodules always grow around a nucleus – a shark's tooth, a volcanic fragment, a fragment of another nodule – the minerals precipitating out of the water or from the sediment below. However, whether the process is entirely a matter of seawater chemistry or whether it is biologically mediated is still contested. Manganese is found in some forams' waste pellets deposited on the nodules, and bacteria isolated from the nodules are also able to precipitate manganese. Sectioning the nodules reveals concentric rings, leading to speculation that nodules are the deep-sea analogue of stromatolites – ancient structures (their fossils date back about 3 billion years) formed by cyanobacteria that also accrete sedimentary material in concentric rings. However, the roles played by chemical versus several proposed biological processes in the growth of the nodules remain shrouded in mystery.[6]

One difficulty is that the nodules grow extraordinarily slowly, only a few millimeters per million years, as measured by several radiometric dating methods. Their growth rate is thus about a thousand times less than the sedimentation rate – a few millimeters per thousand years – even in the central oceanic basins. This is almost certainly why they are predominantly found in unproductive parts of the ocean, where sedimentation rates are lowest. But still, their slow growth compounds the mystery surrounding them. How do they avoid being buried? Drilling and coring shows that some are indeed buried, but mostly they are concentrated at the surface. Do currents winnow away the sediment around them? Do burrowing organisms intermittently push and roll them along, keeping them at the sediment surface? These are the best hypotheses to date.

In the 1950s and 1960s, almost a hundred years after their discovery, manganese nodules were suddenly transformed from a scientific curiosity to the centerpiece of intense global resource speculation. The debate over ownership and access to the nodules continued for decades, culminating in the most substantial revision to the

High density of manganese nodules on the seafloor in the Clarion-Clipperton Fracture Zone of the North Pacific Ocean. (Craig Smith, University of Hawaii.)

international doctrine of the freedom of the high seas since Hugo Grotius developed the concept in the seventeenth century.

The explosion of interest in the nodules can be attributed largely to John Mero, a mining engineer, who became interested in the mineral wealth of the oceans as a graduate student at the University of California at Berkeley in the 1950s. He subsequently took up a position at Scripps, and in 1965 published *The Mineral Resources of the Sea.*[7] Based on data from a number of oceanographic cruises, he concluded that there were well over a trillion tonnes of nodules on the surface of the sediments of the Pacific Ocean alone, representing reserves of manganese, cobalt, and nickel that were more than a thousand times larger than those on land. Even if only 10 percent of the nodules were economic to harvest, the resource was sufficient to provide our global needs of these metals for thousands of years. Indeed, based on his estimate of the nodules' rate of growth, some of the elements were accumulating faster than they would be mined, given projected demand. Renewable mineral resources! A veritable dream!

In the ensuing decade, Mero's book launched, if not a thousand ships, approximately 200 research cruises by the USA, France, Germany, and the Soviet Union alone. Most attention focussed on the Clarion-Clipperton Fracture Zone (CCFZ), a region of the equatorial North Pacific between about 10° and 20° N latitude with a high density of high-grade nodules. (To be of commercial interest, an area must have at least 10 kilograms of nodules per square meter, with a combined nickel, copper, and cobalt content of greater than 2.5 percent.) In 1977 Mero published new resource estimates for this region, revised dramatically downward from his earlier estimate but still substantial: 11 billion tonnes of manganese, 115 million tonnes of nickel and 520 million tonnes of copper. He predicted that nodule mining would be in full swing within 5–10 years.[8]

Almost 30 years on, interest in manganese nodule mining remains active but the prospects for a full-scale commercial operation have receded to a 10–20-year time horizon, due to a combination of technical, economic, and international legal issues. Probably of least consequence is the fact that the nodules do not represent a cornucopia: not only is the CCFZ not quite as rich as Mero believed, with his revised resource estimates scaled back today by between a third and a half again,[9] but his estimates of nodule growth (1 millimeter per 1000 years) were several hundred to a thousand times too high. So the nodules do not represent a magic pudding.

A more significant hurdle is the technical challenge of extracting the nodules. There is no deep seabed mining system in existence. (The only test system to be built was lost off the stern of its vessel after retrieving some 800 tonnes of nodules in 1978 – a testament to the hazards of working on the open ocean.) Most contemporary designs are based on an airlift system. To be economically viable, it would need to retrieve some 3 million tonnes of nodules annually from depths of 4–6 kilometers below the sea surface, equivalent to clearing a square kilometer, or lifting 10 000 tonnes, daily (assuming a density of 10 kilograms per square meter).

While nodule mining is generally

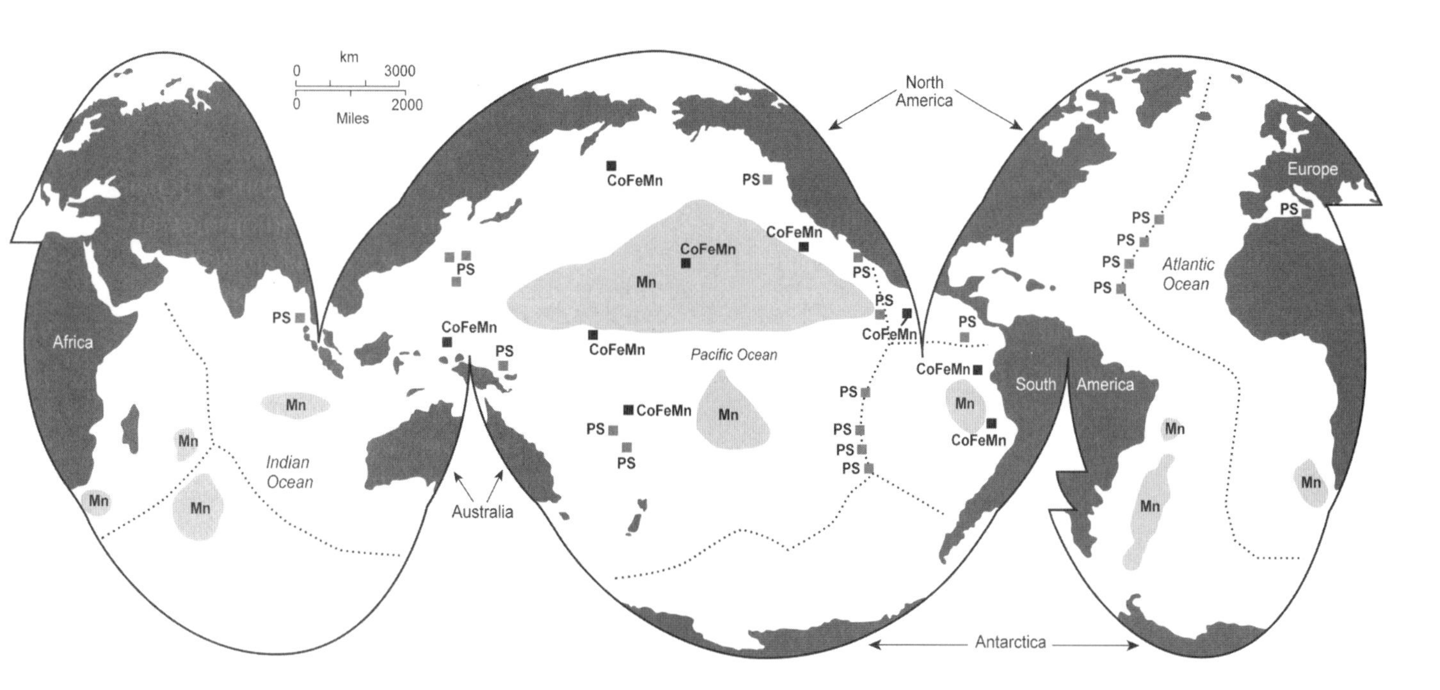

The global distribution of deep-sea mineral resources: manganese nodules (Mn), cobalt-enriched ferromanganese crusts (CoFeMn), and polymetallic sulfides (PS). (Modified from Rona, 2003.)[5]

considered technically feasible, it has not been economically viable, largely due to depressed global metal prices since the early 1980s. Cobalt is a strategic mineral, due to its use in jet aircraft engines, but the discovery of major new land-based deposits of nickel, copper, and cobalt in Canada (Voisey's Bay, Labrador) and Australia has dampened enthusiasm for risky new sea-based ventures.

On the other hand, a number of developments have considerably enhanced the technical feasibility of deep seabed mining: multibeam broad-swath acoustics to map large areas of the deep seafloor quickly and precisely; the global positioning system (GPS) that provides resolution on scales of less than a meter; dynamic positioning, automated GPS systems to maintain precise vessel position; robotics and tracked deepwater ROVs. There is considerable synergy among offshore industries, so technologies developed by the US$100 billion offshore oil and gas industry or for seafloor cable-laying can be readily transferred to the fledgling offshore minerals industry.

Yet another hurdle to manganese nodule mining has been legal and jurisdictional. In the heady days of early nodule exploration, the major industrial nations – primarily the USA, France, Germany, and the Soviet Union – considered that only they had the capital and technology required to undertake such an enterprise. They therefore initially assumed that the enormous potential wealth lying on the seafloor in international waters, beyond the reach of any nation's sovereignty, was theirs for the taking. But the prospect of virtually limitless mineral wealth captured the imagination of the developing world as well. In 1967 the ambassador to the UN from Malta, Arvid Pardo, delivered an historic address to the General Assembly in which he argued that the seabed on the high seas should be considered the 'common heritage' of humankind, to be used and exploited for the benefit of 'humanity as a whole, and not subject to national appropriation.'[10] Pardo's address galvanized the Group of 77 non-aligned and developing nations in the UN, and led to passage later that year of a General Assembly resolution to that effect, over the opposition of the traditional maritime powers.

Over the next quarter century, intense debate and negotiation continued between developing and developed nations over these issues. A UN conference on the Law of the Sea was convened in 1973, but the resulting UN Convention on the Law of the Sea (UNCLOS) was not formulated and signed until 1982 and, due to insufficient signatories, did not come into force until 1994. The G77 nations had the numbers and held sway, but the major maritime powers refused to sign until some aspects of Part XI, having to do with mineral resources on the high seas, were renegotiated. The last major holdout, as in so many international agreements, is the United States, where the treaty has been long blocked in the Senate, but is considered likely to be adopted in the near future. Key results of UNCLOS include recognition of national sovereignty over an Exclusive Economic Zone (EEZ) that extends to 200 miles offshore, and the establishment of a UN agency, the International Seabed Authority (ISA), to regulate the extraction of mineral resources on the seabed in areas beyond national jurisdiction.

In 2001 the ISA granted 15-year licenses to seven nationally based mining consortia to explore the mineral resources and test mining techniques in claim areas, each extending over 75 000 square kilometers. Six of the consortia (Japan, South Korea, France, Russia, a group of former eastern bloc nations, and China) staked their claims in the CCFZ in the northeastern equatorial Pacific; the remaining claim, that of India, is in the central Indian Ocean basin.[11] More recently, Germany staked a claim in the CCFZ as well.

Deep seabed mining remains a form of high-stakes poker game. To stay in the game, each consortium is required to conduct research to obtain baseline environmental data as well as to map out the resources within its claim area. Oceanographic cruises to the region are expensive – a month-long cruise is typically undertaken on a vessel that costs US$25 000–60 000 per day to operate – as are the teams of specialists and their equipment that must be maintained. And like poker, there are likely to be only one or two winners: global resource economics dictate that only one or two claims could be profitably developed before the market is saturated. So the player that makes the first big move is likely to end up the big winner – or loser. To add to the complexity of the game, the player who eventually makes his move for the jackpot may not be playing strictly by the rules of rational resource economics: the Asian players in particular (Japan, China, South Korea, and India) have strategic as well as economic interests, based on guaranteeing their access to vital mineral resources.

Environmental impacts of seabed mining

If deep seabed mining goes ahead, its ecological impact will be enormous. The fauna associated with the sediment, as well as the nodules, will be removed and destroyed within the area from which the nodules are directly removed (an area of approximately 300 square kilometers per year). The suspension and resedimentation of sediments will likely affect an area several times larger, in which the fauna will be buried and disturbed. Over the projected 20-year span of such an operation, one or two mining operations would have a severe impact over an area of at least 10 000–20 000 square kilometers, and more likely several times larger.

The fauna of the abyssal plain has evolved within the most stable and least productive environment on earth. The organisms are fragile and extremely slow-growing, with low reproductive capacity and often limited dispersal capability; in brief, they have not evolved to deal with catastrophic disturbance, so recovery is likely to be extremely slow. Several scientific groups have undertaken to assess the potential impact of deep seabed mining. In the late 1970s when the first mining tests were undertaken, the US DOMES project (Deep Ocean Mining Environmental Study) monitored a site for five years, by which time the benthic fauna had largely recovered. However there was concern that the scale of the disturbance was too small to be realistic, so in the late 1980s Thiel and colleagues ploughed up an area of 11 square kilometers within a German claim off Peru. The abundance of major macrofaunal groups was reduced between 50 and 90 percent following the disturbance.[12] However, after seven years, although the plough marks were still evident, the mega-, macro-, and meiofauna had largely returned to the levels observed in undisturbed sediment.[13]

The intensity of any such experimental disturbances remains far less, and their impacts far more localized, than an actual deep seabed mining operation.[14] Given that humankind appears to accept the clear-felling of its tropical and temperate forests, only the foreseeable extinction of a significant portion of the seabed fauna could likely prevent the development of mining in such a remote part of the planet. So, would deep-sea mining lead to faunal extinctions?

In beginning to resolve this question, the key issue is the range of deep seafloor species. But this issue has been largely ignored in recent decades, at least for the Lilliputian macrofauna; instead, deep-sea ecologists focussed mostly on understanding the fauna's exceptional and still-puzzling species diversity (see Chapter 4). The question of species' ranges is one of the first that ecologists usually pose about ecological communities. Certainly it is the question that deep-sea ecologists from Murray to Ekman and Bruun debated with regard to the megafauna. However, the dispersal capabilities of the Lilliputian benthic macrofauna are so different that its range is likely to differ considerably from that of the megafauna. This is a research issue to which deep-sea ecologists have only recently returned (see Chapter 4).

Seamount mining

The long-standing jurisdictional uncertainty over mineral resources on the high seas, and the current requirement to deal with a UN agency under international scrutiny, has led some players to examine the mineral resource potential in waters under national jurisdiction. This interest was heightened following the invasion of the cobalt-rich Shaba province of Zaire in 1977, which caused cobalt prices to skyrocket approximately six-fold. Concerns about stable access to cobalt led to exploration of the potential of cobalt-rich manganese crusts on seamounts. Manganese crusts, like the nodules, were first encountered on the *Challenger* expedition; and they form much like seafloor manganese nodules, except they precipitate out on the hard rock substrate of seamount summits and slopes, where enhanced currents winnow away the sediments. However, whereas the cobalt content of nodules is generally no more than 0.3 percent, it is as high as 1.7 percent in some crusts. Since 1981 there have been some 48 cruises to explore these crusts, undertaken by the USA, Germany, Japan, and the USSR, representing an investment of about US$90 million.[15] The most promising deposits are within the EEZ of Johnston Island (USA) and the Marshall Islands. Global cobalt consumption is presently about 37 000 tonnes a year. Projected mining from a single seamount could yield 700 000 tonnes of crust per year; at a 1 percent cobalt content, this equates to about 7000 tonnes or 20 percent of the global cobalt requirement.[16] Several companies have expressed interest in seamount crust mining, and Marshall Islands has leased its EEZ for mineral exploration, primarily for its cobalt-rich crusts, but the prospects for these operations are still considered distant.[17] The operation of removing crust from the underlying rock is technically more difficult than nodule mining, despite the shallower depth: generally between 800 and 2500 meters. Furthermore, although about two-thirds of global cobalt is currently supplied by the relatively unstable Zaire (57 percent) and Zambia (11 percent), production from the Voisey's Bay deposits in Labrador is soon to commence, and deposits are also being explored in Australia.[18] If seamount mining were to proceed, its impacts would be more localized than nodule mining, but it would remove the seamount-based fauna, which often has highly localized distributions and endemic species (see Chapter 6).

Hydrothermal deposits

More than a decade after Mero published *The Mineral Resources of the Sea*, an entirely new form of deep-sea mineral deposit was discovered, which today appears much closer to commercial development than manganese nodules and crusts. This type of deposit, which comes in several forms, is formed by hydrothermal processes in regions of seafloor spreading and the subduction of one plate beneath another. In these tectonic regions, magma chambers rise within 1.5–3.5 kilometers of the seafloor, driving a hydrothermal convection system within the earth's crust that leaches various metals out of the deep volcanic rock – mostly iron, manganese, copper, and zinc, but also silver and gold – and precipitates them out, largely in the form of sulfides, at the seafloor in chimneys and other deposits (see Chapter 5). An average 'smoker' deposits approximately 250 tonnes of ore each day. Large sulfide deposits, such as in the TAG (Trans-Atlantic Geotraverse) hydrothermal field along the mid-Atlantic Ridge – 200 meters in diameter, 40 meters thick and containing some 5 million tonnes of ore both above and below the seafloor – are estimated to have formed intermittently over the last 50 000 years.[19]

As recounted in Chapter 5, these deposits, termed massive sulfide or polymetallic sulfide deposits by mineral geologists, were first discovered in the late 1970s along the East Pacific Rise. Today, more than 100 sites of deep-sea hydrothermal mineralization are known, at least 25 with high-temperature (300–400°C) black smoker vent formation. Most are in the Pacific, but several have been located along the mid-Atlantic Ridge, and one in the Indian Ocean (p. 167). However, only about 5 percent of the ocean's 60 000 kilometers of ridges formed by seafloor spreading have been surveyed, so the ultimate distribution of these features is still poorly known.[20]

Only a few of these deposits have been drilled and surveyed sufficiently to estimate the size of the deposit. Most contain in the range of 1–10 million tonnes of ore. In 1979, the Saudi-Sudanese Red Sea Joint Commission carried out the first and most quantitative assessment of a deep-sea hydrothermal deposit, the metalliferous muds found over a 40–60-square-kilometer area of the Atlantis II Deep, 2000 meters below the surface of the Red Sea. The largest hydrothermal deposit discovered thus far, the Red Sea deposit contains about 100 million tonnes, comparable to the largest mineral deposits associated with volcanic processes on land. The deposit, which they demonstrated could be successfully mined, was worth about US$5 billion in 1985 dollars, based on its zinc, copper, cobalt, silver and gold. It is estimated that a commercial operation would mine about 100 000 tonnes of mud per day and would have a lifetime of at least 16 years. However, this operation remains on hold, presumably until economic conditions are more favorable.[21]

The most active current deep-sea mining interest lies with massive sulfide deposits formed in the convergence zone where plates collide, and one is subducted beneath the other. Arcs of volcanic activity occur in front of or behind these subduction zones; many are associated with islands in the western Pacific. In 1997, Papua New Guinea issued licenses to an Australian company, Nautilus Minerals, to explore hydrothermal sulfide deposits within an area of 5000 square kilometers in its central and eastern Manus Basin. Another promising prospect is Conical Seamount, also off the coast of PNG. These deposits are relatively shallow (generally at depths of 1000–2000 meters), contain particularly high gold concentrations, and are relatively close to shore, so the strict licensing requirements of the International Seabed Authority can be circumvented. Similar deposits are being considered off Fiji and New Zealand.

What will be the likely environmental impact of these projects, if one or more should proceed? Unlike nodule fields, vent sites are small and discrete, so the mining operation and its impacts would be localized. The most likely mining sites, such as the vents off PNG, are inactive – attempting to mine within the corrosive, super-heated environment of an active vent would be far more difficult – so the iconic hydrothermal vent fauna is probably not at risk: the giant clams, tubeworms, and other organisms so characteristic of the vent environment quickly die off once a vent becomes inactive. However, to date so little consideration has been given to the potential impacts of mining massive sulfides that we do not even know what organisms may inhabit these deposits. Deep-sea ecologists have such limited opportunities to explore hydrothermal vents in submersibles that, as vent ecologist Van Dover put it, they have simply passed by the inactive deposits on their way to the more exciting active vent sites. The inactive deposits still contain minerals that could support chemosynthetic organisms, albeit not in the profusion found at the active sites, but at this time we simply don't know what is there.

Energy resources of the deep sea

If the mining industry's attitude toward the mineral resources of the deep sea can be likened to someone perched at a great height and unable (for decades!) to summon the courage to dive for the prize hidden in the waters below, the approach to offshore energy resources has been more like a bather making his or her way out from the shore, progressively wading into ever deeper water. It was not a great step in moving from drilling in the sea-marshes of the Louisiana delta country to drilling within similar sedimentary structures on the continental shelf just offshore. By the early-to-mid-1960s, about the time when Mero published *The Mineral Resources of the Sea*, drilling for oil had begun on the continental shelf off Louisiana, in the North Sea, and elsewhere.[22] The offshore oil and gas industry is now worth US$100 billion per year and since the 1990s has expanded into waters deeper than 500 meters. The Brazilian oil company Petrobras now works at depths greater than 2000 meters, and there are roughly 50 drilling rigs in the Gulf of Mexico working at depths greater than 500 meters. Other deep reserves are known off western Africa.[23] The human appetite for oil appears insatiable, and as the more accessible energy resources are exhausted, the industry has gradually moved to extract those that are more technically challenging and costly.

Although catastrophic oil spills are always a possibility, the greatest impact of oil exploration on the ocean environment has proven to be the chronic effect of drill cuttings and drilling muds, which are disposed of during drilling operations. These are complex mixtures, whose composition varies considerably. Generally, however, they contain a weighting agent (usually barite), viscosity agents, swelling and pH (acidity) controllers, dispersants, defoamers, lubricants, and biocides; some drill cuttings are dissolved in water and some in oil, which is preferred for drilling but more toxic. Barite is used in large quantities (around 1500 tonnes per well), and typically contains heavy metal contaminants, in particular zinc, copper, cadmium, and lead. If the chemical mix is complex, so too will be the impacts, ranging from direct smothering of benthic organisms around the well-head to the toxic effects of the hydrocarbons and heavy metal contaminants and the effects on the sediment community of organic hydrocarbon enrichment.[24]

The oil industry is generally required to monitor the chemical and biological impacts of its operations, but its studies are rarely open to review by the public or scientific community. Simple, rather crude, mathematical indices of species diversity are used to monitor ecological impacts on the benthic community. The consensus arising from such studies by the UN Group of Experts on the Scientific Aspects of Marine Pollution (GESAMP) was that major biological impacts – changes in species diversity indices – of oil and gas drill operations were quite limited, extending to a radius only 1–1.5 kilometers from platforms off the coast of the UK and Norway.[25]

However, in Norway, where this information is open to public scrutiny, two scientists from the University of Oslo, Frode Olsgard and John Gray, reviewed the data from Norwegian offshore oil-fields going back to the early 1970s. Using contemporary computer-based methods of community analysis, they showed that the contaminants continued to disperse even several years after drilling operations ceased, eventually covering an area of at least 100 square kilometers around each platform, with impacts on the biological community discerned over an area of approximately 10–60 square kilometers. Whereas simple diversity indices only take account of numbers of species, the faunal analyses of Olsgard and Gray showed a change in the *types* of species, from those living on the surface of the sediments, such as brittle stars, which are a principal prey item for cod, haddock, and flat fishes, to small polychaete worms that live within the sediments and are not so available.[26] It is expected that the impacts of these disturbances will be greater and longer-lasting as oil development moves into deeper water, given the reduced productivity of deepwater communities, their slower rates of recolonization, and their generally reduced ability to adapt to disturbance. Some deepwater oil exploration threatens highly sensitive deepwater coral communities, such as the *Lophelia* reefs off Norway and Scotland, which has led to high-profile deepwater baseline surveys being conducted in these areas and in the Gulf of Mexico; this may lead to protection of these particularly unique and vulnerable habitats.[27]

Fire from ice – methane hydrates

The wild card in offshore energy development is the recent discovery of massive reserves of methane hydrate locked up within the continental slope sediments (and also the Arctic permafrost), containing perhaps twice as much carbon as all other known fossil-fuel deposits.[28] Two areas off the coast of North and South Carolina alone contain more than 37 trillion cubic meters of methane gas, more than 70 times the US gas consumption in 1989.[29]

Methane hydrate has been called ‘fire from ice,’ because the gas is trapped within the lattices of a distinct crystalline ice structure

– but the crystals are cubic, rather than hexagonal as with normal ice. Methane hydrate only forms where there are high concentrations of methane, but its stability properties – a function of pressure as well as temperature – enable it to remain frozen at temperatures well above 0°C. It is generally found at depths below about 250 meters. The methane packing within the hydrates enables a cubic meter of hydrate to contain about 160 cubic meters of methane gas; its high energy density, about 2–5 times that of conventional natural gas, enhances its commercial attractiveness as an energy source.[30] Several countries – Japan, India, the USA, and Canada – have started gas-extraction feasibility studies, and drilling has been carried out in a permafrost region of northern Canada and in Japan's Nankai Trough. It is estimated, however, that commercial gas production is at least 15 years – and perhaps several decades – off, once production of more accessible fossil fuels declines.

The potential environmental risks from methane hydrate extraction are as poorly understood as the methods that will be required to extract the gas commercially from its matrix of sediment and ice. Methane is a powerful greenhouse gas – molecule for molecule it is 25 times more powerful than carbon dioxide in retaining solar radiation – so extensive losses to the atmosphere during drilling operations could considerably exacerbate global warming. Sudden gas releases pose additional, more immediate risks, having been implicated in the loss of industry vessels during conventional offshore drilling operations. Another concern lies in the role gas hydrates play in maintaining seafloor stability. Removing the hydrates could trigger massive underwater landslides, posing the threat of both tsunamis and enormous uncontrolled releases of methane due to the sudden release of pressure, events which are observed in the geological record.[31]

Resource extraction at the last frontier – the long view

Taking the long view – the view over the next half century or so – the conclusion that the deep sea represents the last and greatest of the planet's frontiers for mineral and energy development is unavoidable. Particular projects may come sooner or later depending upon technological developments and political and economic circumstance. But the long-term trajectory for oil and gas production from the land – its peaking and subsequent decline – is reasonably certain; if industrial society stays on its current course, it will develop more remote resources. And the first steps toward development of the most accessible deep-sea mineral and energy resources are already apparent in the offshore oil and gas industry and the leasing of vent sites off western Pacific islands.

The discovery of new mineral and energy resources in the deep ocean continues unabated. The era was ushered in with the discovery of the Red Sea metalliferous muds and the publication of Mero's *The Mineral Resources of the Sea* only 40 years ago. Exploration of seamount crusts and the discovery of hydrothermal vents followed more than ten years later. And exploration of back-arc vents and methane hydrates is more recent still, occurring mostly within the last ten years.

All major deep-sea habitats may be affected by these potential developments: slope environments mostly by oil and gas development, both of the conventional variety and methane hydrates; seamounts for their cobalt-rich manganese crusts; and vast expanses of the abyssal plains for their nodules. Given the many types of resources and the many sectors potentially involved in the deep sea – even without considering the impacts of fishing and climate change – it is apparent that deep-sea development must be monitored, and care taken to maintain a broad overview of total impacts.

One of the most positive developments, from an environmental perspective, has been the establishment of the ISA in 1994, with authority under the United Nations Law of the Sea Convention (UNCLOS) to monitor and regulate resource development on the high seas, at the very dawn of deep-sea industrial development. Over the last decade, the ISA has successfully walked the fine line of an international regulatory agency charged with both fostering the development of deep seabed mining and protecting the deep-sea environment. Wielding more moral authority than either funding or enforcement clout, it has thus far managed to work successfully with both mining consortia and the world's leading deepwater scientists to develop guidelines for the exploration and eventual mining of manganese nodules on the abyssal plain, of cobalt-enriched crusts on seamounts, and of polymetallic sulfide deposits. It has also used its moral authority to assist scientists in obtaining the funding to address critical environmental research questions, such as the biogeography and distributional range of species associated with these deposits, to ensure that mining does not lead to excessive loss of species. However, the offshore mineral industry has lain dormant for this past half-century, mostly awaiting an improved economic outlook. This has engendered a lack of urgency and a certain complacency in the scientific and environmental communities. But if the slumbering dragon should awake, perhaps driven by the rising resource hunger of the emerging Asian market, how long will it pause to consider the fate of the diminutive worms and crustaceans comprising the Lilliputian deep-sea fauna?

CHAPTER 9

Climate Change

In a landmark paper published in 1957, two Scripps scientists, Roger Revelle and Hans Suess, noted that 'human beings are now carrying out a large scale geophysical experiment of a kind that could not have happened in the past nor be reproduced in the future. Within a few centuries we are returning to the atmosphere and oceans the concentrated organic carbon stored in sedimentary rocks over hundreds of millions of years.'[1]

When we burn fossil fuels (oil, coal, and natural gas), whether to generate electricity, power our transport or warm our homes, the carbon, long sequestered from the planet's biogeochemical carbon cycle, re-enters it by combining with oxygen and producing carbon dioxide (CO_2). CO_2, water vapor, and several trace gases, such as methane and nitrous oxide, constitute the planet's so-called greenhouse gases: like the glass in a greenhouse, they are transparent to the incoming radiation from the sun, mostly in the visible and near-ultraviolet range, but absorb and partly radiate back to Earth the heat in the infrared range that the planet radiates into space. Without our atmosphere's greenhouse gases, the mean temperature on the Earth would plunge from an equable 15°C (59°F) to about –20°C (–4°F).

As Revelle recognized, the hypothesis linking industrial activity with increased atmospheric concentrations of CO_2 and global warming had been proposed and developed by eminent scientists over the preceding hundred years.[2] John Tyndall, a mid-19th century British physicist, first measured the heat-retaining properties of water vapor and CO_2 and suggested that reduced CO_2 levels may have been responsible for the Ice Ages. By the end of the 1800s, one of the first Nobel Prize winners, the eminent Swedish chemist Svante Arrhenius, had calculated that halving the atmosphere's CO_2 would lower the earth's temperature by 4°C, sufficient to bring on an Ice Age, and doubling it would increase the temperature by 5–6°C – estimates remarkably close to contemporary calculations.[3] Arrhenius also recognized the effect of positive feedback between CO_2 and water vapor on climate: higher CO_2 levels, by heating the planet, increase evaporation rates, thereby increasing the water vapor in the atmosphere, which further enhances greenhouse warming. Although water vapor concentrations in the atmosphere are higher and contribute about 60 percent of the atmosphere's greenhouse effect, CO_2, which directly causes only about 25 percent of greenhouse warming, is the primary climate driver.[4] As Arrhenius wrote, by

> evaporating our coal mines into the air … the slight percentage of carbonic acid in the atmosphere may, by the advances of industry, be changed to a noticeable degree in a few centuries … [even though] the sea by adsorbing carbonic acid acts a regulator of huge capacity.

Arrhenius, together with one of the leading American geologists of the time, Thomas Chamberlin,[5] realized that the CO_2 in the atmosphere was but a small part of the global carbon cycle, in which carbon and CO_2 are exchanged between the solid earth, the oceans, the atmosphere, and the biosphere. Over geologic time, volcanoes have emitted vast quantities of CO_2, almost all of which has been either absorbed by the oceans, precipitated and buried within the sediments, and eventually transformed into limestone and dolomites (calcium and magnesium carbonates), or taken up by plants through photosynthesis and eventually buried as organic matter in marine sediments or within the soils. At present there are an estimated 50 million gigatonnes (or billions of tonnes) of carbon sequestered in sedimentary carbonates and about 20 million gigatonnes within the sediments, about 100 000 times more than resides in the atmosphere (700 gigatonnes). The oceans and biosphere contain some 42 000 gigatonnes of carbon, about 60 times more than the atmosphere. Chamberlin confronted the paradoxes at the heart of the Earth's climatic system: the equability of our climate for the past billion years (never too hot or too cold for life) despite its apparent sensitivity to CO_2 levels; and the implied stability of CO_2 levels over geological time, despite the very small amount within the atmosphere (only 0.3 percent of the atmospheric gas mixture) and its reactivity with vastly larger components.

Chamberlin recognized that only exquisitely tuned negative feedback controls could maintain atmospheric CO_2 within relatively narrow confines over the long history of life on Earth. He proposed that the vast amounts of CO_2 emitted by volcanoes were regulated by the temperature-dependent weathering of the Earth's silicate and carbonate rocks. CO_2 in the atmosphere dissolves in rainwater, forming carbonic acid (H_2CO_3), which reacts with these rocks, releasing bicarbonate ions (HCO_3^-). The bicarbonate flows into the ocean with the rivers and groundwater, where eventually it is taken up by organisms that form calcareous shells – various plankton and forams, corals and mollusks – and much of it is then buried in the sediments. This drawdown of atmospheric CO_2 provides negative feedback to climate change, since higher temperatures enhance evaporation and rainfall, as well as the rates of chemical reactions, and hence the weathering process, which puts the brakes on warming; cooler conditions, conversely, slow the weathering process, allowing CO_2 levels to build again. Eventually the cycle is closed as the buried carbonate is metamorphosed back into silicate and carbonate rock, which releases the CO_2, making it available to be released through hydrothermal vents and volcanoes.

Chamberlin's hypothesis, which has been somewhat further developed over the past century,[6] is central to our understanding of climate. One of the most striking aspects of the Earth's climate is not its variability but its stability. Over the past 3.8 billion years, our planet has passed many times between greenhouse and icehouse states, yet it has never gone beyond the relatively narrow window where life is able to flourish – it has never either frozen over (the 'snowball Earth' scenario) or gotten so hot that the oceans boiled away. Such scenarios may seem fanciful, but that is in fact what happened on our neighboring terrestrial planets, Mars and Venus. Both once had atmospheres with water vapor and carbon dioxide – indeed, the deep channels on Mars are testament to that planet's once considerable stores of liquid water – but today the water on Mars (which has a mean temperature of –60°C) is all frozen, and on Venus (mean temperature 460°C) it has boiled away. Although the Goldilocks explanation accounts for this in part – Venus is too close to the sun and Mars too far away, but the Earth is just right – the planets' ability to cycle CO_2 provides a fuller explanation.[7] Venus' furnace-like conditions are due more to its runaway greenhouse atmosphere, composed almost entirely of CO_2, than to its position nearer the sun, just as Mars' now-frozen state derives from its negligible CO_2 levels, enough to provide only 6°C of greenhouse warming. As noted above, without its present atmosphere, the Earth would have a mean temperature of only about –20°C. To add a further dimension to the remarkable long-term equability of the Earth's climate, our atmosphere must once have had far higher CO_2 levels, in order to have supported life during the planet's early history. The sun's luminosity (its power or intensity) was about 25–30 percent less when the earth first formed about 4.6 billion years ago and has been slowly increasing ever since. Carl Sagan and George H. Mullen of Cornell University recognized that, with the Earth's present atmosphere, the planet would have frozen over for its first two billion years. Life evolved, however, after about the first billion years of planetary history, and all indications are that the earth was warmer during its early history, rather than colder. Higher atmospheric CO_2 levels were the most likely cause.[8] And the inability of our neighboring planets to cycle CO_2 between atmosphere, oceans, and solid earth – the cycle of volcanic outgassing, rock weathering, and ocean precipitation – underlies the evolution of their present climate, so unsuitable for sustaining life.

Interestingly, although the greenhouse-warming hypothesis was revived in the first half of the 20th century, it was generally not treated seriously, despite its illustrious lineage. Most scientists believed that the present equilibrium between the atmosphere and oceans, with about 60 times more carbon residing in the oceans than the atmosphere, would be maintained: man's CO_2 production would be readily taken up by the oceans, with only a trivial fraction (one 60th) remaining within the atmosphere.[9] Measurements of CO_2 levels in the atmosphere to that time were taken close to population centers and were too variable to show a clear trend.

Revelle and Suess's key insight – the reason their paper finally roused the world to take the greenhouse-warming issue seriously – was their recognition that ocean uptake of CO_2 occurs on geological rather than societal time-scales. Ocean–atmosphere CO_2 exchange is restricted to the upper mixed layer, which on average is about 80 meters deep. The rest of the ocean is effectively isolated from the atmosphere, except through the relatively slow processes of downward mixing and winter deepwater formation in the extreme North Atlantic and Southern oceans. Renewal of the deep ocean takes place on the time-scale of a millennium – a blink of an eye geologically but a very long time for human society. Revelle and Suess also pointed out that the rate of CO_2 uptake is limited by the ocean's dissolved inorganic carbon chemistry, essentially a global marine antacid system. The rate of CO_2 dissolution is limited by the rate at which it reacts with dissolved carbonate (CO_3^-) to form bicarbonate (HCO_3^-).

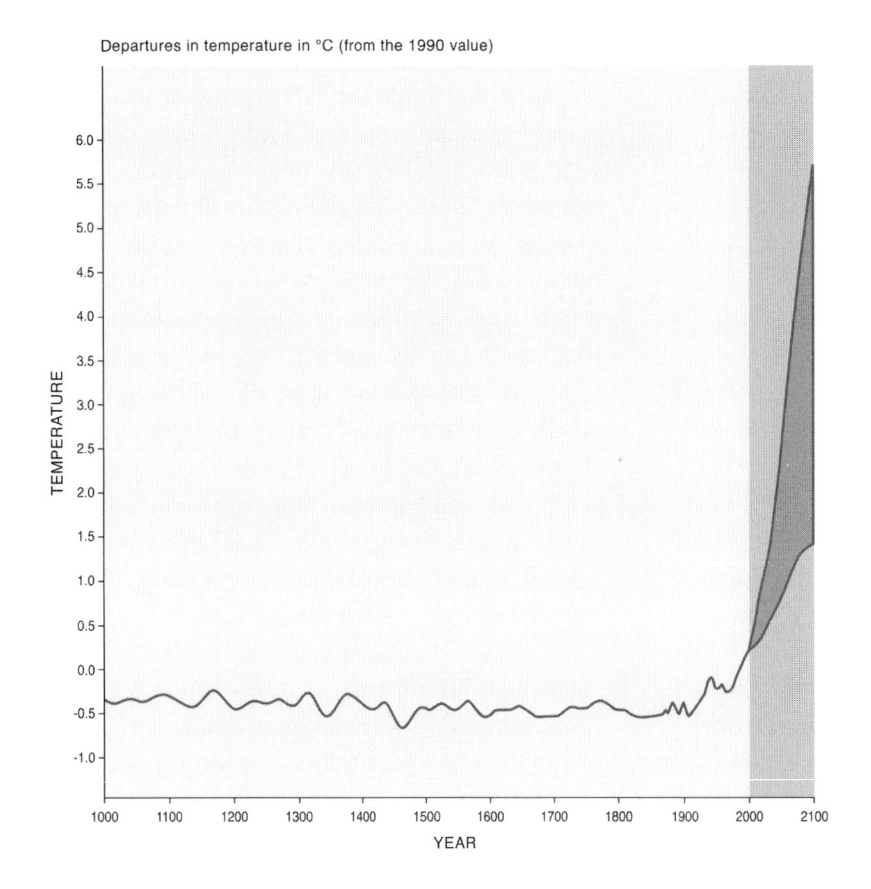

The so-called 'hockey-stick' graph, showing temperature trends over the past millennium, including projected temperature increases over the next century based on the leading climate models. Temperatures since 1861 are based on instrument records; previous temperatures are estimated, based on proxy records from tree rings, corals, and ice cores. (From IPCC, 2001.)[t]

As a result of these factors, about half of our CO_2 emissions each year remain in the atmosphere, and half are taken up by the oceans and terrestrial biosphere. Due to the limitations of ocean mixing, about 25 percent of our CO_2 emissions will remain in the atmosphere even centuries later.[10]

Consummate scientists, Revelle and Suess were not being facetious in bringing attention to mankind's unique geophysical experiment. In 1957, little was known about the Earth's climate system and the role of CO_2 within it. That atmospheric CO_2 levels were increasing was still a hypothesis, not yet clearly demonstrated. Once we had embarked on this experiment, it was incumbent upon the scientific community to learn its lessons. Their paper was published during the International Geophysical Year, and they proposed an immediate program to monitor CO_2 levels in the atmosphere in regions far removed from industrial activity. Within a year, monitoring stations were established at Mauna Loa in Hawaii and in the Antarctic, and the inexorable rise of greenhouse gases in the atmosphere was soon confirmed.[11]

Since then, the global scientific community has undertaken one of the world's largest and most vital research programs, covering all aspects of the earth's climate system:

- monitoring greenhouse gases, temperature, and other parameters critical to climate in the atmosphere, within the oceans, and on land;
- conducting studies on the global carbon cycle and other systems critical to climate regulation;
- reconstructing the earth's past climate history in relation to potential driving mechanisms; and
- developing sophisticated computer models of global climate, building on our growing understanding of climate processes.

The results of this research have been regularly reviewed by an international group of leading scientists known as the Intergovernmental Panel on Climate Change (IPCC). Their third review, issued in 2001, involved delegations from 99 countries and was prepared by 122 lead authors and 515 contributing authors, as well as 21 review editors and 337 expert reviewers.[12]

The 2001 IPCC review is a scientific consensus document: it is not speculative. It focusses on findings that have been reviewed and published in the leading journals, that are consistent with a growing body of data from widely varying sources. As the review notes, global temperature has increased by 0.6°C over the past century. The 1990s represented the warmest decade since the 1860s, when temperature records were first systematically maintained; it was also the warmest decade, and our past century the warmest century, for at least the past thousand years, based on proxy temperature records derived from tree rings, coral banding, and ice cores. The rapid rise in temperature over the 20th century relative to the past millennium, which produces the so-called 'hockey-stick' graph, closely matches the rapid rise of CO_2 and other greenhouse gases over the past century. Current levels of CO_2 in the atmosphere, about 370 ppm (parts per million), are a third higher than during the pre-industrial era (280 ppm) and are already higher than they have been for at least the past 420 000 years – encompassing four major glacial cycles – for which there are direct data from Antarctic ice cores.[13] In fact, the current CO_2 levels are likely higher than they have been for the past 20 million years. There are now several climate models that incorporate the effects of greenhouse gas concentrations and various other natural forcing factors, such as changes in volcanic activity and solar irradiance. These models closely predict the observed rise in temperature when both natural and man-induced factors are included, but provide a poor match to the data when only natural forcing factors are included, leading to the strong conclusion that the observed warming is primarily due to greenhouse gas emissions.

The focus of the 2001 IPCC report, however, was not on the present but the future. The observed increases in CO_2 and temperature over the past century are but a small fraction of the increases expected over the century to come. Based on a range of scenarios regarding future global development, emissions, and land-use policies, atmospheric CO_2 levels will likely increase to between 540 and 970 ppm by 2100, or about 2–3.5 times pre-industrial levels. Globally averaged temperature is projected to rise a further 1.4–5.8°C – temperature increases 2.3–9.7 times higher than those experienced during the twentieth century. Under most scenarios, CO_2 levels and global temperature will continue to increase beyond 2100. Allowing CO_2 levels to increase to four times higher than pre-industrial levels (about 1000 ppm) could lead to a profound restructuring of the earth's climate system, with a shutdown or severe slowing of the North Atlantic and/or Southern Ocean thermohaline circulation – the sinking of cold, salty water in winter that fills and rejuvenates the world's deep ocean basins – and a return to temperature conditions not experienced for many millions of years, possibly not since the greenhouse world of the Age of Dinosaurs.[14]

Implications of climate change for the deep sea

What are the implications of global climate change for the deep sea? Although the deep sea might seem the least of our concerns when considering global climate change, understanding its impacts on this remote ecosystem provides critical insight into the extent and nature of the changes that potentially lie ahead. Indeed, one of the ironies of our current situation is that, despite our extremely limited understanding of deep-sea ecology, we know more about the impacts of climate change on the deep sea than probably any other environment, because our best record of climate change is laid down in the deep-sea sediments. As a result, we know as much or more about how the deep-sea responded to climate events thousands or millions of years ago as we do about its current variability. So there is a great deal to be learned from looking back at how the deep sea has responded to past climate change. Some key questions to focus on are:

- How does the scale of potential future climate change compare with past changes?
- How has past climate change affected the deep-sea environment and deep-sea communities?
- Are there any close analogues to our present geophysical 'experiment' that we can learn from?
- What, in the end, can we say about the likely impacts on the deep sea of climate change over the next century and beyond?

Time-scales of climate change

Climate has many characteristic time-scales, each with its own suite of dominant climate drivers. In this way, the long history of the Earth's climate is like an exquisitely complex symphony. A geological era, like a symphonic movement, has certain dominant motifs, typically driven by major tectonic events: the opening and closing of oceans, the separation and collision of continents. Our own Cenozoic Era (meaning 'recent life'), often referred to as the Age of Mammals, extends over the past 65 million years and might be called the Icehouse Movement in contrast to the Greenhouse Movement of the previous Mesozoic Era, the Age of Dinosaurs. Driving the progressive Cenozoic chilling of the deep ocean and of the world as a whole has been the increasing isolation of Antarctica, which led eventually to permanent Antarctic glaciation, as Australia separated and started its slow drift northward about 37 million years ago. Previously, warm currents had extended south to Antarctica, maintaining a temperate climate there, such that this continent was forested and inhabited by dinosaurs, and the deep ocean was filled with warm (10–15°C) salty water.[15]

Playing over the top of these long, slow motifs is the beat of the periodic variations in the eccentricity, obliquity, and precession of the Earth's orbit: 'wobbles' in the Earth's rotation about its axis that affect the seasonal distribution of solar radiation at high latitudes.[16] The ebb and flow of ice sheets over the Northern Hemisphere over the past 500 000 years closely followed these periodic variations in the earth's orbit, producing cycles of approximately 20 000, 40 000, and 100 000 years, and similar periodicities date back to the earliest glaciations during our Cenozoic period, some 34 million years ago.[17] These orbital variations are probably not sufficient to drive the glacial cycles on their own. But like a percussion section, they provide the underlying beat, which drives variation in greenhouse gases, such as CO_2 and methane, fundamental changes in the ocean's deep thermohaline circulation, and changes in the Earth's albedo, which amplify the orbital effects. (Glaciation increases the amount of radiation reflected back into space, further cooling the planet once glaciation sets in.)

Most of what we know today about the Earth's past climate history is derived from long cores of sediment drilled from the deep sea, and ice cores obtained from the Greenland and Antarctic ice sheets. The sediment cores, obtained from different regions, provide a record of conditions in the different oceans and at various latitudes, extending back tens of millions to over a hundred million years. Reconstructing the Earth's past climate history depends on proxy records, such as variation in the ratio of stable isotopes of oxygen and carbon, preserved within the calcium carbonate ($CaCO_3$) shells of forams. The oxygen isotopes provide a sensitive indicator of seawater temperature when the foram shells formed during non-glacial periods and a mixed signal of ice volume and temperature during glacial periods.[18] Because some species of forams are planktonic and others benthic, temperature records can be obtained for both the sea surface and seafloor. The carbon isotopes, on the other hand, provide insight into the global carbon cycle, because different carbon sources (for example, volcanic outgassing or the dissolution of methane hydrates) have distinct isotopic signatures, and photosynthesis itself removes the lighter ^{12}C preferentially over ^{13}C. However, the sediment record is generally not very fine-scaled, because deposit-feeding and burrowing organisms rework the surface sediments, blurring the temporal resolution. Fortunately high-resolution climate records can be obtained for the past several hundred thousand years from cores through the Greenland and Antarctic ice sheets. The annual layering in these cores can be counted back about 14 500 years, and the air bubbles trapped within the ice provide direct samples of ancient atmospheres that have now been extended back 420 000 years from Antarctic ice cores. The stable oxygen isotope ratio here mostly provides a record of the original ice volume in the glaciers, due to the fractionation of water containing lighter and heavier oxygen when water evaporates. CO_2 levels at the time of ice formation can be directly measured within the ice bubbles.

Before the ice-core records became available in the early 1980s, the paradigm for climate change was that it occurred on geological time-scales – in the absence of a cataclysmic event, such as an asteroid slamming into the Earth, which dramatically closed the Age of the Dinosaurs. The climate system seemed slow and ponderous, with large-scale change damped by the considerable heat capacity of the deep ocean, which requires approximately a millennium to turn over.

A key discovery from the finely resolved ice-core record has been the abruptness of climate change. Indeed, the only period in this climate record not characterized by abrupt change has been our own Recent Epoch – the past 11 000 years has been the longest and most stable warm period of the past 420 000 years. Thus although from our limited perspective we often view our climate as marked by droughts, floods, and extremes of temperature, mankind has in fact enjoyed an extended stretch (over 10 000 years) of exceptionally stable climate that has enabled settled agriculture-based civilizations to develop and prosper.[19] During the previous Ice Age, which extended over about 100 000 years, the climate underwent 22 abrupt shifts, in which climate dramatically warmed over just a few years or decades, particularly at high latitudes in the North Atlantic, followed by intense cooling over the following centuries. These millennial-scale events, often referred to as Dansgaard-Oeschger events for the glaciologists who led the teams that discovered them in Greenland's ice cores, add yet another voice in the climate music score.[20]

What can be responsible for such abrupt shifts in the Earth's climate? Broecker, one of the most fecund minds in the field, hypothesizes that the oceans' deepwater or thermohaline circulation, a key flywheel of the planetary climate system, has several stable modes of operation that shift abruptly from one to another. At present, there are two fairly equal sources of ocean deepwater: the Antarctic, where brine released as sea ice forms in winter creates the coldest, densest water in the ocean; and the North Atlantic, where the relatively salty water of the North Atlantic Current, the extension of the Gulf Stream, cooled in winter, forms a deepwater only marginally less cold and dense. But this North Atlantic branch of the thermohaline system is particularly vulnerable to perturbation. All it takes is a sudden addition of freshwater, such as from accelerated glacial melting that releases a flotilla of icebergs into the far North Atlantic, and deepwater convection shuts down. Broecker has called the North Atlantic the Achilles' heel of the climate system, and studies of the ice cores now provide ample evidence of such sudden shifts in the North Atlantic.[21] This is the scenario so fancifully presented in the film *The Day after Tomorrow*. Of course, glaciation doesn't set in overnight, but neither is climate change necessarily 'glacial' in its mode and tempo.

Because the climate system is linked through the atmosphere as well as the oceans, a fundamental shift in climate mode can be communicated virtually instantaneously around the globe. During the last glacial cycle, for which the best geological data are available, rapid shifts in North Atlantic thermohaline circulation often led to changes in: the trade winds and associated upwelling across the equatorial Atlantic and Pacific; the Asian monsoon; sea surface temperatures of the Pacific warm pool, which drives the El Niño cycle; upwelling, productivity, and deepwater circulation along the margin of the Pacific; and accompanying changes in the Southern Hemisphere as well.[22]

Interactions between the atmosphere and ocean circulation drive the very highest voices in the climate symphony, variations on time-scales of a few years, such as the El Niño–Southern Oscillation cycle, or a few decades, such as the North Atlantic or Pacific Decadal Oscillations. Until the advent of our present era of global warming, these rapid trilling figures were all that we ever experienced of the great climatic symphony, living on our time-scale of three-score years and ten.

In general, the most profound climatic shifts require the greatest expanse of time, insofar as they involve the drift of continents and the like. Global warming over the last century amounted to a little more than half a degree Celsius. The difference between our present climate and that experienced 15 000–20 000 years ago during the most recent glaciation is ten times greater, about 5°C – on the order of some projections for climate change for the 21st century. In other words, the difference between our present climate and the projected climate 100 years from now may be as great as the difference between the present and the climate of the Ice Age 15 000 years ago – the climate change of millennia compressed into a century, one of the most rapid changes in planetary greenhouse gases and temperature that the planet has ever experienced. The temperature difference in the deep ocean, however, between the present and the last Ice Age is only 1.5°C. In the mid-Miocene, about 15 million years ago, just before permanent Antarctic glaciation set in and Antarctic Bottom Water formation took on its modern character, the deep ocean was about 5°C warmer; in the early Cenozoic, about 60 million years ago when the Earth was still a Greenhouse world, it was about 11°C – approximately 10 degrees warmer than today.[23] There is a limit to how much the deep ocean can cool during an Ice Age – it is already only a few degrees above freezing. But there is far more potential for warming, depending on how the deep ocean thermohaline circulation is affected by global warming. What, then, are the likely implications for deep-sea ecology?

Climate impacts on the deep sea

Climate change will affect the deep sea through its impact on deepwater circulation patterns, which determine the temperature and oxygenation, within the deep sea, as well as through its impact on the region's overlying productivity. The increased infusion of CO_2 into the ocean will also directly acidify the ocean, a factor largely neglected until recently.

Climate models predict that global warming will lead to a general weakening of deepwater circulation, most notably of North Atlantic Deepwater formation, which may shut down altogether, but global climate models also predict a slowing or even a shutdown of Antarctic Bottom Water formation if CO_2 levels climb to 1000 ppm, about four times higher than pre-industrial levels.[24]

The first signs of such change are already apparent in the global ocean, extending to considerable depth. The oceans are the planet's major heat sink, and ocean observations since 1955 indicate that the heat content, mostly of the upper 300 meters of the worlds' oceans but extending down to 3000 meters, has increased, consistent with estimates of greenhouse-induced warming.[25] Climate models predict that global warming should lead to enhanced precipitation at high latitudes and hence decreased salinity. This has now been observed in all the major

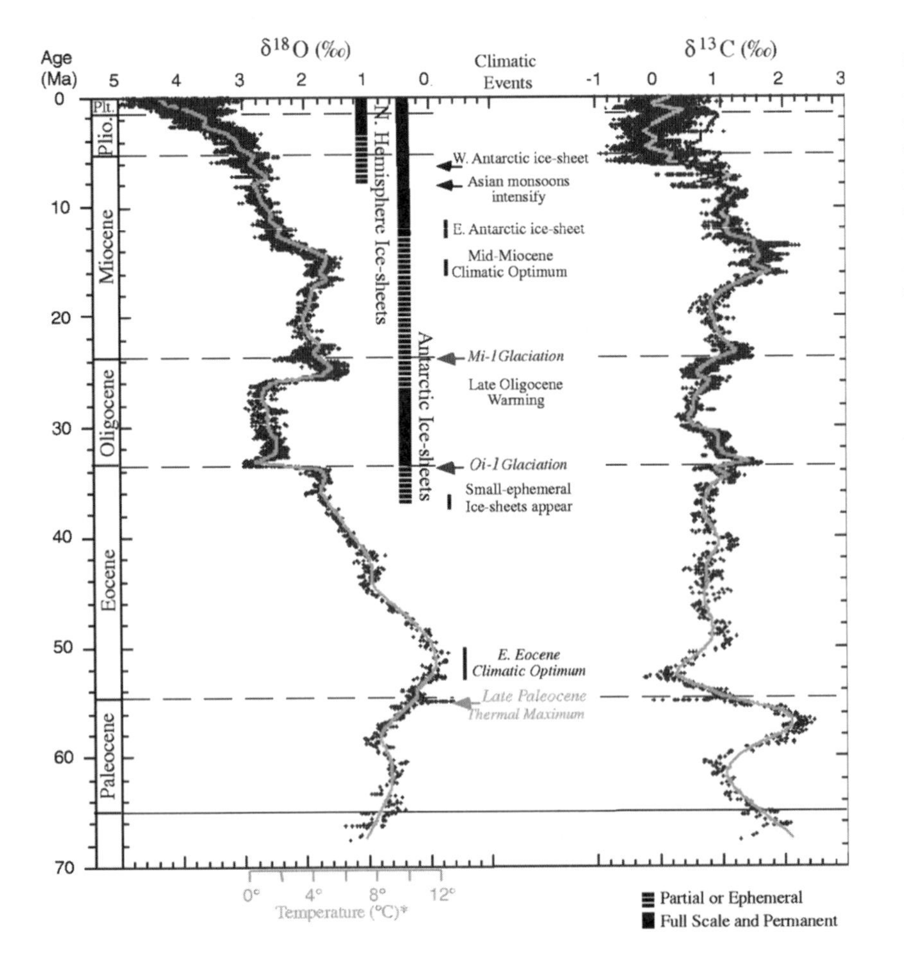

Global deep-sea oxygen and carbon isotope data extending back to the dawn of our Cenozoic Era. A temperature scale is shown, based on the oxygen isotope data in an unglaciated world. (The oxygen data reflects changes in both temperature and ice cover.) Major climatic and tectonic events are shown, along with the onset of Antarctic and Northern Hemisphere glaciation. (From Zachos et al., 2001.)[u]

oceans, extending down into the ocean's major intermediate-depth water masses in the North Atlantic, North Pacific, and Southern Oceans.[26] Freshening and warming serve to both enhance the stability of the upper waters and inhibit convective overturning. And there are already the first tentative signs that the global deepwater circulation may be weakening.[27] Enhanced stability also inhibits nutrient mixing into the upper waters, which global models predict will lead to decreased productivity.[28]

At what point will these changes begin to affect the ecology of the deep sea? The answer is, sooner than we once thought. Deep-sea ecologists extending all the way back to Murray and the *Challenger* expedition recognized that the deep sea during greenhouse periods was far warmer than it is today, and hence poorly oxygenated (oxygen solubility decreases at higher temperatures) and probably unsuited for most present forms of life. But the last greenhouse period was a *very* long time ago, and the deep sea was generally believed to be stable on shorter time-scales – even over extended geological time. This was, after all, the fundamental assumption underlying Sanders' stability–time hypothesis to explain the exceptional diversity of the deep-sea benthic fauna (Chapter 4). That assumption had considerable intuitive appeal, but that was all it was based upon.

Over the past decade, a very different picture has emerged regarding the tempo of ecological change in the deep sea, indicating that it is subject to substantial change even on decadal time-scales, those rapid shifts in marine climate that I likened to the highest trilling in the climate symphony. At present, there are only two sites in the worlds' oceans – one in the North Pacific and the other in the North Atlantic – where deep-sea observations have been carried out sufficiently long – ten years or so – to examine susceptibility to such change.

As noted in Chapter 4, in 1989 Ken Smith and a number of collaborators initiated a time-series study at a site 220 kilometers west of central California at 4100 meters depth.[29] Smith had pioneered *in situ* measurements of benthic community metabolism in the deep sea – the use of a respiration chamber to measure the oxygen consumption of the overall community living within the sediments. Over the first seven years of the study, Smith and Ron Kaufmann, from the University of San Diego, found that the oxygen consumption of the sediment community, a measure of its energy use, varied seasonally – highest in summer and lowest in winter – but otherwise showed no consistent trend or pattern. The food input to the deep-sea site, on the other hand, which they measured with sediment traps that sampled the rain of detritus to the seafloor, matched the benthic energy requirements during the first two years of the study but subsequently declined. By the end of the study, the rain of sediment to the seafloor met only 21 percent of the benthic community's apparent energy requirements.[30] Since the mid- to late-1970s, the North Pacific had been undergoing large-scale oceanographic changes, leading to warming and decreased productivity over extended regions, including the California Current.[31] Smith and Kaufman hypothesized that the declining food input to the deep sea was most likely linked to these large-scale changes in

North Pacific oceanography and productivity.

This decline in the flux of organic material to the seafloor reversed at the end of the 1990s, with changes in upwelling and surface productivity leading to higher inputs of organic matter to the deep sea than previously observed. Photographic surveys showed that a suite of echinoderm species – holothurians (sea cucumbers), sea urchins, and brittle stars – underwent dramatic changes in abundance, often by a factor of 10–100. The sea cucumber *Elpidia minutissima*, which increased in abundance through the early 1990s, eventually attaining densities of one per square meter, disappeared altogether in 2001 and 2002. Other species increased by comparable amounts.[32] Thus one suite of species dominated under conditions of low food input and another when the detrital flux was high.

Comparable changes were observed in the northeast Atlantic at depths of about 4800 meters in the Porcupine abyssal plain to the southwest of Ireland. Trawl sampling had been carried out in this region between 1989 and 1994 and then between 1996 and 1999. Between those two periods, there were significant changes in abundance for many groups of organisms: sea anemones, polychaete worms, decapod crustaceans, sea spiders (pycnogonids), bivalve mollusks, tunicates, and holothurians. Two holothurians increased more than a hundred-fold between the two sampling periods. One species, *Amperima rosea*, had been quite rare in sampling that extended back to 1977, with a mean abundance of 2–6 individuals per hectare. In 1996 it had increased to about 75 per hectare, and from 1997 through 1999, its abundance was consistently 100–230 per hectare; this dramatic change came to be known as the '*Amperima* event.' Its cause remains unknown, although the holothurians that increased in abundance all feed primarily on the phytoplankton detritus that sediments to the bottom following blooms, leading to speculation that there was a change in surface production.[33]

In both instances, we know too little about the ecology of deep-sea organisms to ascribe the shifts to any particular cause with certainty, although changes in surface productivity were suspected. Most important was simply the repeated observation of substantial change in deep-sea communities, which indicates that they are far more susceptible to change than previously recognized – and hence probably more resilient.

Our next example is of the response of the deep-sea environment to Dansgaard-Oeschger events, those dramatic millennial-scale events occurring during the Ice Ages, when the climate abruptly warms, followed by centuries of cooling. One of the most remarkable records illustrating the trans-oceanic nature of these climate shifts and their impacts on deep-sea circulation and fauna is from the 600-meter-deep Santa Barbara Basin, off the coast of California. The North Pacific does not form its own deep water – its surface waters are too fresh – so the North Pacific is filled with old, oxygen-depleted deep water, originally from the North Atlantic and Antarctic, that upwells to intermediate depths. Deepwater circulation within the Santa Barbara Basin is restricted and its oxygen is entirely depleted, due to the high sedimentation and subsequent degradation of organic matter from the productive overlying waters of the California Current. Higher organisms, notably burrowing worms and the like, cannot live there, so the sediments are undisturbed. In a situation virtually unique in the world's oceans, the sediments are preserved in thin annual layers, due to the alternation of dark material from terrestrial runoff during the wet season and whitish carbonates from marine production during the dry, upwelling season. These conditions have prevailed during the whole of our Recent Epoch. However, during the last Ice Age, North Atlantic Deepwater production was much reduced and the North Pacific was more vigorous in creating its own intermediate water, leading to well-oxygenated conditions in the Santa Barbara Basin. A recent study of a long core of sediment from the basin, lain down over the past 60 000 years, revealed that 16 of the last 17 Dansgaard-Oeschger events recorded in Greenland's ice cores were precisely matched in the Santa Barbara Basin, with a rapid transition from disturbed sediments characterized by an assemblage of benthic forams typical of well-oxygenated conditions to laminated sediment conditions dominated by a few species able to tolerate oxygen-depleted conditions, and then back again. Remarkably these transitions from one set of conditions and one fauna to another sometimes occurred within decades, though more typically over about 130–140 years.[34]

The record from the Santa Barbara Basin is unique in the world's oceans in allowing environmental and ecological change in the deep sea to be examined over the last glacial cycle on a virtually annual basis. Over longer time-scales, we need to examine sediment cores from elsewhere in the world's oceans to document how the deep-sea benthic fauna tracked glacial cycles over time-scales of tens to hundreds of thousands of years. The North Atlantic is a particularly interesting region to study, because North Atlantic Deepwater formation has been a focal point of climate change for the past several million years. Several studies now document how the diversity and abundance of benthic deep-sea forams and small benthic crustaceans known as ostracods tracked the ebb and flow of the glacial cycles, the modes of deepwater circulation, and near-surface productivity.[35]

The Earth swings both ways: deep-sea extinctions during Greenhouse–Icehouse oscillations

Taken together, these studies indicate that over the past several million years the deep-sea environment has varied significantly on all known climatic time-scales, from the orbital beat of tens to hundreds of thousands of years, which sets the major cycles of glaciation, down to decadal scale shifts. There were no significant waves of extinction during this period. Rather, the assemblages appear to have advanced and retreated in their latitudinal or depth distribution as conditions became more or less favorable – a process comparable to the advance and retreat of the tundra and spruce forests with the waxing and waning of the Northern Hemisphere ice sheets.[36] The deep sea is more resilient to this variability than deep-sea ecologists, such as Sanders, might have initially predicted.[37]

However, the shifts and swings of global climate within the past several million years have all been within the framework of an icehouse world. The glaciers advanced and retreated but never disappeared altogether from either hemisphere. And while the mode of North Atlantic Deepwater formation has shifted, Antarctic Bottom Water formation has not faltered, so the deep basins of the world ocean have fluctuated only between cold (under present inter-glacial conditions) and even colder, by about 1.5°C, during the Ice Ages when the relatively warm North Atlantic Deepwater largely shut down. But deepwater circulation has remained vigorous, keeping the deep ocean well oxygenated, except for portions of the Pacific and northern Indian oceans. How would the deep sea respond to more fundamental change?

If we look further back in the geological record, major deep-sea extinction events have followed more profound changes in global climate and ocean deepwater circulation in either direction – whether warming or cooling. The last major extinction event took place in the middle of the Miocene Epoch, about 15 million years ago, when the permanent Antarctic ice-sheet first set in, along with a shift to the modern thermohaline circulation pattern: deep water forming predominantly around the Antarctic and, intermittently, in the North Atlantic. The early Miocene, from about 24–15 million years ago, experienced some of the warmest conditions since the Age of the Dinosaurs. Well-developed coral reefs extended to the North Island of New Zealand; and in the Northern Hemisphere tropical mangroves and reefs were found off central Japan, and a subtropical flora extended to Kamchatka and eastern Siberia.[38] A tropical sea, known as the Tethys, reached from the northern Indian Ocean through what is today the Mediterranean, bathing Eurasia with warm water. The northern Indian Ocean and the eastern end of the Tethys Sea provided a warm, highly saline source of deep water, which flowed south to the Antarctic, warming both that continent and the bottom water it produced, known as Southern Cold Water. The North Atlantic also produced a deepwater, which further moderated the Southern Cold Water.[39] The deep ocean was therefore several degrees warmer than it is today. Benthic foram species today associated with relatively warm, low-oxygen conditions were found deeper and relatively uniformly with depth, indicating a warmer, less stratified, and more sluggish deep ocean environment.[40] But around 15 million years ago, Africa pressed northward against Eurasia, creating the Mediterranean but shutting off the Tethys and its production of warm saline deep water.[41] Intensified latitudinal temperature gradients enhanced global wind systems, leading to enhanced upwelling and marine productivity. Organic carbon–rich sediments were laid down around the North Pacific, producing a massive formation known in California as the Monterey Shale. The resulting drawdown of atmospheric CO_2 amplified the impact of the tectonic plate movements.[42] As the deep ocean cooled, species that had formerly lived within a 'corridor' of Antarctic Bottom Water expanded their range to fill the deep ocean basins, and other species shoaled in their distribution to obtain warmer conditions. Eleven foram species living at intermediate and bottom water depths went extinct, and three new species evolved, leading to the deepwater assemblage of forams that has persisted for the most part to the present day[43] – one suited to the deep ocean of our present icehouse world.

Has our great experiment already been done? The Paleocene–Eocene extinction event

Is there a climate analogue to the dramatic change in CO_2 levels envisioned for the coming centuries – a two- to four-fold increase in CO_2 concentrations above pre-industrial levels – and accompanying change in global temperature on the order of 5°C?

In fact, Revelle and Suess may have been wrong – our geophysical climate experiment may not be unique in the geological record. An earlier extinction event, one of the largest of the past 90 million years – far more significant to the deep sea than the extinction event that marked the close of the Age of the Dinosaurs – provides a remarkable analogy to our species' current atmospheric CO_2 dump. Interest in this event, which marks the transition between the Paleocene and Eocene Epochs about 55 million years ago, dates to 1991, when a landmark paper was published in *Nature* by two marine geologists and paleoceanographers, James Kennett and L.D. Stott of the University of California in Santa Barbara and Los Angeles.[44] They found that the close of the Paleocene Epoch, which saw the extinction of between a third and a half of all benthic foram species living beyond the continental shelf, was marked by a dramatic carbon isotope signature, a sudden drop of about 2.5 percent in the ratio of ^{13}C to ^{12}C, indicating a massive infusion of isotopically light carbon into the ocean and atmosphere. There are few sources of light carbon that could be naturally injected so quickly – within a few thousand years – into the global carbon cycle. Indeed, the sudden release of frozen methane hydrates from the marine sediments seems the only reasonable candidate.

In the previous chapter we introduced the methane hydrates, a vast reservoir of hydrocarbons – about twice as much carbon is estimated to be locked up within methane hydrates (predominantly within ocean sediments) as there is in all known reserves of oil, coal, and natural gas. The stability of frozen methane hydrate is a function of temperature and pressure, such that in today's ocean, for example, the methane hydrate is stable below about 250 meters. A warming of, say, 5°C would cause the methane hydrate to dissociate down to about 400 meters, releasing the methane as gas.[45]

The most likely scenario at the end of the Paleocene is that some initial warming, such as from volcanic outgassing of CO_2 and/or changes in deepwater circulation, warmed intermediate water masses sufficiently to melt and destabilize methane hydrates within the sediments, leading to a massive methane release.[46] The sediment record indicates an initial release of about two-thirds of the 1200–

2000 gigatonnes of carbon eventually injected into the atmosphere over a few thousand years at most, followed by successive smaller waves of methane release, as the greenhouse warming from one release triggered further methane hydrate meltdowns and releases at successively greater depths.[47] Evidence of slope 'failure' – the collapse of the sediments associated with release of methane gas – at this precise period in the sediment record on the Nose of the Blake Plateau off the southeastern United States provides direct support for this hypothesis.[48] The massive spike in the carbon isotope record is matched by a coincident spike in the oxygen isotope record, indicating a sudden warming of Antarctic surface and bottom waters by about 5–7°C, leading, for a brief moment geologically speaking – about 200 000 years – to the warmest conditions observed over the past 65 million years of our entire Cenozoic Era.

Nothing in the geological record so closely approximates our species' present manipulation of the planet's atmosphere and climate. It is estimated that over the coming centuries mankind will inject 2000–4000 gigatonnes of carbon into the atmosphere, an amount comparable to, or perhaps several times larger than, the late Paleocene release. Some have speculated that anthropogenic global warming may trigger the release of methane hydrates, substantially increasing greenhouse warming. The IPCC panel did not factor this potential into their scenarios, and many scientists consider the potential for significant methane release from the sediments remote, though not impossible.[49] The Earth 55 million years ago – just 10 million years after the extinction of the dinosaurs – was still very much a greenhouse world. Ocean bottom temperatures were around 11°C. Under those greenhouse conditions, hydrate would have only been stable below about 920 meters depth, and a 4°C warming of the ocean's bottom waters would have dissociated all the methane hydrate down to about 1460 meters. Jerry Dickens and his collaborators at the University of Michigan have shown that if marine hydrate reserves at the end of the Paleocene were comparable to those today, the release of the methane between those depths would be sufficient to account for the carbon isotope signal.[50]

In today's relative icehouse world, methane hydrates are stable at much shallower depths. Kennett and collaborators recently showed that methane hydrates within the Santa Barbara Basin at depths of about 400–600 meters have regularly dissociated during inter-glacial periods, as changes in intermediate water circulation led to a 2–3.5°C warming. Major departures in the carbon isotope signal also indicated several brief but large methane releases within the past 60 000 years.[51] A contemporary, albeit small-scale, release of methane has been monitored on the continental slope of the Gulf of Mexico.[52] There is considerable methane hydrate within the sediments around the continental margins at intermediate water depths – 400–1000 meters – that are susceptible to rapid warming and dissociation in response to a climate-induced shift in intermediate or deepwater circulation. There are also methane seeps in high-latitude seas and considerable methane hydrate locked up within the permafrost, available for release as the permafrost warms.

Several geochemists have proposed that methane cycling in and out of methane hydrates may have contributed significantly to glacial cycling over the past several million years. Ice-core records show that atmospheric CO_2 and methane cycle in close synchrony with the advance and retreat of the glaciers. Methane may both limit and amplify these cycles. As glaciation advances, a fall in the sea level – around 100 meters at the peak of the last glacial cycle – will decrease the pressure on shallow hydrates, causing them to dissociate, making greenhouse gas levels rise again and limiting glacial advance. As warming sets in, the sea level rises and some arctic permafrost regions are flooded, warming them and releasing further methane. Eventually the rising sea level enables methane hydrates to reform within relatively shallow ocean depths, drawing down carbon from the atmosphere, and completing the cycle.[53]

Anthropogenic global warming may trigger the release of methane from hydrates buried within the permafrost by both direct warming and flooding of low-lying permafrost regions. Arctic seas will warm most dramatically, and shallow hydrates there are most susceptible to dissociation. The greatest methane stores by far, however, are within the ocean sediments – an estimated 10 000 gigatonnes, compared to 400 gigatonnes within the permafrost[54] – and the major concern here is that a shift in intermediate or deepwater circulation could warm and destabilize the hydrates over large sections of the seafloor, as occurred at the end of the Paleocene.[55] Such triggering of further greenhouse gas releases is potentially the worst-case doomsday scenario for our own present course, the very opposite to that depicted in *The Day after Tomorrow*.[56] Interest in the role of methane in global warming and past extinction events has increased substantially in recent years, and there is now evidence of massive methane releases associated with several earlier extinction events during the Mesozoic.[57] At this time, however, too little is known to evaluate its potential for release under current global warming scenarios.

There are some sobering lessons to be learnt from this 'blast from the past,'[58] as we come to realize that our great geophysical experiment has been carried out at least once and perhaps several times before, with catastrophic results. The initial conditions were considerably warmer in the Mesozoic or at the end of the Paleocene, but the magnitude of the temperature change – a 5–7°C increase in high-latitude sea-surface and deepwater temperatures – is most impressive and implies a massive change in ocean circulation.

It is also notable that about 120 000 years were required for the carbon and temperature spikes to return to former levels, which corresponds closely to present estimates of the cycling time for carbon.[59] In other words, if mankind burns up the planet's hydrocarbon reserves in the coming century or so, more than 100 000 years may be required to get the genie back in the bottle – for the carbon to be sequestered again as carbonates or as organic carbon buried within the sediments.

The late Paleocene thermal maximum, though geologically brief, had an enormous impact. On the land, the modern mammals – primates, artiodactyls (camels, sheep, goats, pigs), and perissodactyls (the odd-toed hoofed mammals, such as the horse) dispersed from Asia, where they had evolved, into Europe and North America, rendering extinct the previous mammalian fauna. In the deep ocean, the dramatic warming, possibly combined with more sluggish deepwater circulation, led to poorly oxygenated or virtually anoxic conditions over much of the deep ocean. Laminated sediments, much like those now restricted to the Santa Barbara Basin, appeared widely in the sediment record, indicating a lack of disturbance because higher organisms were no longer able to live in such oxygen-depleted conditions. The infusion of CO_2 into the ocean, forming carbonic acid, reduced pH, so organisms had greater difficulty forming their calcium carbonate shells and skeletons. Surface productivity likely declined as well. These factors, whether singly or in combination, resulted in the rapid extinction of 35–50 percent of benthic foram species, the fauna adequately sampled within the sediment cores.

Global climate change – looking into the future

It is tempting to assume that raising atmospheric CO_2 levels to around 1000 ppm, the levels experienced in the Eocene, approximately 40 million years ago,[60] will cause the planet to return to the climatic conditions experienced at that time. But the Earth has changed since then – Antarctica has become further isolated and the Isthmus of Panama between the Atlantic and Pacific has closed, which has in turn enhanced the Gulf Stream and set up the present North Atlantic thermohaline circulation. Mankind has set the planet's climate on a course into uncharted waters.

Unfortunately there are wild cards in the climate deck. The rapid fluctuations in climate records indicate instability: the potential for abrupt change to occur when the climate system crosses a threshold and reaches a new equilibrium, often in response to relatively small forcing.[61] Climate models are poor at predicting such transitions; by their very nature such instabilities are virtually unpredictable. Although most climate models do not predict a complete shutdown of North Atlantic Deepwater production during the 21st century – a profound restructuring of the oceans' thermohaline circulation – it becomes increasingly likely as CO_2 rises above twice pre-Industrial levels.[62] Such shutdowns have occurred in the past,[63] where they have generally led to the sudden onset of global cooling (the *Day after Tomorrow* scenario), but with the Earth's new greenhouse atmosphere, the impact will be quite different. Once such a transition occurs, global warming and freshwater inputs will likely enhance the stability of the near-surface layer, further sealing the Earth's climate system into its new state. Decreased deepwater formation will further limit the sequestration of atmospheric CO_2 in the deep ocean, amplifying its build-up in the atmosphere. The wind systems that drive ocean upwelling will decline with a decreased temperature gradient between low and high latitudes, and the ocean's enhanced stability will further limit nutrient input to near-surface waters. Global models predict a general decline in ocean productivity, which will further limit CO_2 uptake by the oceans, yet again amplifying the greenhouse effect.[64]

Global climate change and its potential impacts are still highly uncertain, not least because the scale of the human impact remains uncertain – it is still possible to limit emissions. But predictions based on the business-as-usual scenario are dire. If accessible hydrocarbon resources continue to be consumed following current trends, or even at somewhat reduced rates, CO_2 levels in the atmosphere will rise two to four times above those in the pre-Industrial era. Global warming in the coming centuries could take the world back to a climate not seen since the Miocene or the Dinosaur Age, 55–90 million years ago, with a warming of 7–10°C over the continents. With a four-fold increase in CO_2 levels, the global deep thermohaline circulation may shut down altogether; a doubling of CO_2 levels might lead to just the North Atlantic limb of the deepwater circulation system shutting down.[65] As the deep ocean warms and its circulation grows more sluggish, oxygen levels will fall, and the increased infusion of CO_2 into the ocean as carbonic acid will increase its acidity. Global marine productivity will decline due to reduced nutrient input to near-surface waters as upwelling declines and stratification increases. These impacts, predicted by current global ocean models, are all seen in the paleoceanographic record.

Rising CO_2 and ocean acidification

Climatologists have largely focussed on the impacts of rising temperatures from increased greenhouses gases, but ocean-ographers have recently begun to examine the direct environmental impacts of increased CO_2 on ocean acidity. Based on IPCC projections that atmospheric CO_2 concentrations may reach 800 ppm by the end of the century, the pH of ocean surface waters is predicted to drop by 0.4 units – an increase in acidity (the concentration of hydrogen (H^+) ions) of about 150 percent. By 2300, the continued depletion of fossil fuel reserves could lead to a drop in ocean pH of 0.77 units, the largest change in ocean pH experienced over the past 300 million years. This could have a dramatic impact on organisms that build calcium carbonate ($CaCO_3$) skeletons, particularly those groups, including corals, with skeletons made from aragonite, a particularly soluble form of $CaCO_3$. In Chapter 6, I noted that deepwater reefs created by stony corals are found extensively around the rim of the North Atlantic but not the North Pacific, possibly related to the considerable difference in the aragonite saturation horizon in those two oceans: 1500–2600 meters in the North Atlantic compared with 120–580 meters in the North Pacific. The aragonite saturation horizon is estimated to have already shoaled 30–100 meters in the North Pacific since the dawn of the Industrial era.[66] Based upon the IPCC projection, undersaturation in aragonite could extend to the very surface of the subarctic Pacific by the end of this century, and the saturation horizon in the North

Atlantic could rise from 2600 meters to 115 meters north of 50° N latitude.[67] No experiments have yet been done with deepwater corals, but experiments with shallow tropical corals indicate a decline in calcification rates of up to 50 percent with changes in acidity of around 0.4 pH units. Deepwater corals may prove even more sensitive, given the reduced variability experienced in deep waters. Might this spell the end of deepwater coral reefs around the world? Nothing is yet certain, but this is a subject of increasing concern. Deepwater corals may be as threatened by climate change as shallow tropical reefs.

The impacts of global warming on the deep sea may seem like the least of mankind's concerns in considering strategies to deal with global greenhouse emissions. However, the deep-ocean record from the end of the Paleocene provides the clearest example of the massive, long-standing implications of returning vast quantities of carbon to the global biogeochemical cycle. The quantities of carbon returned to the atmosphere are uncannily similar between then and now, and the observed impacts thus far are sufficiently close to the predictions of current models to cause reasoning policy-makers to pause before repeating that experiment.

CHAPTER 10

Deepwater Fisheries: No More an Oxymoron

In the decades immediately following World War II, the prevailing paradigm for the deep sea was of a vast realm so depauperate that deepwater fisheries seemed an oxymoron. Tows with plankton nets through the open water or with dredges along the seafloor indicated that biomass declined exponentially with depth in all the world's oceans (p. 50).[1] Phytoplankton production at the base of ocean food webs is restricted to the well-lit near-surface waters – approximately the upper 100 meters or so – and it is there consumed by a variety of protists, small crustaceans, and gelatinous filter feeders. Little but the detritus from this feast filters down to the deep ocean: feces, unconsumed phytoplankton from large blooms, crustacean molts and discarded mucous larvacean 'houses.' It seemed a truism that substantial fisheries would not be found in the deep ocean.

The Paradigm of the Depauperate Deep remained entrenched into the 1980s, despite mounting evidence over several decades, such as Isaacs' pioneering study of fish concentrations sustained on offshore banks (see Chapter 6) and from the fisheries themselves. However, this paradigm, a modern reworking of the Azoic Hypothesis, lingered long past its 'use-by date,' partly because deep-sea biologists were isolated from fisheries developments and partly because it seemed to make such good sense – like the Azoic Hypothesis, in its time.

Much of the deep sea is indeed depauperate. But it depends where you look, and most deep-sea biologists were looking where it was easiest – and least productive – on the open sediment-covered seafloor.[2] The vast majority of the deep sea does consist of these open plains; but fishers have always understood that most of the ocean, even the upper waters, is a very dilute broth, and that the action is to be found at certain hotspots defined by the currents and underlying topography.

To most deep-sea biologists, however, it came as a complete surprise when, in the 1980s, New Zealand and Australian fishers working the hard-scrabble slopes of seamounts trawled up massive quantities of orange roughy from 700–1000 meters depth, sometimes as much as 50 tonnes at a shot, with the trawl sometimes on the bottom for only a few minutes (Plate 16). Lesser, but still sizeable, quantities of cardinal fish (*Epigonus telescopus*) and black and smooth oreos (*Allocyttus niger* and *Pseudocyttus maculatus*) were also trawled from these seamounts. And a decade earlier, almost no one had noticed when Soviet and Japanese trawlers cleared more than 900 000 tonnes of pelagic armorhead (*Pseudopentaceros wheeleri*) from North Pacific seamounts. Western fishery scientists and deep-sea biologists alike failed to anticipate the magnitude of the new seamount fisheries. Even the species themselves came as a surprise: biologists had thought of them until then as rather obscure taxa, minor players in the ecology of the deep sea. So these species – and the general topic of deepwater fisheries altogether – received virtually no mention in the leading books on fisheries and in scientific reviews of deep-sea biology well into the 1980s.[3]

Traditional deepwater fisheries

In fact, deepwater fisheries have a long history, extending into the pre-modern era among traditional fishing cultures in the North Atlantic and South Pacific. The most remarkable of these artisanal deepwater fisheries developed on volcanic islands, whose small size, steep terrain, and characteristically sharp drop-offs provided limited potential for agriculture and shallow-water fisheries. Several large predatory fishes, not dissimilar to those found around seamounts and canyons, live at midwater depths around such islands, preying on the smaller fishes that impinge on the steep slopes or that are attracted to the heightened productivity around islands.

In the South Pacific, Polynesians from a number of islands have traditionally fished for the oilfish (*Ruvettus pretiosus*), a snake mackerel or gempylid that attains a length of 1.8 meters and a weight of almost 70 kilograms. It continued to be fished with handlines off the steep island slopes at depths of 150–750 meters well into the 20th century. A sophisticated but unbarbed V-shaped wooden hook was used. A baitfish was lashed at the base of the V along the main shaft, which was 20–35 centimeters long, and the unbarbed spur was designed to catch in the fish's gill as it worked its way onto the shaft to feed on the bait.[4]

In the North Atlantic, the black scabbard fish (*Aphanopus carbo*), growing to 145 centimeters, has been one of the most important fisheries around Madeira for centuries, even being mentioned in a poem dating back to 1635. It is a cutlass fish or trichiurid – a family closely related to the

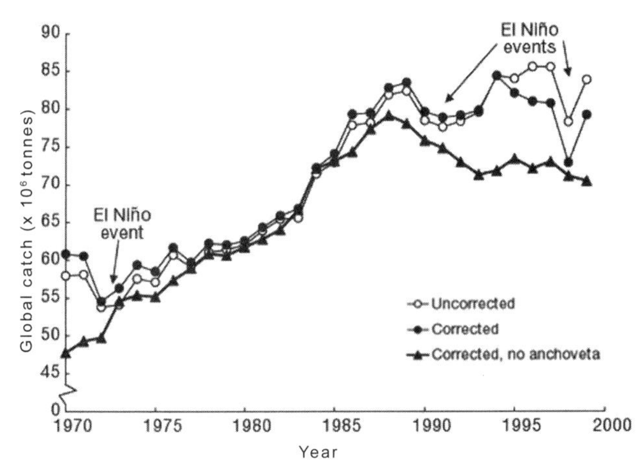

Global marine fishery landings, corrected for systematic misreporting by China, and presented with and without Peruvian anchoveta landings. (From Watson and Pauly, 2001; reprinted from *Nature,* Macmillan Publishers Ltd.)[V]

gempylids. The traditional fishery, maintained until the 1980s, was carried out by setting vertical droplines through the night, with a number of barbed hooks set at depths of 550–1100 meters.[5] Artisanal deepwater line fisheries also developed off the Azores for alfonsino (*Beryx* spp.) and red seabream (*Pagellus bogaraveo*).

The Inuit living in certain deep West Greenland fjords jigged for Greenland halibut (*Reinhardtius hippoglossoides*) through the ice in mid-winter to depths of 350–700 meters with handlines woven from whale baleen. By the early 1900s, a commercial longline fishery had developed in West Greenland for the halibut, replacing the Inuit subsistence fishery.[6] In the northeast Atlantic, Norwegian longlining for Greenland halibut extends back at least to the 1760s, when Norway began to supply the fish to Russia; this trade continued until the Russian Revolution.[7] By the mid-19th century, Scandinavian fishermen had also developed longline fisheries for two deepwater relatives of the Atlantic cod – ling (*Molva molva*) and tusk (*Brosme brosme*) – in the waters off Iceland, Norway, and the Faroes.[8]

These artisanal deepwater fisheries, carried out with simple technology, are among the few deepwater fisheries that have proved sustainable, although their future is now in jeopardy as more efficient fishing methods replace traditional practices.

The rise of modern deepwater fisheries

Marine fisheries entered an era of exponential growth following the Second World War. Between 1948 and 1970, landings increased more than 3.5-fold, from 17 million to 61 million tonnes, growing at about 6 percent per annum.[9]

Fishers have always been competitive, inventive and independent, quick to adopt new technologies and untrammeled by theoretical constructs. The remarkable post-war expansion of world fisheries was fostered by a suite of new technologies, many transferred from the military, combined with largely unregulated access to the high seas. Following the war, fishers quickly adopted radar, enabling them to work in the fog and at night; echo-sounders to locate fish through the water column; synthetics, such as nylon, to build stronger nets; electronic navigational systems, such as LORAN (Long-Range Navigation) and more recently, a satellite-based global positioning system (GPS); and the capacity to freeze their product at sea, thereby massively extending both the time they could spend away from port and their range of operations.

The world's traditional fisheries and fishing grounds – salmon in the North Pacific, cod, haddock, and herring in European waters, the fisheries of Georges Bank, the Grand Banks, and the seas around Japan – had already been fished to near capacity. But these new technologies enabled fishers to exploit new species and new grounds. The Soviet Union, Japan, Poland, Spain, Taiwan, South Korea, and other nations built distant-water fishing fleets that ranged the globe in search of new fishing grounds.

Until about 1977, when most coastal states adopted a 200-mile jurisdiction over their fishery and other marine resources, most of the world ocean beyond 12 nautical miles from shore was a global commons. Of course it suffered the 'tragedy of the commons,' so aptly described by Garrett Hardin in 1968.[10] Without private ownership or an authority to conserve and manage the fisheries, there was no incentive for any fishing interest to limit its effort – any catch foregone would simply be taken up by others.

The distant-water trawling fleets exploited the most valued, most accessible fish resources first. A pattern described variously as *pulse fishing*, *serial depletion*, and *fishing-up* ensued. Massive unregulated fishing on a target *stock* (or local population) led to its depletion, at which time the fleet moved on to the next. The preferred species in the most accessible fishing grounds were progressively overfished, forcing the fleets to range further and further afield, generally into deeper and less easily fished grounds, and to return with less and less favorable species.[11]

Most of the expansion of global marine fisheries in the decades following World War II occurred on the open ocean (for tunas) and in distant continental shelf and upwelling regions – the Bering Sea, Labrador, the Patagonian and other continental shelves, the Peruvian and Benguela upwelling zones.[12] The movement into deeper waters was gradual, first involving

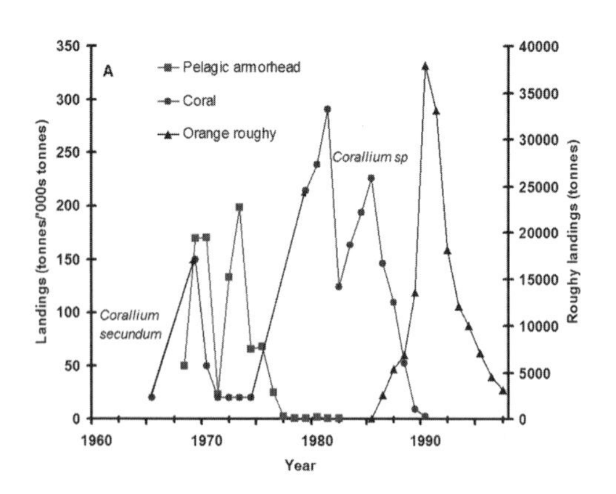

Pelagic armorhead (*Pseudopentaceros wheeleri*) and red coral (two species: *Corallium secundum* and C. sp.) from the North Pacific (left-hand y-axis, measured in thousands of tonnes and tonnes, respectively) and orange roughy (*Hoplostethus atlanticus*) from Tasmania (right-hand y-axis).

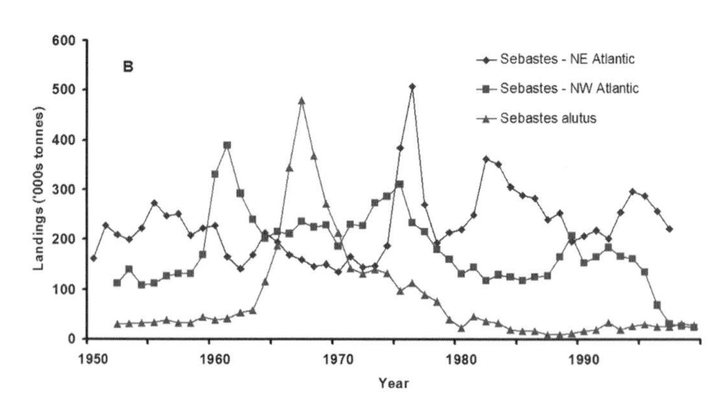

Pacific Ocean perch (*Sebastes alutus*) and redfish (*Sebastes* spp.) from the northeast and northwest Atlantic.

species at the shelf break and upper slope – hakes and species of *Sebastes*, such as Pacific Ocean perch in the northeast Pacific, and redfish (other than deep-sea redfish, *S. mentella*) in the North Atlantic. These are generally not considered true deepwater species: deepwater fisheries are held to start at 400–500 meters depth.

The first modern deepwater fisheries developed toward the end of the 1960s in the North Atlantic and Pacific. The Soviet distant-water trawling fleets initiated these fisheries in both oceans, but they evolved uniquely due to differences between the oceanic regions themselves (Plate 15).

Pacific seamount fisheries – the pelagic armorhead

The Pacific Ocean contains hundreds of times more seamounts than the North Atlantic (see p. 118) so, not surprisingly, seamount fisheries first developed in the North Pacific. The first seamount fisheries, developed by the Japanese shortly after the war, were for albacore. It is still not well understood why, but these fish were found concentrated over seamounts in the Emperor Seamount chain and elsewhere between Japan and Hawaii.

Japanese interest in seamounts focussed on pelagic fisheries for decades, but in 1967 a Soviet trawler discovered large aggregations of a demersal fish, the pelagic armorhead (*Pseudopentaceros wheeleri*), on seamounts in the southern Emperor Seamount chain, some 3000 kilometers northwest of the Hawaiian Islands, at depths between 260 and 600 meters.

Until then, the pelagic armorhead was an obscure species, known from isolated specimens across the North Pacific – so obscure that until its unique life history came to light, scientists believed that the seamount population consisted of two distinct species. Prior to sexual maturity, the armorhead ranges widely in the upper waters of the North Pacific, where it feeds and

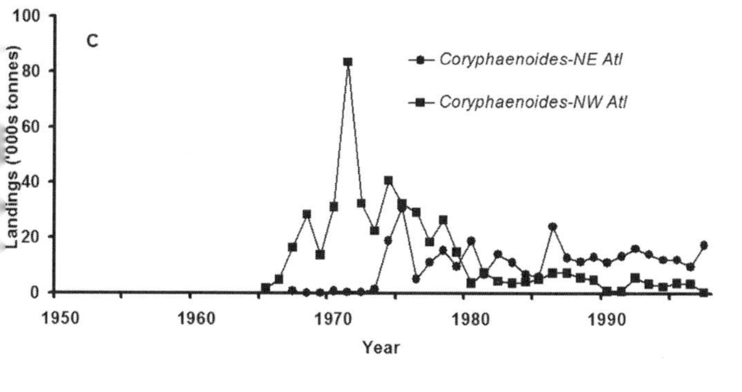

Roundnose grenadier (*Coryphaenoides rupestris*) from the northeast and northwest Atlantic.

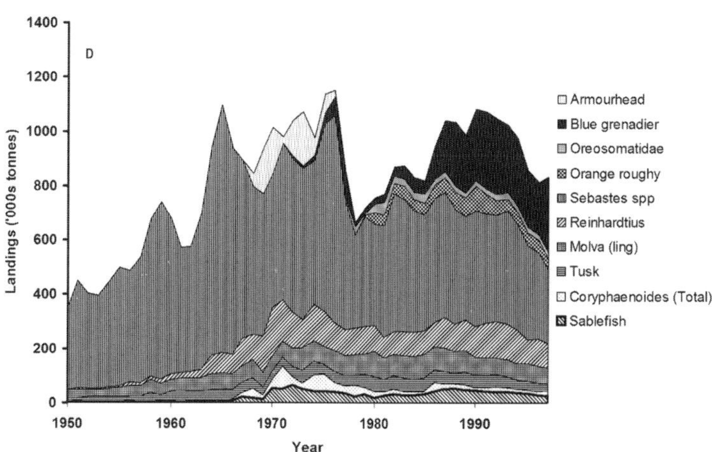

The total landings from the main deepwater fisheries. (Data from FAO, modified from Koslow et al. 2000.)[W]

grows rapidly, much like the Pacific salmon. But whereas salmon return to their natal river systems in Asia and North America to spawn, the armorhead converge on a few seamounts in the central North Pacific.[13] They do not spawn immediately upon reaching the seamounts – they remain there for a year first – and they do not die immediately afterward. Like the Atlantic salmon, they spawn several times before dying. And just as massive salmon runs cannot be supported by the lakes and streams to which they return (salmon must build up their fat reserves at sea to fuel their spawning migration), the armorhead are too densely aggregated to support themselves on these seamounts in the unproductive central Pacific. Extremely fat when they first arrive on the spawning ground, the armorhead progressively lose weight and condition – to the extent that scientists believed at first that the newly arrived armorhead were a different species from the emaciated fish that had spawned several times and were near the end of their life.

Fish species that spawn in large aggregations are most vulnerable to overfishing at that time, and none more so than the pelagic armorhead, which is restricted to a few seamounts. The underlying principles for managing a fishery on a spawning aggregation are straightforward. They do not require elaborate population models or a modern scientific infrastructure, although the Soviets had both. The aboriginal peoples of Asia and North America sustainably harvested salmon fisheries for millennia without formal science. Although they had the ability to block off rivers and fish out entire spawning runs, they knew to leave a reasonable proportion to reproduce.[14]

But the pelagic armorhead fishery was in international waters, and the Soviets gave no consideration to regulating or conserving it. The intensity of fishing on these localized topographic features, with areas of 3–450 square kilometers, was staggering.[15] Soviet fishing effort from 1969 to 1975 totaled 18 000 trawler days.[16] The Soviet trawlers were later joined by Japanese trawlers, and together they landed 50 000–200 000 tonnes of pelagic armorhead each year from the seamounts, a total of over 900 000 tonnes within ten years. The first major scientific symposium on seamount fisheries was held in 1984, and was co-sponsored by Japanese and American fisheries institutions.[17] The Soviets did not attend. The pelagic armorhead had already been fished to commercial extinction.

The boom-and-bust pattern of the pelagic armorhead fishery would continue to be played out, with minor variations, for much of the remainder of the century in other deepwater fisheries. The open ocean was virgin territory to both science and the fishing industry, and for several decades fishers depleted one stock after another – sometimes entire species – while science and management played catch-up, almost always a step or so behind. Indeed, seamount science remained dormant for 20 years following Isaacs' pioneering study, published in 1965, that first explored the processes sustaining fish aggregations over submarine rises (Chapter 6). As noted in a comprehensive review published in 1987, although over 100 seamounts had been sampled by that time, a mere handful – five – had been sampled at all adequately, and these accounted for 70 percent of the invertebrate species recorded from seamounts worldwide.[18]

Precious coral fisheries in the Pacific

The iconic benthic invertebrates on many seamounts are the corals, and no group better illustrates the gap between fishers and scientists in the early period of seamount exploration. A great variety of corals across three orders are harvested for jewelry: gorgonians, such as the red and pink corals (*Corallium* spp.) and bamboo corals (*Lepidisis* spp. and *Keratoisis* spp.); gold corals (*Gerardia* spp.) in the zoanthid order; and the black corals (*Antipathes* spp. and *Cirrhipathes* spp.) in the order of horny corals or Antipatharia.[19] In the Mediterranean, precious corals have been harvested for millennia.

The relationship between coral abundance and the accelerated currents on seamounts was first described in 1986, in the pre-eminent scientific journal *Nature*. The paper began:

> Few biological data have been collected for deep seamounts. Here we present some of the first quantitative observations of hard-bottom (non-hydrothermal) fauna in the deep sea.[20]

There is no small irony here. In 1965, fully 20 years before 'some of the first quantitative observations' were published, Japanese fishermen discovered a huge bed of red coral (*Corallium secundum*) at 400 meters depth in the Emperor Seamount chain. A veritable 'coral rush' ensued. Records for the fishery are poor, but it apparently peaked at about 150 tonnes per annum around 1970 and then precipitously declined, until a second species of *Corallium*, unknown to science, was discovered at depths of 1000–1500 meters. At the peak of this fishery in 1981, more than 100 boats from Taiwan and Japan were involved, and almost 300 tonnes were landed across the Pacific. By 1990, just a few years after the landmark seamount coral study was published, red precious corals had been depleted across the North Pacific: no more coral came out of the Emperor Seamount chain, and the total harvest from the North Pacific was about 3 tonnes per annum.[21]

The collapse of the fishery is not surprising, given that deepwater corals are long-lived, and the fishery was in international waters and unregulated. *Corallium secundum* lives to about 75 years.[22] The age of the deeper *Corallium* species was never determined, but similar corals living at those depths can be exceptionally long-lived: yet another species of *Corallium* from 600 meters depth lives to 180 years, bamboo corals (*Keratoisis* spp.) off Tasmania and New Zealand have been aged to 300–500 years, and a specimen of gold 'coral' (*Gerardia*) (actually a colonial zoanthid sea anemone) collected from 620 meters depth off the Florida coast appears to be the great Methuselah, aged between 1300 and 2700 years.[23] Precious coral fisheries within Hawaiian waters have been carried out with small manned submersibles, so only corals of a specified size were harvested, but the Emperor Seamount coral fishery, carried out largely with tangle nets, was hugely destructive and wasteful. Tangle nets consist of concrete blocks, stones, or iron bars that are dragged up and down seamount slopes, with netting attached behind that entangles and recovers a fraction of the broken coral.

The orange roughy: a case study

Following the demise of the pelagic armorhead fishery, Soviet trawlers scouring the Pacific for further seamount resources discovered orange roughy off New Zealand in the mid-1970s. But a sea change was underway in global fisheries, and the orange roughy fishery was to develop very differently.

By 1977, many coastal states had declared exclusive economic zones out to 200 nautical miles from their coasts, and this was recognized by the 1982 United Nations Convention on the Law of the Sea (or UNCLOS, as it is generally known). This was the first step – necessary but not sufficient, as it turned out – in ending the rapidly unfolding tragedy of the commons in global fisheries: for each nation to take control of at least the resources off its own coast. UNCLOS was the instrument whereby the international community wrested control of most of the world's fisheries from the depredations of the high-seas fleets. Unfortunately, many deepwater fisheries remained outside the fold, still on the high seas and unregulated.

The Soviets apparently first fished orange roughy (*Hoplostethus atlanticus*) on the extensive submarine banks and seamounts around New Zealand in the years 1972–77, just before New Zealand took control of the fisheries out to its 200-mile limit. Soviet landings were small – a maximum of 3500 tonnes in 1977 – and it is unlikely that they discovered the major spawning aggregations. The species was not even clearly identified – the only record is of a 'berycoid' fish (orange roughy is in the order Beryciformes). It was in fact a rather obscure fish then, a member of a family originally known as the 'slimeheads' (the Trachichthyidae), due to the copious amounts of wax ester that liquefy and ooze from canals in their heads when they are brought to the surface.

Within a few years of New Zealand taking over the fishery and rebadging the species as 'orange roughy' for the export market, and with a further large spawning aggregation discovered off Tasmania, Australia in 1989, the orange roughy had become the largest, deepest and most valuable deepwater fishery in the world. In many respects, the New Zealand and Australian orange roughy fisheries provided a best-case scenario for deepwater fishery management. The stocks were entirely within their jurisdictions, their scientific and management infrastructures were well developed, and sufficient resources were allocated to conduct the requisite research. There were even some lucky breaks scientifically. Despite their great depth, orange roughy proved readily amenable to conventional survey methods, because each stock conveniently and predictably gathered for several weeks each year at a localized spawning ground – often a single small seamount. Yet 15–20 years after the onset of these fisheries, not one of the more than a dozen orange roughy stocks in New Zealand and Australian waters was above the management target – 30 percent of the unfished stock size – set for a sustainable fishery. Most were severely depleted. So what went wrong – and what was likely to go wrong in other deepwater fisheries around the world?

As the fisheries developed, Australia and New Zealand instituted scientific programs to estimate the size and productivity of their respective orange roughy stocks. I myself came to Australia in 1989 to lead research on the orange roughy at the Commonwealth Scientific and Industrial Research Organization (CSIRO). I focussed initially on developing methods to survey the orange roughy population, choosing two independent methods: acoustic surveys and egg surveys. These were standard fishery techniques, but they had never been successfully applied to such deep-living fish populations.[24] Colleagues at other institutions worked to determine the age of orange roughy specimens from their otoliths. This was the other key piece of information needed for fishery management, since to estimate productivity one needs to assess the growth rate (the increase in length and weight with age), the number of young fish entering the population, and the natural mortality rate (the decline in abundance with age).

All this might seem rather pedestrian – and applied science often is – but there were many pitfalls for the unwary, and many technical difficulties to overcome. The orange roughy research had an abundance of such problems. Many issues with enormous management implications took years to resolve, such as determining the species' longevity. In 1980 a Russian fishery scientist, A.N. Kotlyar, estimated from the otoliths that orange roughy lived to about 20–25 years, a reasonable-sounding age for a fish that attained a length of 45–50 centimeters.[25] New Zealand set its fishery quotas accordingly, but the rapid decline of its orange roughy stocks and the lack of young fish coming into the fishery indicated that they might be greatly over-estimating the fish's productivity. Alternative radiometric methods to age long-lived fish – similar to the carbon-14 dating method, but based on radium-226, which is incorporated into otoliths in trace amounts as an analogue of calcium – had been worked out in the 1980s for several northeast Pacific deepwater fishes: sablefish (*Anoplopoma fimbria*), Pacific Ocean perch (*Sebastes alutus*) and other species of *Sebastes*.[26]

In 1991 the radiometric method was successfully applied to the orange roughy. Remarkably, it indicated that the species lived to 100–150 years: after about 20–25 years, the orange roughy otoliths, like those of many long-lived fish, grew thicker with age rather than wider, with the rings piled up on one another. However, when the otoliths were ground down appropriately or cracked in half and examined side-on, the ring counts matched the radiometric estimates of age. Such longevity, never encountered previously in a fish only 45–50 centimeters in length, indicated extremely low rates of growth and population turnover – meaning low natural mortality and a low input of young fish into the population. Sustainable catches, which are based on harvesting a portion of the natural production, were thus only a fraction of previous estimates.[27] Subsequently it was shown that smooth and black oreos and deepwater cardinal fish, also fished on southwest Pacific seamounts, all lived to 100 years or more.[28]

By the early 1990s, Australia's acoustic and egg surveys both pointed to a similar stock size, and the scientific community generally accepted that the orange roughy lived to 100 years or more. All the key parameters required to estimate sustainable catch limits were now available, and a mood of optimism briefly prevailed that deepwater fisheries could be maintained sustainably.[29] By the end of the millennium, however, the Australian orange roughy fishery was sorely depleted, with the main stock assessed at about 10 percent of its original size.[30] New Zealand, with its extensive plateaux and seamounts at the appropriate depth and latitude, had more than a dozen orange roughy stocks and fisheries. The apparent stability of their landings – they had ranged from about 30 000 to somewhat more than 50 000 tonnes per year during the 1980s and were still only somewhat less than 20 000 tonnes by 2001 – had in fact been maintained by serially depleting one stock after another. But by 2000 virtually all were depleted, and there were no new discoveries.[31] That year, New Zealand closed its Challenger Plateau orange roughy fishery, one of its most longstanding. Despite quotas having been reduced since 1992, the stock had continued to decline, to the point that it was then estimated to be only 3 percent of its original size. The quotas for other depleted orange roughy fisheries were slashed as well.[32] The only exception to this trend was the largest New Zealand stock, the Chatham Rise stock, which notably did not spawn on seamounts.

Why did fisheries science and management fail for the orange roughy? A number of factors appear to be responsible, and they are worth exploring, if only because deepwater fisheries continue to collapse around the world, despite a growing chorus of protest from concerned fishery scientists and environmental organizations.

The first strategic error that doomed the Australian orange roughy fishery, among others, was violation of what I'll call the Scott Joplin principle. Joplin often marked the tempo of his pieces, 'Not fast.' Later in life, he seems to have lost patience, writing above one of his scores: 'Notice: It is never right to play "Ragtime" fast. Author.'[33]

For 'ragtime,' substitute 'fisheries.' It is an elementary principle of managing any new fishery – simple in theory but much more difficult to put into practice than to slow a reckless ragtime pianist – to play it slow: to allow sufficient time to assess the sustainable potential of the fishery before ramping it up. It is always easy to open the door wider, to increase catch levels and to allow in more fishermen. But once you let them in, it is never easy to push them back out and close the door.

When the Tasmanian orange roughy spawning aggregation was first discovered, there was enormous pressure from the region's trawlers, licensed to fish the continental shelf and upper slope, for deepwater fishing licenses. The traditional fisheries were largely depleted and the industry economically depressed, yet each day orange roughy trawlers were returning to port in Hobart with 100 tonnes of orange roughy in their holds – at $10 per kilo, each load was worth about a million dollars. The Australian Fisheries Management Authority buckled under the pressure, and granted deepwater licenses to the mass of trawl operators. By the second year of the fishery, 66 vessels were involved. The reported landings of orange roughy soared to 40 000 tonnes, and at least 30 percent more is estimated to have been lost due to the inexperience and greed of the fishers, who too often trawled up more than they could land, causing them to spill dead fish back into the water. 'Burst bags' were often reported: nets filled to the point that they burst, losing the entire catch.[34] How long could such a fishery last when the total population was estimated at less than 200 000 tonnes?

The second key strategic error was to allow fishing to continue on the spawning aggregation. The population was exceptionally vulnerable for the several weeks that it spawned over a single small seamount off the northeast coast of Tasmania, and fishing pressure was enormous. Because of the small size of the seamount and the many trawlers in the fishery, they had to queue for their shot, such that one trawler after another shot away at the aggregation, 24 hours a day, for the duration of the spawning season. Over the first four years of this fishery, about 7700 trawl shots were directed at the spawning fish. Another 11 200 trawl shots were directed at the fish in their feeding grounds: small seamounts that dotted the continental slope off the south coast of Tasmania. Within a few years the spawning aggregation had virtually disappeared. In theory, a sustainable quota can be fished from a spawning aggregation as well as during some other part of the fish's life cycle. In practice, though, it violates the Joplin principle. Intensively fishing a population when it is massed and most vulnerable leaves no margin for error. And there was ample opportunity for error when the methods to assess the fishery – the deepest in the world to date, on a previously obscure species – were still under development.

The third strategic management error might be crudely termed lack of backbone – the failure to follow scientific advice and the agency's own management objectives. Early in the development of the fishery, it was agreed that the stock would be maintained above 30 percent of its original size. If it slipped below, quotas would be reduced to allow the stock to rebuild. If the stock fell below 20 percent of its original size, the fishery should be closed. By 1994, only five years after the fishery had gotten underway, the stock was assessed to be below 30 percent of its original size. A quota of 2000 tonnes on the spawning ground was then set, based on the results of a population model which indicated this would allow the stock to re-build.

However, an acoustic survey in 1996 showed that the stock was continuing to decline and that only a substantial reduction in the quota would allow the 30 percent objective to be achieved over a seven-year time horizon, assuming reasonable recruitment of young fish to the population. The industry disputed the survey's validity, and the managers chose not to stand behind the scientific evidence. In countries like the USA and Australia, where the industry is given a strong voice in managing the fisheries, there is considerable pressure to achieve consensus between the industry and managers. (Proponents of these co-management arrangements claim that it fosters industry responsibility and a sense of ownership; others believe it gives the foxes charge of the chicken coop.) I did not join in this consensus and consequently was eased out of the stock assessment arena. The quotas remained unchanged until 2000, when an assessment indicated that the stock was likely less than 20 percent of its original size. The model, which had thus far failed to predict the continued decline of the stock, now indicated that even cutting the quota in half would not allow it to rebuild in a reasonable time frame. The quota was actually cut by just 10 percent. Then in 2002, an independent review by overseas fishery scientists concluded that the stock was severely depleted, probably 10–20 percent of its initial size. The quota, which should have been set to zero, was only cut in half. In 2006, Australia's governmental Threatened Species Scientific Committee recommended that orange roughy in Australia be listed as endangered. If the government accepts the recommendation, the fishery may finally be closed.

Of course, much of the responsibility for the failure of the fishery lies with the industry. Its short-sighted self-interest, combined with its power and influence, proved virtually unstoppable in running down populations as valuable, vulnerable, and unproductive as the orange roughy. A sustainable yield for this species was estimated to be only a few percent of the original biomass. So, an industry consultant asked at a meeting of the scientists, industry, and management, why not just 'mine' it out? Whether explicitly stated or not, that became the industry objective, and it used every means at its disposal to delay the inevitable quota reductions required to sustain the fishery. A time-honored tactic was to exploit uncertainty in the scientific assessments – and there is always uncertainty in this science – to argue that severe quota reductions were not yet justified. And the strategy succeeded: each year the management agency acquiesced and compromised, setting the quota higher than the scientists recommended.

When all other avenues failed, the industry under-reported its catches. One year, however, Tasmanian marine police secretly videotaped the trawlers in port. Orange roughy landings were estimated at approximately twice the amount officially reported. The police implicated virtually everyone in the industry, but only one operator was prosecuted. In addition, catches from the spawning ground were mis-reported as coming from other areas where quotas were more difficult to fill.[35] But cheating is only one more facet of the tragedy of the commons: once it becomes rampant, any individual who does not participate appears foolhardy. When everyone's hands are in the bag grabbing for the lollies, it's clear they will soon be gone.

Some responsibility for the failure resides with the scientists as well, particularly those who proposed and placed their faith in management models that had little margin for error – and the human elements in the management process. The models used to manage orange roughy fisheries indicated that the populations could be fished down to 30 percent of their original biomass without ill effect.[36] But in a world with the usual greed, mendacity, weak managers, and strong political and economic pressures, it was virtually inevitable that the fishery would not be adequately restricted until the stock was well below that target.[37] A target of 50 percent of the original stock size would have been more realistic.

The models also assumed a more-or-less constant input of young fish to the spawning stock, which would enable the population to rebuild once fishing quotas were reduced.[38] However, recruitment to long-lived deepwater fish populations appears to be highly episodic, such that decades may pass before significant numbers of young fish enter the adult stock.[39] As a result, several orange roughy stocks continued to decline even after quotas were severely reduced.

The largest orange roughy resources proved to be around New Zealand and Tasmania, where that species dominates the fish biomass at mid-slope depths (700–1200 meters), within a water mass known as Antarctic Intermediate Water. The distribution of orange roughy and a number of other fish species with which it co-occurs largely follows the circulation pattern of this water mass around the rim of the Southern Ocean, extending northward into the North Atlantic.[40] Smaller orange roughy fisheries developed in the 1990s on seamounts and ridges off Namibia and Chile, in the southwestern Indian Ocean, on the Louisville Ridge in international waters to the east of New Zealand, and in the northeastern Atlantic.[41] These fisheries followed a similar boom-and-bust pattern: catches quickly peaking following discovery of a new ground, and then plummeting almost as rapidly, as the stock was depleted. There was little attempt to manage these fisheries, many of which were in international waters, except for the fishery off Namibia, where there was an attempt to learn from the Australian and New Zealand experience. The fishery was limited to five vessels and a more precautionary fishdown target was set – 50 percent of the initial stock biomass. However, catches were not limited in the first years of the fishery due to an initial overly optimistic assessment of the population size. Within six years, catches had dropped dramatically, and most spawning aggregations were assessed as being already well below 30 percent of their original size. The quota was slashed by 84 percent, but it was too late. Setting an apparently conservative target is not sufficient, if an adequate assessment is not carried out *before* the fishery is allowed to expand.[42]

Patagonian toothfish

No fishery so dramatically illustrates the problems of the unholy triad of illegal, unreported, and unregulated fishing (known collectively as IUU fishing) as the fishery for Patagonian toothfish (*Dissostichus eleginoides*). This large predatory fish, which grows to over 2 meters in length and about 130 kilograms, belongs to the Nototheniidae, or cod icefish family, which evolved and diversified in the waters around Antarctica. As its name implies, the Patagonian toothfish is found around the southern end of South America, but it is also the key predator in deep water right around the sub-Antarctic, wherever suitable habitat is available: seamounts and the canyons, deep plateaux, and ridges associated with such islands as South Georgia, Heard and Macdonald, Crozet, Kerguelen, and Macquarie. Ecologically it fills a niche similar to the oilfish and black scabbard fish around tropical and sub-tropical islands in the South Pacific and North Atlantic. And like these other large predatory fish, it is relatively fast growing, maturing at about 70–95 centimeters when it is 6–9 years of age – although the larger fish are estimated to be at least 50 years old.[43] The older fish live deeper, down to 2000 meters and beyond, and are generally found over rough bottoms.[44]

The fishery for Patagonian toothfish dates back to the 1970s, when it was initially carried out with trawls. However, the fishery began to intensify with the introduction of longlining in the mid-1980s; this targeted the larger, older fish that were inaccessible to trawling. In the 1990s, following the collapse of the austral hake (*Merluccius australis*) and golden kingclip (*Genypterus blacodes*) fisheries in Chilean waters and the decline of many northern hemisphere fisheries, pressure on the toothfish ramped up dramatically, with reported catches in the order of 30 000–40 000 tonnes and illegal and unreported catches several times that.[45] The toothfish is an oily but highly esteemed fish with a good market value in Japan, North America, and the EU. In cases like this where there is a ready market, new fisheries can take off far faster than the management apparatus.

The Patagonian toothfish fishery represents in microcosm many of the complexities and difficulties associated with managing multinational distant-water fisheries. The stocks within the EEZ of Chile or Argentina are managed by those countries, but most stocks lie within the Southern Ocean and are managed by CCAMLR (Commission for the Conservation of Antarctic Marine Living Resources; pronounced 'Camel-R'), established in 1982 as part of the Antarctic Treaty to manage Southern Ocean resources. Twenty-four nations are Contracting Parties to CCAMLR, and they meet annually to review the status of Southern Ocean fisheries and establish quotas and other management measures. Other nations are sometimes invited to the meetings because of their relevance to the issues: for example, Belize and Panama, the principal flag-of-convenience states used by illegal fishers to avoid their own countries' regulations, and Mauritius, where much of the IUU toothfish is landed and trans-shipped.

In 1997, CCAMLR acknowledged the magnitude and potential impact of IUU fishing for Patagonian toothfish, when it estimated that the IUU catch that year (68 000 tonnes) was more than double the legal catch (32 736 tonnes). Chile and Argentina cracked down on the use of their ports for unloading IUU toothfish; Chile, for example, required vessels to use VMS, the satellite-based vessel monitoring system, whereby a ship's position can be automatically monitored by a regulatory agency. New trans-shipment ports then emerged: Vigo (Spain), Durban (South Africa), Montevideo (Uruguay), Port Louis (Mauritius), and Walvis Bay (Namibia). All but Mauritius were Contracting Parties to CCAMLR. Over the next two years CCAMLR adopted measures to require Contracting Parties to mandate use of VMS on all toothfish vessels, inspection of vessels licensed to fish in the CCAMLR area, and adoption of a Catch Documentation Scheme to halt trans-shipment of IUU toothfish. Some illegal fishers switched their registration, using flags of convenience. Subsequently Panama and Belize, the two countries most guilty of enabling this practice, were invited to CCAMLR's meeting and agreed to take steps to eliminate it. The cat and mouse game goes on; unfortunately it usually ends only when the fish are all caught and the fishers move on to new grounds.

Of course the problem of eliminating IUU toothfish fishing is compounded by the remoteness of the fishing grounds and the complicity of host fishing nations and large multinationals. To discourage poaching on its lucrative toothfish grounds on the Kerguelen Plateau around Heard Island, situated at 53° S latitude in the southern Indian Ocean, some 4000 kilometers from Perth, Australia mounted several epic chases across the high seas. The most spectacular involved an unarmed Australian patrol vessel pursuing the *Viarsa* for 21 days through mountainous seas, in and out of pack ice for 7200 kilometers, most of the way back to its home port of Montevideo, before it was apprehended by a posse of armed South African and British vessels. It turned out that a Uruguayan fisheries department observer was on board the *Viarsa*.

Apprehension of several such vessels has revealed that these modern-day pirates, though manned by ragged, poorly-paid Third World crews, are typically owned and operated by European nationals, such as the Spanish vessel *South Tomi*, apprehended after a 14-day chase from Heard Island to South Africa, or by shadowy networks of multinational companies. The Hong Kong–based global fish-trading company Pacific Andes, through its Indonesian subsidiary Sun Hope Investments, operated two Russian-flagged vessels, the *Lena* and the *Volga*, which were apprehended with about 200 tonnes of toothfish worth about $2.5 million. The toothfish were caught illegally from Australian waters around Heard Island, and as reported by *Four Corners*, an Australian investigative television program, these vessels were just the tip of the iceberg. A large illegal Sun Hope/Pacific Andes fleet was operating in these waters.[46]

By 2000, CCAMLR estimated that the IUU catch of toothfish by member states had been eliminated and that it was reduced overall to 25 percent of the total catch. But a report by TRAFFIC, which monitors the ecologically threatening trade in wildlife, analyzed the trade figures for toothfish – imports into North America, the EU, and Japan – and estimated that IUU fishing still supplied 57 percent of the global trade in toothfish and was about four times higher than CCAMLR estimated.[47] CCAMLR assessments indicate that Patagonian toothfish stocks are being depleted and that continued IUU fishing severely threatens their sustainability.

Hoki

Amid the spreading disaster of southern hemisphere deepwater fisheries, the fishery for blue grenadier (*Macruronus novaezelandiae*), or hoki (its Maori name) as it is known in New Zealand, stands out as a welcome bright exception. The New Zealand hoki fishery was only the second fishery in the world to be accredited as ecologically sustainable by the Marine Stewardship Council. The reason for its success, however, lies as much in the biology of the species as in its stewardship.

In a good example of the confusion arising from fishes' common names, the blue grenadier is in fact a hake and only distantly related to grenadiers, which are mostly sluggish, long-lived deep-sea fishes living in close association with the bottom. The blue grenadier, in contrast, is an active, fast-growing fish, living mostly at upper slope depths (300–600 meters). Although it is often near the bottom in the daytime, it feeds in midwater on fish, squid, and crustaceans and is known to follow its myctophid prey up into near-surface waters at night.[48] It grows to 120–130 centimeters (about 1.5 kilograms), maturing at about 5 years of age and generally living to 15–20 years.[49] Its largest populations are in New Zealand waters, where it is mostly fished when it forms large spawning aggregations. Its fisheries have been assessed and monitored in Australia and New Zealand with egg and acoustic surveys.[50] Catches and quotas in New Zealand have ranged between about 200 000 and 250 000 tonnes since 1987. It is a productive fishery, and the New Zealand fishing industry appears committed to its long-term sustainability – and to maintaining its Marine Stewardship Council cachet, which provides preferred access to certain export markets.

North Atlantic deepwater fisheries

North Atlantic deepwater fisheries predate those in the North Pacific, and from the outset they had an altogether different character. The Pacific is notable for its trenches, which severely limit the extent of the continental slope, as well as for its tens of thousands of seamounts. Pacific deepwater fisheries developed largely on the seamounts, which tend to be dominated by a relatively few species uniquely adapted to that habitat: pelagic armorhead on central North Pacific seamounts, orange roughy in the temperate South Pacific and Indian Oceans, alfonsino (species of *Beryx*) in the tropics, and Patagonian toothfish in the sub-Antarctic. The North Atlantic, on the other hand, including the Norwegian, Greenland, and Labrador seas, has relatively few seamounts but considerable continental slope area – in the Atlantic more than twice as much of the area (7.1 percent) is between 200 and 1000 meters depth, compared to the Pacific or Indian Oceans (3.1 percent).[51] The North Atlantic also features offshore banks and other complex topography, including the Mid-Atlantic Ridge, all suitable as deepwater fish habitats. It is also noted for its complex hydrography: Arctic outflows, the Gulf Stream, and the outflow from the Mediterranean are all in relatively close proximity and each has its characteristic complex of deepwater fish species.[52] Add to this the range and diversity of the region's fishing cultures – every nation bordering on the Atlantic and the Baltic is a major fishing nation with its own longstanding fishing history and distinctive fleet characteristics – and one can appreciate the complexity of the deepwater fisheries that developed there.

As noted earlier, longline fisheries have an extended history of targeting deepwater species: Greenland halibut from Greenland to Norway; ling and tusk, deepwater relatives of the Atlantic cod, in the waters off Norway, Iceland and the Faroes; black scabbard fish, alfonsino, and red seabream off Madeira and the Azores. These fisheries have all been modernized, particularly in Scandinavia, where a fleet of much larger longliners (30 meters and more) now operates, equipped with automated baiting systems and facilities to process the catch at sea. There are 50 or 60 such vessels in Norway alone. This fleet now fishes all the northeast Atlantic slopes, targeting redfish (*Sebastes*) and Greenland halibut as well as ling and tusk over a much wider area. In European waters farther south, the longline fleet now sometimes targets deepwater sharks for their livers and a deepwater hake, the greater forkbeard (*Phycis blennoides*).[53]

Trawl fisheries on both sides of the North Atlantic traditionally targeted cod, haddock, and flatfishes, species distributed primarily over the region's broad continental shelves: the North Sea, the Grand Banks, and Georges Bank, among other productive shelf areas off Europe and North America. In the initial postwar decades, North Atlantic fisheries – predominantly the trawl fisheries – participated in the global boom, with landings virtually doubling, from 7.45 to 14.4 million tonnes, between 1950 and 1970. Much of this expansion was not sustainable, however, particularly in the northwest Atlantic, where cod, haddock, redfish, and other fisheries subsequently collapsed across much of the region: from West Greenland and the Labrador shelf, extending south across the Grand Banks to Georges Bank off the northeastern United States. These fisheries still show little sign of recovery: marine fishery landings in the northwest Atlantic in 2001 were less than half their levels in 1968, when the fishery peaked. In the northeast Atlantic the fishery peaked in 1976 and subsequently declined by a third during the 1980s.[54]

As the traditional North Atlantic trawl fisheries began to decline, the Soviets, among the first to develop a distant-water trawl fleet, pioneered deepwater trawling of the North Atlantic slope waters, just as they were later the first to trawl North Pacific seamounts. In 1967, they initiated commercial deepwater trawling in the waters off Newfoundland, targeting Greenland halibut and roundnose grenadier (*Coryphaenoides rupestris*), a macrourid initially found in high concentrations at mid-slope depths (600–800 meters) and subsequently fished to 1500 meters. From there, the Soviet deepwater trawl fisheries extended eastward, targeting seamounts and ridges of the mid-Atlantic Ridge and the slope and deep banks of the northeast Atlantic.[55] They were soon joined by other Eastern bloc nations – Poland and the German Democratic Republic, in particular. These were nations without extensive traditional trawl grounds, whose large domestic markets had limited consumer choice and were the first to accept the soft-bodied deepwater fishes. As noted earlier, the Russian market for Greenland halibut extended back hundreds of years. Eventually other nations extended their trawl operations into deep water, and the fishery spread to other species.

Roundnose grenadier

Species such as the roundnose grenadier, Greenland halibut, and deepwater redfish (*Sebastes mentella*) are fished at mid-slope depths right around the rim of the North Atlantic, from the Barents Sea and Norway to Canada. The roundnose grenadier lives to about 70 years, although most individuals in the fishery are 20–30 years old.[56] The early fishery was unregulated, with little research conducted to assess its sustainability. As a result, even after decades of fishing, such basic questions as the number of stocks across the North Atlantic, their initial abundance, where they spawn, and their distribution as juveniles remain unresolved.[57] Most of the roundnose grenadier trawl fisheries peaked in the 1970s and declined in the 1980s and 1990s. In Canadian waters, standardized catch rates from research trawl surveys carried out between 1978 and 1994, with supplementary data to 2003, indicate that this species has declined 99.6 percent and should now be considered critically endangered based on standard IUCN (International Union for the Conservation of Nature and Natural Resources) criteria.[58] Four other deepwater species from the region, mostly taken as by-catch in the Greenland halibut and redfish fisheries, also declined precipitously over the period and similarly qualify as being critically endangered.

Greenland halibut

Major concentrations of Greenland halibut are found off Norway and in the Barents Sea in the northeast Atlantic and from West Greenland and the Labrador Slope to the slope off the Grand Banks in the northwest. The species has an extensive depth range, with the younger fish found along the upper slope and the larger, older fish occurring much deeper: the distant-water trawl fleet operating in the 1990s in international waters – in the Flemish Pass off the Grand Banks, for example – was working in depths of 1200–1800 meters. Although not as prized as the cod, Greenland halibut is often fished in its absence.[59]

Unlike many deepwater fishes – and most flatfishes as well – Greenland halibut is an active, voracious fish, often found feeding well up into the water column; at times it has even been caught in salmon gillnets set near the surface. To facilitate its pelagic habits, the Greenland halibut is the only flounder that is pigmented on both sides of its body, with its left eye only partially migrated to the other side. It is the second-fastest-growing flatfish in the northwest Atlantic, increasing by 6–8 centimeters per year for the first several years of its life. The females live longer than the males and grow somewhat larger. Although Greenland halibut as old as 30 years have been reported, by 1988 the oldest females in the northwest Atlantic were 20 years old and the oldest males only 12; by 1996, heavy fishing had further reduced the spawning stock, such that the oldest fish were only 14 and 11 years old, respectively.[60]

Greenland halibut do not aggregate like the orange roughy, so the larger fish often evade trawls. As a result of their behavior, wide distribution, and greater productivity, Greenland halibut fisheries have proven more sustainable than most deepwater fisheries. Landings since the mid- to late-1960s on each side of the North Atlantic have mostly fluctuated between 30 000 and 60 000 tonnes, with brief periods of unsustainable higher landings. This was most worrisome in the early 1990s, when fishing pressure on Greenland halibut increased substantially on both sides of the Atlantic, following the collapse of Canadian cod and flatfish stocks, along with the decline of European shelf fisheries. Since 1978, fishery scientists have advised on annual catch levels in European waters, but the fishery remained unregulated until 1992. In the years immediately prior to regulation, Greenland halibut landings in the northeast Atlantic rose to between 60 000 and 80 000 tonnes annually, with fleets from the former Soviet Union, the two Germanys, Poland, and the UK all working the fishery. In 1995, the continued decline of the spawning stock and the apparent failure of several year classes led to a recommended zero quota. However Norway still permits a small fishery, and the species continues to be caught as a by-catch of other fisheries.

In the northwest Atlantic, the distribution of Greenland halibut contracted to the Flemish Pass area in the early 1990s, similar to cod and flatfish stocks prior to their collapse. This area was outside Canada's 200-mile EEZ and an intense unregulated fishery developed there, carried out largely by Spain, which increased its catch from a few thousand to 35 000–40 000 tonnes per year, causing overall landings to skyrocket to 75 000–90 000 tonnes. There appears to be only a single stock over much of the region, and Canada, which had managed the Greenland halibut fishery within its waters since 1974, found its stock plummeting. In 1995 this led to an incident known as the 'turbot war' between Canada and the EU. As a consequence, jurisdiction for most of the fishery was turned over to the Northwest Atlantic Fisheries Organization (NAFO), which imposed a 27 000-tonne quota that year. There is now evidence of recovery of the stock.[61]

Redfish

Redfish, also known as ocean perch, comprises the third major longstanding deepwater fishery found around the rim of the North Atlantic. The fishery in fact exploits a complex of similar species: golden redfish (*Sebastes marinus*) and deepwater redfish (*S. mentella*) right around the northern North Atlantic; Acadian redfish (*S. fasciatus*) in the northwest Atlantic; and a somewhat smaller species, *S. viviparous*, in the European Atlantic. These species are difficult to distinguish – in fact, *S. mentella* was only described as a distinct species in 1951, and *S. fasciatus* was not universally accepted as a distinct species for many years after that.[62] Landings are generally reported simply as 'redfish,' which greatly complicates understanding the dynamics of the fishery. It is believed, for example, that the fishery initially focussed on *S. marinus* and *S. fasciatus*, which are found predominantly along the upper slope. As these were depleted, the fishery increasingly targeted deepwater redfish (*S. mentella*), somewhat deeper and further offshore, predominantly between 350 and 700 meters.[63] To further complicate the picture, a distinct and previously unknown stock of *S. mentella*, generally referred to as 'oceanic' redfish, has been increasingly fished since 1982 in the open waters of the Irminger Sea between Iceland and Greenland. Yet another redfish, known as 'giant' redfish, which is either a genetically distinct population of *S. marinus* or perhaps yet another species,[64] was discovered in 1996 at 400–1000 meters depth over seamounts and coral on the Reykjanes Ridge, an extension of the mid-Atlantic Ridge south of Iceland.[65]

Redfish are a long-lived, relatively slow-growing species, requiring between 8 and 12 years to attain maturity, at which time they are about 30 centimeters in length, and living to an estimated 75 years.[66] They are more pelagic than most other deepwater commercial species. Although they are often found over the bottom, where they are fished with bottom trawls, redfish feed up in the water column on krill and other plankton and small fish. They are often concentrated in midwater, where they are increasingly fished with pelagic trawls. The oceanic redfish in the Irminger Sea can be fished only with pelagic trawls. The giant redfish, on the other hand, is generally 70–80 centimeters in length and, being found predominantly over seamounts, is fished mostly with longlines.

Exploitation of redfish began in the 1920s and was already substantial before World War II, peaking at 105 000 tonnes in 1938. The fishery resumed on both sides of the North Atlantic following the war. Unlike the fishery for Pacific Ocean perch (*S. alutus*) along the shelf edge and upper slope off the west coast of North America, which followed the now-familiar boom-and-bust cycle, redfish landings remained between about 100 000 and 400 000 tonnes on each side of the Atlantic, at least until about 1990. Virtually every fishing nation bordering the Atlantic has been involved in the fishery – the former Soviet Union and other Eastern bloc countries, Spain, Portugal, the USA, Norway – but the fishery today is dominated by Canadian trawlers working off the coast of Newfoundland and Labrador, and Icelandic vessels in the waters around East Greenland, Iceland, and the Faroes.[67]

Redfish declined dramatically in the northwest Atlantic from the mid-1980s until the mid-1990s, when the fishery was sharply curtailed. In the 1990s the catch per unit of effort (CPUE) from scientific surveys, the main index of stock size, had dropped to less than 10 percent of its levels in 1978–1985, and the mean size of the fish had declined to about half that in the 1980s; these were sure signs that the spawning stock was depleted.[68] In the northeast Atlantic, the apparent stability of the fishery was achieved by a shift to deepwater and oceanic redfish (*S. mentella*) as golden redfish (*S. marinus*) was depleted. The larger redfish stocks are now believed to be over-exploited.[69] Good year-classes of redfish appear relatively infrequently – only two were observed, for example, in 15 years of surveys in Icelandic waters between 1985 and 2000[70] – a phenomenon also observed in other long-lived fishes, such as orange roughy and Pacific Ocean perch. This could make the process of rebuilding redfish stocks more uncertain, as it has in these other fisheries.

The present state of North Atlantic deepwater fisheries can only be described as extremely precarious, ranging from the severe decline of redfish and Greenland halibut to the virtual disappearance of roundnose grenadier. Nonetheless, the relative stability of North Atlantic deepwater fisheries for several decades, in marked contrast to the 5 to 10 year boom-and-bust cycle observed in most deepwater fisheries in the Pacific and southern hemisphere, seems enigmatic. The Greenland halibut (like the blue grenadier) may be considered a deepwater anomaly: a fast growing, productive, and not particularly long-lived species that often feeds high in the water column. And the roundnose grenadier seems to have undergone only a slightly more extended boom-and-bust cycle. But this still leaves the long-standing fishery for the redfish complex to be explained.

The productivity of all these species appears to be sustained largely by their feeding on the deep scattering layers as they impinge on the slope, a process not so different from those sustaining seamount species.[71] However, a high proportion of the North Atlantic consists of this productive slope habitat. These species have a virtually unbroken distribution around the rim of the North Atlantic from northern Europe, across Iceland and Greenland, and down along the shelf edge of Canada to the northeastern United States. Contrast this with the isolated seamounts and banks available to pelagic armorhead, orange roughy, and Patagonian toothfish.

The North Atlantic is also notable for its intense spring bloom, which extends right across the ocean from North America to Europe. This is one of the most striking phenomena seen in satellite images of ocean color that show the intensity and distribution of plant life over the world's oceans. The blooms consist mainly of the larger diatoms, which support a short food chain readily utilized by fish. These diatoms later aggregate and sink to the seafloor to sustain deepwater food chains. Blooms of this intensity are not observed in the open North Pacific and Southern oceans, which are starved of iron, an essential micro-nutrient.

Another factor that may have sustained the redfish, particularly the deep-sea or ocean species (*S. mentella*), is its pelagic habit, which provides a refuge from demersal (or bottom) trawlers. Historically, pelagic trawls have been far less effective than demersal trawls, which close off a fish's key escape response – to dive beneath the gear. However, advances in gear technology have now overtaken this refuge. In 1989 Iceland developed the Gloria trawl, a super-sized midwater trawl specifically designed for catching ocean redfish. The circumference of Gloria trawls ranges from 544 to 3072 meters – making for a mouth opening almost a full kilometer in diameter! These nets have proven enormously successful at catching oceanic redfish in the open waters over the Reykjanes Ridge, the portion of the mid-Atlantic Ridge that extends south of Iceland, and in the Irminger Sea.[72] Clearly this fishery must now be managed carefully if it is to be maintained. The giant redfish, the exception that proves the rule, are restricted to seamounts. Like other seamount-associated fisheries, this one underwent a rapid boom-and-bust cycle, being largely depleted after a single year of intensive fishing.[73]

European deepwater trawlers have also targeted several species found only in the northeast Atlantic. German trawlers first ventured into deep water in the 1970s to target spawning aggregations of blue ling (*Molva dypterygia*) around banks in the Rockall Trough to the west of Scotland. Icelandic fishers targeted another blue ling spawning aggregation around Iceland in the early 1980s, and French trawlers subsequently discovered yet another on the southern Reykjanes Ridge.[74] Blue ling aggregate to spawn on small topographic features at 700–900 meters depth, and like orange roughy spawning aggregations, these are quickly fished out. As a result, German fishers were soon largely replaced by the French, who were able to market roundnose grenadier and developed a more diversified deepwater fishery.

A number of further trawl fisheries have developed in European waters in recent years. Angler fish or monk fish (*Lophius* spp.) is now fished on the upper slope. Semipelagic trawl fisheries target spawning aggregations of blue whiting (*Micromesistius poutassou*) and greater silver smelt (*Argentina silus*) during their spawning seasons. France and other nations target orange roughy on steep slopes, seamounts and the mid-Atlantic Ridge.[75] But orange roughy is not the dominant species at mid-slope depths in the North Atlantic as it is around New Zealand and southeastern Australia. Not surprisingly the spawning concentrations have been relatively small and quickly depleted.

Europe has had a fisheries science infrastructure since 1902, when the International Council for the Exploration of the Sea (ICES, pronounced 'I-seas') was founded to collect fishery statistics and assess its fishery resources. However, the management of European fisheries has proven difficult, due to the large number of fishing nations and jurisdictions. European fisheries are notorious for being chronically overfished; only the high productivity and resilience of many North Sea and northeast Atlantic fish populations has thus far prevented their collapse. However, the deepwater fisheries represent an extreme case of European neglect, and unfortunately many do not share the productivity and resilience of their shallower counterparts. A recent review by leading deepwater fishery scientists from England, Scotland, Germany, Norway, and France noted that:

> a feature of deep-water fisheries (whether they be longstanding or new, artisanal or mechanized) is that almost all have developed without programs in place to collect biological and fisheries data.[76]

Without baseline data on catches and effort, levels of catch or effort cannot be determined scientifically, nor can quotas be allocated equitably among the various EU fishing nations. ICES recently concluded in its management advice for the northeast Atlantic deepwater fisheries:

> Most exploited deepwater species are, at present, considered to be harvested outside safe biological limits. ICES recommends immediate reduction in fisheries that cannot be shown to be sustainable. New fisheries should be permitted only when they expand very slowly and are accompanied by programs to collect data which allow evaluation of stock status.[77]

ICES specifically recommended a halt to directed fishing for blue ling, a 30 percent reduction in fishing effort on ling and tusk, and a 50 percent reduction in several areas for black scabbard fish and roundnose grenadier. To date the deep-sea fishing nations, led by France, have resisted these pressures: EU quotas for 2005 were reduced by only 10–15 percent.[78]

An overview of deepwater fisheries

Global deepwater fishery landings have fluctuated between about 600 000 and somewhat over 1 million tonnes per year since the 1960s. Few are sustainable, and most have undergone a depressingly familiar boom-and-bust cycle, due to a combination of ecological and life-history characteristics (extreme longevity and concomitant low productivity), vulnerability (aggregation on seamounts or other small topographic features), and IUU fishing. The productive upper and mid-slope region of the deep ocean is now a far emptier place: the pelagic armorhead and corals of North Pacific seamounts; the orange roughy of New Zealand, Tasmania, Namibia, and the southwestern Indian and North Atlantic oceans; the roundnose grenadier, Greenland halibut, and redfish of the northwestern Atlantic and the blue ling of the northeastern Atlantic – all have largely disappeared. When, or even whether, these species will return is unknown: why some species bounce back and others fail to is one of the great mysteries of fisheries science today.[79]

To place this in perspective, deepwater fisheries account for approximately 1 percent of global marine fishery landings. The least sustainable fisheries – those carried out on the high seas – involve only 11 countries and account for about 200 000 tonnes or 0.25 percent of global marine fishery landings. Their value – an estimated US$300–400 million – is equivalent to about 0.5 percent of the value of global marine fisheries; the number of vessels involved is only 100–200, out of the 3.1 million fishing vessels in operation.[80]

So what does the future hold for deepwater fisheries? Economists often confidently predict that the unregulated 'commons' will tend toward a bio-economic equilibrium: that fishing will stop when it is no longer profitable, and that this will occur before the resource is entirely depleted. Unfortunately, these economic models are based on idealized fish populations that are more or less randomly distributed over a large area; as the population becomes too 'dilute,' the fishery stops, in theory leaving behind a still substantial population. But when a stock is aggregated on small topographic features, such as seamounts, the fishery may not stop so long as schooling fish remain. Species may continue to be caught as a by-catch even when they are not specifically targeted. Furthermore, world fisheries are heavily subsidized. Global marine fish landings worth about $70 billion in total cost about $124 billion to catch; the $54 billion deficit is covered by government subsidies, such as for fuel and for building boats.[81] So unregulated fisheries are often driven below the putative bio-economic equilibrium. The main hope for deepwater fisheries – other than a change in global marine fisheries management, discussed in the next chapter – may be the dramatic increase in oil prices that many geologists predict will set in over the next decade or so, and an end to fuel subsidies, which would render many distant-water fisheries uneconomical.[82]

Impacts of trawling

The impact of many deepwater fisheries extends well beyond the fish populations. Trawling is the most destructive fishery method, particularly for life-forms on the seafloor. Concern about trawling impacts has a very long history, with a petition presented to the British parliament as early as 1376 complaining about the effects of the so-called 'wondyrchoum,' whose small meshes removed vast quantities of small fishes 'to the great damage of the whole commons of the kingdom.'[83] But until the age of diesel, the size of trawls was quite limited, and until recently the gear was restricted to soft-bottom habitats. However, modern deepwater trawls are particularly large and heavy, enabling them to sink rapidly to depth, and are often outfitted with large steel or rubber bobbins or discs up to 60 centimeters in diameter, enabling them to roll or bounce over boulders and other rough ground. The footrope of a net like this is between 50 and more than 100 meters long and weighs up to 4800 kilograms, including the bobbins and other weights along its length. Metal doors weighing about four tonnes each help sink the net and hold it open as it is towed.[84] Whatever is not dragged up by such a behemoth is generally left crushed in its wake, and a modern trawler working around the clock may have its gear on the ocean floor for 18 hours a day. Given the 100-meter footrope and a trawl speed of 3 knots, a single trawler will disturb approximately 10 square kilometers of seafloor each day.

Concern about the potential impacts of deepwater trawling on seafloor communities arose in the 1990s, well after many North Pacific seamounts had been swept clear of Pacific armorhead, precious corals, and the benthic fauna associated with them. However, the impacts of trawling have now been examined on seamounts around New Zealand and Tasmania, as well as in the vicinity of the *Lophelia pertusa* reefs off Norway and the Darwin Mounds in the northeast Atlantic, in the Canadian Atlantic, on *Oculina varicosa* reefs off the Florida coast, and within the coral and sponge gardens around the Aleutian Islands in the North Pacific. The benthic communities in all these regions were discovered, and their exceptional diversity, antiquity, and sheer magnificence recognized, only well after trawling got underway.

The seamounts around New Zealand and Tasmania, the center of the orange roughy fishery, characteristically contain dense coldwater coral reefs, with species of stony corals, such as *Solenosmilia variabilis* or *Goniocorella dumosa*, forming the reef matrix and many hundreds of fish and invertebrate species living in association with it (Chapter 6). When the orange roughy fishery first developed in 1997 on the South Tasman Rise, a ridge system extending south of Tasmania, observers were placed on New Zealand vessels. They provided the first good data on by-catch of coral and other species, from the outset of a deepwater fishery. The South Tasman Rise fishery proved relatively small: the total orange roughy catch through about half of the 2001–2002 fishing season was 11 000 tonnes, with landings of about 4000 tonnes per year in the first two years. Coral by-catch in the first year of the fishery was 1.6 tonnes per hour of towing, for a total of 1762 tonnes – about 44 percent of the orange roughy landings that year. The coral

Mark Lewis, CSIRO's 2 meter tall technician, with a black coral collected from a seamount south of Tasmania.

by-catch consisted predominantly of the reef-forming coral *Solenosmilia variabilis*, although many other species were obtained as well. Two years later the coral by-catch rate had dropped 76 percent to 428 kilograms per hour of towing – still sizeable, but an indication that the deepwater reef was being removed from the fishing ground.[85]

Not surprisingly, surveys that compared fished and unfished seamounts off New Zealand and Tasmania found virtually no intact reef-forming coral on heavily fished seamounts, whereas coral cover on unfished or lightly fished seamounts was as high as 50–100 percent.[86] In both regions, the most heavily fished seamounts had been towed across several thousand times, based on fishery logbook records. Given the size of the trawls and the small extent of the seamounts, typically only several hundred meters high and a few kilometers across at their base, it is little wonder that heavily-fished seamounts were scraped clean. In 1999 and 2001, when the diversity and uniqueness of the deepwater reefs, along with the extent of fishery impacts, became known, Australia and New Zealand established protected seamount areas: a 370-square-kilometer reserve on the slope south of Tasmania that contained 12 relatively pristine seamounts at the edge of the fishing ground, and some 19 seamounts protected in New Zealand waters, spread around the fishery region.

Fishers on both sides of the North Atlantic who use passive gear – longlines and gillnets – have long known about deepwater corals: because fish aggregate over deepwater reefs, these fishers traditionally set their gear near these features. When rockhopper trawl gear was developed in the late 1980s, enabling trawlers to venture into coral-covered ground, the longline and gillnet fishermen were the first to recognize the loss of deepwater corals and to alert scientists and government agencies to the issue. A subsequent extensive survey of the *Lophelia* reefs off the coast of Norway (the most extensive deepwater reefs in Europe) showed that 30–50 percent had already been impacted or destroyed by fishing.[87] Since 1999 Norway has protected six reef sites, including Sula and Røst, the largest *Lophelia* reefs known.[88]

Lophelia has also been found growing on recently discovered mounds at depths of 500–1200 meters in the Rockall Trough and Porcupine Seabight, extending from northwest of Scotland to southwest of Ireland. The Darwin Mounds, about 185 kilometers northwest of Scotland, were discovered in 1998. Only 5 meters in height, their influence on the ambient currents still appears sufficient to promote the growth of *Lophelia*. The mounds also have a long downstream 'tail,' colonized by xenophyophores, 'giant' single-celled protists up to a centimeter across and distantly related to amoebae. Unlike the carbonate mounds to the west of Ireland, which can be 300 meters high – the size of some small seamounts fished off Australia and New Zealand – and quite rugged, the Darwin Mounds are highly vulnerable to trawling. Following a Greenpeace lawsuit in 1999, in which the High Court ruled that the EU Habitats Directive applied to offshore environments, the UK made a commitment to protect the Darwin Mounds as its first offshore Special Area of Conservation. But the trawling industry strongly opposed fishing restrictions, and the EU only announced emergency fishing restrictions in 2003, following a study that provided photographic evidence of extensive trawl damage to corals on the mounds.[89] Ireland is also seeking to protect deepwater coral sites off its coast, as are the Azores, Madeira and the Canary Islands.[90]

In the northwest Atlantic, concern over reports of fishery by-catch of corals led to extensive surveys in the Canadian Atlantic by the Department of Fisheries and Oceans. As a result, three areas off the Scotian Shelf are now protected from fishing: the Gully, the largest canyon on the eastern Canadian continental margin, where particularly diverse octocoral gardens were discovered; parts of the Northeast Channel, leading into the Gulf of Maine, which also contains abundant octocorals but evidence of significant trawl damage; and an area at the mouth of the Laurentian Channel, leading into the Gulf of St Lawrence, where the first *Lophelia* reef off Atlantic Canada was discovered.[91]

Octocoral and sponge gardens are particularly diverse around the Aleutian Islands, and several commercial fish species, such as Atka mackerel and several rock fishes (*Sebastes* spp.) are associated with them. Submersible and video surveys have found areas damaged by trawling, and more than 2000 tonnes of coral and sponge were removed as by-catch between 1990 and 2002, based on fishery observer data.[92] For several years, the trawl fishing industry and the National Marine Fisheries Service (NMFS) resisted pressure to protect the corals and sponges. Eventually, the threat of a lawsuit from conservation groups such as Oceana, based on the Magnuson-Stevens Act, which provides for protection of essential fish habitat, and strong condemnation by an international scientific review (which I participated in), led to a dramatic turnaround in US fishery policy. In February 2005, the North Pacific Fishery Management Council limited the trawling footprint to those areas already being trawled, an area of about 25 000 square miles (65 000 square kilometers), and closed a vast area of some half a million square miles (1.3 million square kilometers) of coral and sponge habitat around the Aleutians – an area about nine times the size of Washington State.[93] The Pacific Fishery Management Council shortly followed suit, closing to trawling the previously untrawled areas off Washington, Oregon, and California, including the continental slope beyond 1280 meters depth (700 fathoms), as well as seamounts, canyons, various banks, and other sensitive habitats.[94] These actions, following upon the New England Fishery Management Council's protection of deep-sea coral habitats in New England and mid-Atlantic waters in 2003, signal a sea change in the protection of sensitive ecological habitats in American waters.

The last several years have witnessed a growing global recognition of the need to protect deepwater habitats from the impacts of trawling. In 2003, four nations (the USA, Australia, New Zealand, and Norway) had closed about 4 million square kilometers to deepwater bottom trawling. By May 2006, the area closed or proposed for closure had more than doubled to over 10 million square kilometers, involving the EEZs of 11 nations, with 49 nations bound to the closures. Significantly, the EU had agreed to close the areas around the Canaries, the Azores and Madeira to bottom trawling (1.2 million square kilometers), and the countries around the Mediterranean had had prohibited bottom trawling in waters deeper than 1000 meters (1.6 million square kilometers).[95]

Deepwater fisheries – where are we now?

Deepwater corals appear to be distributed throughout the world's oceans at the appropriate depths and where bottom currents are sufficiently enhanced. But deepwater reefs and octocoral 'gardens' have mostly been discovered only within the past ten years. More detailed sampling, with cameras, video, or submersibles, is required to ascertain the extent and diversity of these habitats – whether the corals are isolated or form reefs or extensive 'gardens' – and so far this has only been carried out in the waters around Europe, North America, and parts of Oceania. The Indian, Southern, and South Atlantic oceans are virtually unexplored, as are large parts of the Pacific and virtually all of the high seas – the areas beyond national jurisdiction. Developed nations have now begun to protect some of the more outstanding deepwater features within their jurisdictions. In 2002 the UN General Assembly passed a resolution calling upon nations to protect the biodiversity of the high seas as well. But without concerted effort, deepwater trawling will continue to crush and remove these remarkable habitats from the face of the deep.

What is the bottom line? Are we really facing the extinction of deepwater fish or invertebrate species? How much deepwater coral and sponge habitat will have been lost before deepwater trawl activities are reined in?

Unfortunately, such questions cannot yet be answered with confidence. On seamounts, where a large proportion of species are rare and may have a limited distribution – at least so far as our limited sampling indicates – there is a high probability that continued trawling and habitat elimination will lead to extinctions. Of course it is possible that many of these species will turn up elsewhere, or outside the range of trawling, but we don't know that for sure. Species endemic to the Emperor Seamounts may have already been rendered extinct by the successive waves of pelagic armorhead and precious coral fishing, but we may never know.

Of what use are these deepwater species? Again, clear answers cannot yet be given – we are at such an early stage of exploring these environments. Many benthic invertebrates, slow-growing and obviously loaded with chemical defenses, may provide new pharmaceuticals to fight human cancer and infections. This potential has barely begun to be explored. And might tourism be developed, based around submersible trips to seamounts or deepwater coral reefs? Tourists have already been taken to hydrothermal vents at the mid-Atlantic Ridge and to visit the *Titanic*. But do we even need to be able to set down a list of reasons to justify conserving one of the major ecosystems on the planet?

What would the disappearance of these ecosystems mean for the ecology of the oceans? Again, we can only speculate. Ecologists have only recently come to realize the key role that sponges play in filtering the water around coastal reefs. What then is the role of coldwater corals and sponges in filtering and cleansing the oceans? We are at such an early stage in exploring the deep sea that such questions are only now being posed.

It has been said that destructive fishing practices are an evolutionary stage in mankind's use of the oceans, akin to the domestication of the Earth's wild spaces – the leveling of forests to make way for farms; the replacement of buffalo with cattle. But the mining out of the Earth's deepwater fish resources and the destruction of its deepwater coral environments can provide only the most ephemeral gain. That this process continues is testament to the ongoing capitulation of weak managers and corrupted government departments to the greed of a few. How human agencies may be able to rein in these excesses is the subject of the concluding chapter.

CHAPTER 11

The Way Forward: Conservation and Policy Options

written with Kristina Gjerde

Man is defined not by what he creates, but by what he chooses not to destroy.

E.O. Wilson

The evolution of ocean governance

From the earliest days of exploration, when the first evidence of life in the deep was retrieved in the process of surveying for and repairing the telegraph cables that spanned the Atlantic and Mediterranean, scientific discovery and industry have fed upon and reinforced each other. Grassle and Maciolek's realization that the deep seafloor may be home to some millions of species was based on sampling to assess the impacts of oil and gas exploration;[1] the discovery of Norway's vast *Lophelia* reefs arose from Statoil's pipeline surveys for offshore oil fields;[2] and the discovery of the diverse and highly endemic faunas on seamounts arose from concerns about the impacts of deepwater fisheries.[3] The relationship works the other way too, of course, with potential new industries spawned from serendipitous scientific discovery: pharmaceuticals from deep-sea corals and sponges; a new frontier in biotechnology from the discovery of thermophilic microbes spewed from hydrothermal vents; and the marine mining industry, which has its origins in manganese nodules trawled up during the *Challenger* expedition and in the hydrothermal vent fields discovered a century later.

Ocean governance evolves as well, but at an altogether slower pace, to meet sweeping changes in our use of the sea and our understanding of potential human impacts. Until relatively recently, the so-called 'freedom of the seas' was virtually the sole doctrine regulating use of the oceans beyond a few nautical miles from shore. This doctrine itself had its roots in the conflicts that arose from expanded European use of the high seas following the Age of Exploration: Hugo Grotius developed the doctrine in the 17th century as the basis for the Dutch rights to free transit and free trade with the East Indies against Portuguese and Spanish claims to sovereignty over vast areas of ocean.[4] Grotius extended the concept to the right to fish freely on the high seas, as well as to navigate, arguing that natural law forbade ownership of goods that seemed 'to have been created by nature for common use.' The commons, he argued, citing authorities extending back to Ancient Greece and Rome, consist of what cannot be occupied and 'although serving some one person it still suffices for the common use of all other persons,'[5] or as his follower John Locke stated, 'there is enough and as good left in common for others.'

It is now clear that high-seas fisheries no longer leave behind what 'still suffices for the common use of all other persons.' Like the air we breathe, the open ocean comprises a critical part of the global environmental commons. About half the planet consists of ocean waters beyond areas of national jurisdiction. The deep sea is home to wondrously diverse ecosystems found nowhere else on the planet. The open ocean is critical as well to the global biogeochemical cycles that regulate oxygen and carbon dioxide in our atmosphere – and hence global climate and the very furtherance of life on Earth. Humankind once treated the deep ocean as a vast dumping ground for old munitions (including poison gas), mine tailings, sewage, nuclear waste and other discards. Other pollutants are not deliberately dumped into the ocean, yet it remains one of their ultimate sinks: semi-volatile heavy metals such as mercury, chlorinated hydrocarbons such as DDT, PCBs, and related pesticides and industrial compounds are invisibly distributed around the world in the atmosphere and mostly end up in the oceans.

Will the global ocean commons be left as good as we found it for future generations' use? In response to these challenges, the London Dumping Convention largely put an end to ocean dumping. The production and use of DDT and PCBs has been sharply curtailed, and use of mercury is being reduced as well. The Kyoto agreement is a welcome first step toward reducing greenhouse gas emissions, although unfortunately some of the world's largest emitters have not yet agreed to its minimal provisions. And in the most comprehensive and far-reaching change in ocean governance since the time of Grotius, the UN Convention on the Law of the Sea (UNCLOS) and subsequent implementing agreements have dramatically changed the fundamental framework of national and international rights and obligations underlying the governance of the high seas; a process of change the global community is still struggling to get right.

The UN Law of the Sea Convention

UNCLOS is today the chief framework agreement, sometimes called the 'Constitution of the Oceans'; it governs all ocean uses and activities. It had its origins in the conflicting claims coastal states increasingly made on the resources off their coasts after World War II. Negotiated throughout most of the 1970s, it was finalized in 1982 but did not come into force until 1994.[6] Even before it was officially ratified, UNCLOS had a pervasive influence on all maritime activities and laws, to the extent that it was considered binding on all states as customary international law.

Like many framework documents, the UNCLOS is the product of many competing pressures, and to gain acceptance it had to satisfy various competing interests. Coastal states were granted jurisdiction over the natural resources out to 200 nautical miles from their shores in newly recognized 'exclusive economic zones' (EEZs), thereby transferring a third of the oceans to their national domain. In this way, coastal states were protected from the worst depredations of the distant-water fishing fleets, but to appease these fishing interests, UNCLOS also legitimized traditional open access rights to stocks on the high seas beyond the EEZs. And in a nod to the great powers, jurisdiction over EEZs was limited: UNCLOS accepted the practically unrestricted access of shipping, including warships, through a coastal state's EEZ.

At the same time, the environmental fervor of the 1970s was not without effect. UNCLOS incorporated language from the Stockholm Declaration of 1972, the first 'Earth Summit,' which declared that states have the unqualified obligation to protect and preserve the marine environment (Article 192). They also have the obligation to protect rare and fragile ecosystems such as coral reefs, and vulnerable species such as sea turtles, sharks and whales, and their habitats, and other forms of marine life (Article 194.5). Further, the rights of coastal states in their EEZs were to be exercised pursuant to environmental policies and in accordance with their duty to protect the marine environment. In the two thirds of the oceans that remained outside national jurisdiction, states were to cooperate to conserve living marine resources and protect the marine environment, and to develop additional rules and guidelines where necessary for these purposes. Unfortunately, these words have thus far proven too vague to have had much effect.

A major concession was also made to the developing world, which in the 1970s increasingly sought economic equity – a 'New International Economic Order.' Manganese nodules, rich in strategic minerals and ostensibly worth trillions of dollars, were reported to be lying on the flat abyssal plain just waiting to be scooped up. The industrialized countries argued that deep-sea mineral resources on the high seas should remain part of the global commons, available on a first-come, first-served basis. In the end, the developing countries prevailed, and the deep seabed and its mineral resources were designated the 'common heritage of mankind,' (Article 136) to be used 'for the benefit of mankind as a whole with particular consideration for the interests and needs of developing countries' (Article 140) (see Chapter 8). The International Seabed Authority (ISA) was established to act on behalf of mankind in the exercise of rights over seabed resources beyond national jurisdiction; its mandate was not only to promote seabed mining, but also to protect biodiversity and minimize impacts from mining activities.

UNCLOS and the living resources of the high seas

Unfortunately, rather than setting up a system of global stewardship for all the resources of the oceans, living and non-living, beyond national jurisdiction, UNCLOS limited the 'common heritage of mankind' concept to the deep seabed and its mineral resources. Fishing has had greater immediate impacts on deep-sea ecosystems and biodiversity than any other human activity, but UNCLOS provided only minimal rules for governing deep-sea fisheries in the two-thirds of the ocean beyond coastal states' EEZs, or otherwise protecting those deep-sea environments from anything other than deep seabed mining. Deep-sea areas beyond national jurisdiction remained an open access commons for fishing, based on the customary policy of 'first-come, first served' that had dominated governance of the high seas at least since the time of Hugo Grotius.

Even before UNCLOS came into force in 1994, the international arrangements for regulating fisheries on the high seas were widely recognized as inadequate. But re-opening the Convention on the Law of the Sea to develop a new overarching agreement was generally viewed as too difficult, given the decades that had gone into negotiating UNCLOS and the delicate balance of interests underlying it. Instead, a patchwork of regulatory arrangements has developed around deep-sea and open ocean fisheries, such that today some fisheries and regions are covered adequately, others inadequately, and still others not at all.

The primary issue is that of open access. So long as a fishery today remains within the global commons, it will fall prey to the tragedy of the commons, leading to overexploitation and depletion of the resource.[7] Of course, under UNCLOS, many deepwater fisheries now lie within the expanded EEZs of coastal states. As reviewed in the previous chapter, management of these deepwater fisheries has proved particularly difficult because of their vulnerability and low productivity, as well as the impacts of trawling upon sensitive benthic habitats. However, under pressure from a strong international scientific consensus, developed nations around the world – Australia, New Zealand, the United States, Canada, Norway, and to a limited extent the European Union – have now taken steps to improve management and to protect deepwater corals, seamounts, and other sensitive deep-sea habitats. Our main concern, therefore, is for those species and stocks that remain within the global commons.

In the 1980s the most intense high-seas fisheries conflicts were over the tunas and other highly migratory fish stocks – the most valuable of the high-seas fisheries – and over stocks that straddled EEZs and international waters, where the management efforts of coastal states were typically undermined by unregulated fishing in international waters. To remedy this situation, the Agreement for the Implementation of the Provisions of the LOS Convention relating to the Conservation and Management of Straddling Fish Stocks and Highly Migratory Fish Stocks was concluded in 1995. Otherwise known as the UN Fish Stocks Agreement (UNFSA), this so-called 'implementing agreement' to UNCLOS achieved several major advances on the original Convention but has yet to live up to its early promise.

The UN Fish Stocks Agreement

The UNFSA is a far-reaching document that sets out to bring the governance of high-seas fisheries in line with contemporary fisheries management practice, as set down in the 1995 UN FAO Code of Conduct for Responsible Fisheries, a voluntary agreement accepted by all 170 member states of the UN Food and Agricultural Organization.[8] Over the past decade or so there has been a sea change in the global perspective on fisheries management. From the 1950s through at least the 1980s, fisheries management tended to be narrowly focussed on the exploited fish stocks; its primary role was to set quotas to maximize the estimated sustainable yield. Heavily influenced by the fishing industry and its short-term perspective, quotas were generally ratcheted upward and only reduced in the face of strong evidence of overexploitation.[9]

But fisheries science is not yet a precise science. For a start, the data are imperfect: ocean populations and processes are notoriously difficult to measure, and fishery data are often biased due to illegal fishing, misreporting, and the dumping of undersized and unwanted fish at sea. On top of that, the impacts of climate on fisheries and the broader ecosystem impacts of fishing down certain species and damaging benthic habitats are still poorly understood. As a result, management that sails close to the wind is virtually doomed to fail. Fully a quarter of the world's fish stocks are today either overexploited, collapsed, or recovering.[10] Some, such as the Grand Banks cod, had been among the most productive in the world, with a several-hundred-year history as the mainstay of local economies.[11]

Recognizing the inherent uncertainties of fisheries assessments, the FAO Code of Conduct sets down a new ethos for fisheries management, based on a more precautionary, ecosystem-based approach. The UNFSA defines precautionary management in terms of determining reference points more conservative than the theoretical maximum sustainable yield, adopting 'cautious conservation and management measures [for new and exploratory fisheries until] there are sufficient data to allow assessment of the impact of the fisheries on the long-term sustainability of the stocks,' and, more generally, not allowing 'the absence of adequate scientific information [to] be used as a reason for postponing or failing to take conservation and management measures.' The UNFSA also recognizes the need to take into account the ecosystem impacts of fishing: the impacts on non-target species, on special habitats, and on marine biodiversity generally.

Regional fisheries management organizations – a mixed review

Under the UNFSA, management of highly migratory and straddling fish stocks was to be left, as it had been traditionally, with so-called regional fisheries management organizations (RFMOs), comprised of those states with an interest in a particular fishery or set of fisheries within a region. When the UNFSA was concluded in 1995, there was already a patchwork of RFMOs covering certain shared fisheries, such as tunas, salmon, and Pacific halibut, and some parts of the world ocean. Only a few, however, had the legal competence to regulate high-sea bottom-trawl fisheries: NAFO (Northwest Atlantic Fisheries Organization) and NEAFC (North East Atlantic Fisheries Commission) in the North Atlantic, and CCAMLR (the Commission for the Conservation of Antarctic Marine Living Resources) covering the Southern Ocean. And probably only CCAMLR had had any success. Although demersal fish stocks had been intensively fished in international waters across the North Atlantic for almost 30 years, NEAFC had yet to regulate any of these fisheries, and while NAFO, led by Canada, had attempted to do so, it had met with little success (see Chapter 10).

RFMOs had often proved ineffectual, for a variety of reasons. Most operated by consensus, which often leads to standards being set at the lowest common denominator. Most RFMOs were established with an earlier management paradigm, and their management practices were generally not precautionary, nor did they take into account ecosystem impacts. Inadequate enforcement is also a perennial issue in international organizations, in relation to both vessels of member states and 'free riders' – vessels from non-member states, often re-flagged to flag-of-convenience states. In the international arena, enforcement of maritime regulations is mostly left to the flag state – the state where the vessel is registered and whose flag it flies – so there is a strong incentive for vessel owners to fly the flag of countries such as Belize, Honduras, Dominica, and others that do not sign onto – and often flout – international regulations.

Though it was clearly a case of putting new wine into old bottles, those negotiating the UNFSA considered that they had a greater chance of success in strengthening and reforming existing international institutions than in attempting to create new ones. Not only did the UNFSA provide a new set of management principles for many RFMOs – if member states were signatories to the UNFSA, then RFMO operating procedures needed to be consistent with the principles laid down in the UNFSA – it also attempted to reform some aspects of their structure, though certainly not all. Under the UNFSA, access to the relevant fisheries was to be restricted to member states of the RFMO or states that agreed to abide by its rules, but all states with an interest in the fishery were eligible to join. RFMO decisions were to be scientifically based and transparent, such that their activities were open to the public, and organizations, such as non-governmental conservation organizations, could take part in their meetings.

CCAMLR's long and largely successful history in managing fisheries such as krill and Patagonian toothfish, as well as trawl fisheries, in one of the most remote and difficult regions of the ocean indicates that the RFMO model can work effectively. The CAMLR Convention, adopted in 1982 as part of the Antarctic treaty system that regulates human use of the Antarctic and the oceans around it, was one of the first pieces of international legislation to incorporate an ecosystem-based management approach. Established in response to concerns about the potential impacts of increased krill fishing on the recovery of baleen whales, CCAMLR from the outset has examined the requirements of dependent species as well as those directly exploited.[12] CCAMLR also explicitly adopts the precautionary approach, but decisions on 'matters of substance' are made by consensus.[13] Of particular note are its rules for new and emerging fisheries. These restrain the development of new fisheries until the information is in place to enable sustainable use, protect dependent and associated species and prevent long-term changes. Though not without its problems – it is to be expected that decision-making will be somewhat tedious and cumbersome in a large and highly diverse international organization – CCAMLR remains the best model for co-operative management of high-seas resources. Had this approach – essentially the approach adopted by the UNFSA – been in place in other areas of the high seas, deepwater fisheries would not have developed so rapidly, nor collapsed so quickly.

Unfortunately, progress in implementing the UNFSA has been slow. Although considered an 'implementing agreement,' in fact it considerably curtailed the traditional open-access rights to high-seas fisheries granted under UNCLOS. Not surprisingly, many UNCLOS signatories with an interest in open-access high-seas fisheries have held back from signing onto the UNFSA. Six years were required for sufficient countries to sign the agreement for it to come into force. As of May 2006, only 57 nations were parties to the UNFSA, compared with 148 that had signed onto UNCLOS. Most flag-of-convenience states and some key distant water fishing nations – most notably Japan and Korea – have not yet joined the UNFSA. However, at the end of 2003 the EU signed on, so together with other key signatories – Australia, New Zealand, Russia, Norway, Iceland, and Ukraine – the parties to the UNFSA now account for about 90 percent of the high-seas bottom-trawl catch.[14]

Given how recently the EU signed on, it is still too early to assess how the UNFSA will ultimately perform. However, there are some promising signs. In 2002 NEAFC finally agreed to regulate deep bottom-trawl fisheries in the international waters of the northeast Atlantic, following some 30 years of intensive fishing there (see Chapter 10). At its 2004 meeting, it agreed that future fishing should not exceed the highest effort of the past and that fishing in certain fisheries should be reduced by 30 percent in 2005. NEAFC also banned trawling on four seamounts in the Northeast Atlantic and on portions of the mid-Atlantic Ridge for three years.[15] On the western side of the North Atlantic, NAFO, led by Canada, has long attempted with little success to regulate fishing on the overfished deepwater stocks off its coast, such as Greenland halibut and straddling stocks of cod, plaice, and other species. However, at its most recent meeting in 2004, NAFO agreed to adopt a precautionary approach in setting catch limits.[16] While it is too early to tell how successful this will prove, NAFO clearly recognizes the need to reform. There is presently a complete moratorium on fishing straddling stocks of cod and plaice in the northwest Atlantic region, and quotas on Greenland halibut have been reduced. The sea change that has been sweeping global fisheries management – recognition of the need for a precautionary, ecosystem-based approach – appears to be taking hold of the RFMOs as well.

Despite these positive developments, vast regions of the world's oceans remain outside the framework of any RFMO with legal competence to regulate high-seas bottom-trawl fisheries: the entire North Pacific and all Southern Hemisphere oceans apart from the Southern Ocean. Some steps have been taken to remedy this situation. The convention underlying the General Fisheries Commission for the Mediterranean (GFCM) was recently amended, giving it competence to regulate bottom-trawl fisheries, and in 2005 the GFCM banned bottom trawling below 1000 meters in the Mediterranean. This does nothing to protect shallower depths, where most deepwater trawling takes place, but it is nonetheless a step in the right direction. In 2003, a new RFMO came into force with competence to cover bottom-trawl fishing in the southeast Atlantic (SEAFO, or the South East Atlantic Fisheries Organisation), and there have been negotiations to form one for the southeast Pacific.[17]

The international community is taking halting steps to improve the governance of bottom-trawl fisheries on the high seas, but they still largely remain outside any regulatory framework. And the workings of the international community are far too slow and cumbersome to effectively deal with them. To cite the most recent example, when the fishery for orange roughy developed on seamounts in the southwestern Indian Ocean in 1999, intergovernmental consultations began two years later to develop a regional fisheries management organization. The consultations were continued in 2004, with agreement still not reached, by which time the fishery had collapsed.

There is an urgent need to address the issue of high-seas bottom trawling. The locations of the world's major seamount and ridge systems are known, and trawlers are scouring these hotspots for the last unfished stocks.[18] Orange roughy populations, once depleted, will require decades to recover, perhaps longer – no one knows for certain. A commercial bottom trawl with its heavy ground gear works like a bulldozer, cutting a 50–100-meter-wide swath across a seamount or the continental slope. Deepwater reefs built up over the millennia are wiped from the face of the deep in a single season. The time-scale for recovery of these deep coral habitats is probably on the order of centuries to millennia, with the threat of extinction for localized species.

Scientists' proposal for a moratorium on high seas bottom trawling

In what was probably the largest concerted action ever undertaken by the marine scientific community, 1136 marine scientists and conservation biologists from 69 countries released a statement in 2004 calling on governments and the United Nations to adopt a moratorium on high-seas bottom trawling.[19] A broad coalition of more than 50 of the world's leading conservation organizations, including Oceana, WWF, Greenpeace, Conservation International, the Marine Conservation Biology Institute, and the Natural Resources Defense Council have joined together to press the case at the UN. A moratorium is seen as the only viable mechanism to halt the destruction of seamount and deep-sea coral habitats, until effective governance of high-seas bottom-trawl fisheries is put in place. The proposal follows the precedent set by the UN General Assembly's global moratorium on high-seas drift gillnetting, an unsustainable fishing practice that indiscriminately affected a wide range of pelagic by-catch species, including threatened turtles and dolphins. A series of resolutions passed by the UN General Assembly between 1989 and 1991 effectively ended the practice.[20]

In 2003 the UN General Assembly passed a resolution stating that there was an urgent need to address 'the threats and risks to vulnerable and threatened marine ecosystems and biodiversity in areas beyond national jurisdiction.'[21] Whether the General Assembly will follow this up with a moratorium on bottom-trawl fishing remains an open question. At present, the primary opposition to the moratorium is from the EU, whose position has been dominated by Spain, which today accounts for 40 percent of the reported high-seas bottom-trawl catch.[22] On the other hand, pressure for urgent action, based on the scientific consensus, has been building. Since 2004, resolutions have been passed and statements issued by a growing number of the relevant international agencies and fora, including the Committee on Fisheries of the UN Food and Agricultural Organization, the Seventh Conference on the Convention on Biological Diversity, the UN Task Force on Environmental Sustainability, and the IUCN World Conservation Congress, all essentially pointing to the need for urgent action to protect biodiversity on the high seas.[23] Many have called explicitly for a moratorium on high-seas bottom trawling.

High-seas bottom-trawl fisheries today are a small but highly destructive and unsustainable component of global fisheries pursued by relatively few, mostly developed countries.[24] It is unconscionable that the meager short-term interests of so few should stand in the way of urgent globally recognized reform. But whether timely action will be taken remains to be seen.

The way forward[25]

It is remarkable that the existence of life in the deep sea has been known for less than 150 years. Unquestionably, however, a new era in our understanding and utilization of the deep sea, and of our impacts upon it, opened only half a century ago. Each decade since the 1960s has brought a virtual revolution in our understanding of major deep-sea environments and faunas: the discoveries of the biodiversity of the abyssal seafloor; of chemosynthetic-based ecosystems at hydrothermal vents, cold seeps and whale falls; of seamount fisheries and deep-water coral ecosystems; of the thermophilic archaea and their possible role in the very origins of life on Earth.

The modern era has also seen deepwater fisheries sweep across the face of the deep and the energy industry gradually work its way down the continental slope. The mining sector is now taking its first tentative steps into the deep sea. Major issues looming on the horizon include the impacts of deep-sea mining; of extracting less conventional deep ocean energy resources, such as methane hydrates; of CO_2 disposal in the deep sea. Other issues of overarching concern may continue to build insidiously without ever leading to a clear decision-point: Can we contain the impacts of global pollution by heavy metals and persistent organic compounds? What will be the effects of global warming and the acidification of the oceans that accompanies it?

But the most grave and immediate threat to ocean environments of the high seas is unregulated deep-sea bottom trawling. These deep-sea fisheries fell through the cracks of UNCLOS, which addressed exploitation of seabed mineral but not biological resources, and this has been only partly redressed by the UN Fish Stocks Agreement. These unregulated and largely unreported fisheries are exploited in disregard of the FAO Code of Conduct and without regard for sustainability. They have consequently led to the serial depletion of fish stocks throughout the world's oceans, and severely damaged unique, diverse, and threatened hard-bottom benthic communities of the deep sea.

There are several policy and conservation options to consider in addressing the threats to high-seas biodiversity posed by bottom trawling:

1. A global moratorium, based on a UN General Assembly resolution, on demersal trawling on the high seas, to protect seamounts and deepwater coral and sponge habitats until these areas can be identified and assessed.
2. The establishment of RFMOs with the authority to regulate demersal fishing and protect biodiversity on the high seas.
3. A UN-based declaration of a representative system of Marine Protection Areas (MPAs) to protect the high-seas' sensitive and unique environments, with agreement to monitor and police fishing fleets' whereabouts with satellite-based vessel monitoring systems (VMS).

Establishing a global moratorium on bottom trawling on the high seas is a key first step in halting the ongoing damage to deep-sea ecosystems and biodiversity on the high seas, while other options for the governance of the living resources of the high seas are considered and developed. A moratorium would effectively close the gate on the high seas as an unregulated global commons for bottom trawling. It would provide a 'time out' while the global community considers how best to proceed. The UN moratorium on high-seas drift gillnetting and the International Whaling Commission moratorium on whaling provide valuable precedents for such action.

If further demersal fishing is to be carried out on the high seas, it must be done within the framework established in the UNFSA and the FAO Code of Conduct; that is, it must be managed in a precautionary manner within the context of an ecosystem-based approach, overfishing must be prevented, and by-catch and adverse impacts on the marine environment minimized. This should be the basis of any future agreements to regulate high-seas fishing, both pelagic and demersal. New deepwater fisheries should only be permitted if they are sustainable: if they develop in a precautionary manner, accompanied by data collection and scientific assessment programs. Sustainability must be evaluated at the ecosystem level: endurance of targeted fish stocks is not sufficient if destruction of habitat or food web connections threatens other components of the ecosystem.

Many nations have now taken steps to protect their deepwater coral and sponge environments: Norway, Australia, New Zealand, the USA, Canada, and the EU. This should provide an impetus to protect known areas of high biodiversity and critical habitat on the high seas as well, and this can be best undertaken by the UN or by other international agreement. Research may be required to identify such areas, so other policy and conservation options should be set in train first. Nonetheless, we currently have at least a rudimentary understanding of the distribution of representative habitats and of the biogeography of the open ocean. Application of the precautionary principle suggests that we should protect the unique biodiversity in each general habitat type (e.g. seamounts, mid-ocean ridges and hydrothermal vents, the coral and sponge reefs, and submarine canyons along continental margins) within each biogeographic province. This will necessitate the design of a representative network of MPAs distributed throughout the high seas, requiring input from a diverse suite of scientists and other interested parties. Further, effective monitoring and enforcement of MPAs on the high seas requires that an effective international management framework is in place, including responsible regulatory agencies, licensing systems, and use of satellite-based Vessel Monitoring Systems (VMS) for all vessels licensed to fish or conduct other environmentally threatening activities (e.g. mining or bioprospecting) on the high seas. Clearly, the design and management of such a network of high-seas MPAs should be mediated through the authority of the UN. Integration across agencies and organizations responsible for regulating these different ocean sectors is required, so that the cumulative effects of the various sectors' activities can be managed.

The deep sea may be out of sight, but it is no longer out of mind. Discovery of the deep interior of our watery planet is one of humankind's more exciting journeys of scientific exploration, a journey that we have only just begun. Given the mineral wealth in the deep sea, its biodiversity, its energy and biological resources, as well as our growing impacts upon this vast realm, our futures will continue to be intimately linked.

Nor is the deep sea really out of sight today; it is coming into view ever more clearly through the scientific media: books, journals, documentaries, television, and the internet. This provides a rapidly expanding awareness of the wonders of the deep watery portion of the planet, as well as of our growing impact upon it. Without global stewardship, the oceans will not remain 'as good left in common' for future generations. The 17th century premise on which the open access regime of the high seas was based is no longer valid. In the 21st century it is time to revive the proposal of Arvid Pardo (see Chapter 8) and others and once again seek to declare the oceans and all their living creatures part of the common heritage of humankind.[26] To maintain deep-sea fisheries and biodiversity beyond the areas of national jurisdiction for the benefit of future generations, it is clear that business as usual will not suffice. The UNFSA is an important first step, but it must now be extended across the oceans. The open-access regime of the high seas must give way to a system of conditional access, shared stewardship and global responsibility. What is required is a new approach to ocean management: an ecosystem-based approach that is comprehensive, integrated and effective, and that addresses all activities and threats in a precautionary fashion. We also need to improve the ways decisions are made, to ensure that these decisions reflect the concerns of all humankind and not just a limited club of users. In other words, all human activities need to be harmoniously managed for the long-term benefit of the seabed, humanity, and the planet.

Glossary

Abyss The deep seafloor, between about 4000 and 6000 meters depth.

Abyssopelagic The zone within the water column extending from about 2500 meters (the base of the bathypelagic zone) to just above the seafloor over the abyssal plain (see p. 44).

Albedo The fraction of the solar radiation that is reflected back into space.

Azooxanthellates The stony corals, mostly living in deep water, which do not live in symbiotic association with algae. They are entirely dependent on capturing particles, living or detrital, from the water. (Literally: without zooxanthellae – Gk.)

Bathyal The benthic depth zone, extending along the continental slope from about 200 to 2000 meters depth.

Bathymetry Related to the depth of the seafloor and its mapping.

Bathypelagic The zone within the water column below about 1000 meters (the lower limit of the mesopelagic zone), extending to about 2500 meters, the base of the continental slope (see p. 44).

Benthopelagic Refers to the fauna living within the water column in association with the seafloor, often extending some tens of meters above the bottom (see p. 44).

Benthos/benthic The fauna and flora of the seafloor.

Biogeography The science of the geographic distribution of species.

Biological pump The biological mechanisms by which organic matter is relatively rapidly transported from near-surface waters to the deep ocean, such as through the grazing and defecation of fast-sinking fecal pellets by copepods and larger organisms, and the vertical migration of near-surface feeders to mesopelagic depths.

Cenozoic The most recent geological era, extending from 65 million years ago, when the dinosaurs became extinct, to the present. Sometimes referred to as the Age of Mammals. (Literally: recent life – Gk.)

Chemosynthesis The process of carbon fixation carried out by certain microbes, based on the energy in certain reduced compounds, such as methane or hydrogen sulfide, as opposed to the use of light energy, as in photosynthesis.

Chlorinated hydrocarbons (CHCs) Hydrocarbons are compounds composed of carbon and hydrogen. When they are chlorinated, chlorine is substituted for one or more of the hydrogens. Low molecular weight CHCs, such as carbon tetrachloride, are highly volatile and relatively unstable in the environment. Of most concern in this book are the relatively high molecular weight, semi-volatile and highly stable CHCs, such as the PCBs, DDT, and related compounds.

Climax community A biological community at equilibrium.

Congener Species belonging to the same genus.

Copepods An order of small crustaceans on the order of a millimeter in length. They are a diverse group, with benthic and parasitic forms, but the pelagic copepods typically dominate the mesozooplankton.

Cosmopolitan In biogeography, this refers to species that have a global distribution.

Curie The basic unit of radioactivity. Named after Marie Curie, it is equal to the radioactivity emitted by a gram of radium, or 37 billion disintegrations per second. The Three Mile Island nuclear accident released about 50 curies.

Deep scattering layers The layers observed through the water column on echo sounders. The name refers to the way they scatter sound waves. The layers are typically composed of krill, midwater fishes, and siphonophores.

Demersal Occurring on or near the seafloor. Typically used to describe the fish obtained in bottom trawls.

Endemic Restricted to a region or habitat, such as species being restricted to seamounts or to the seamounts in a particular region.

Epifauna Organisms living mostly on the surface, e.g. of the sediment.

Epipelagic Referring to the reasonably well-lit near-surface layer of the ocean, generally the upper 200 meters (see p. 44).

Eukaryotes The group of so-called higher organisms with a distinct nucleus in their cells; includes the plants, animals, protists, and fungi.

Euphausiids An order of pelagic shrimp-like crustaceans, typically up to a few centimeters in length as adults. A key component of deep scattering layers, they are often found in swarms. They are frequently referred to as krill (literally: whale food) as they are an important food for baleen whales.

Euphotic The upper layer of the ocean, where there is sufficient light for net phytoplankton production.

Evenness The equitability in the relative abundance of species in a sample or community.

Fecundity Egg production.

Foraminifera (or forams) Protozoans ubiquitous in the oceans, related to amoebae but having shells.

Forcing A disturbance or force driving change. For example, climate forcing may lead to changes in deep ocean circulation.

Gravid Carrying eggs.

Guyots Flat-topped seamounts formed by wave erosion of former volcanic islands, which subsequently subside beneath the sea surface. First described by Harry Hess and named after the 19th century Swiss geologist Arnold Guyot.

Hadal Referring to the ocean trenches, e.g. to the depth zone or the fauna within the trenches, at depths greater than about 6000 meters.

Ichthyology The study of fishes.

Infauna The organisms living within the sediment on the seafloor.

Larvaceans Gelatinous filter feeders, at times important phytoplankton grazers. They spin a transparent mucilaginous 'house,' in which they live. Their body consists of a trunk and an elongated tail, which they beat to set up a current that draws water into the house and through a fine filter, on which phytoplankton is collected. Larvaceans, along with salps, belong to the invertebrate protochordate subphylum of the phylum Chordata (which is mostly made up of vertebrates).

Macrofauna Refers to pelagic fauna between 200 microns and 2 millimeters. In the shallow benthos, refers to organisms retained on a 1 millimeter mesh but too small to be trawled or observed in photographs (about 1–2 centimeters). In the deep sea, where much of the benthic fauna is reduced in size, the term refers to the same taxa (polychaete worms, small crustaceans, and mollusks) retained on a 0.3–0.5 millimeter mesh.

Megafauna Those pelagic species large enough to be readily trawled or, in the benthos, observed in photographs or from submersibles.

Meiofauna The benthic fauna consisting mostly of nematodes, larger protists such as forams, and small benthic copepods (harpacticoids) that are retained on the finest meshes (about 62 micrometers) and are generally smaller than the macrofauna.

Mesopelagic The layer of the water column between about 200 meters (the lower limit of the epipelagic layer) and 1000 meters (the lower detection limit of sunlight, and the approximate lower limit for the vertical migration of fishes that feed in near-surface waters) (see p. 44).

Mesoscale The spatial scale in the ocean on the order of tens to hundreds of kilometers.

Mesozoic The geological era, often referred to as the Age of Dinosaurs, extending from the end of the Permian Period 245 million years ago to the end of the Cretaceous 65 million years ago. The era comprises three periods, the Triassic, Jurassic, and Cretaceous. (Literally: middle life – Gk.)

Microbial loop The pelagic food web consisting of small phytoplankton, microzooplankton, and microbes, which efficiently grows, grazes, and recycles nutrients.

Micronekton The larger crustaceans, such as pelagic shrimps, and small fishes and squids within the water column. (Literally: small swimmers – Gk.)

Mixed layer The upper layer of the ocean that is reasonably well mixed due to the action of the wind, tides or winter convective cooling. This layer is reasonably homogenous in terms of its temperature, salinity, and nutrient concentrations; this is also the layer where most primary production occurs.

Mysids Small planktonic crustaceans, sometimes known as opossum shrimp, on the order of a centimeter in size.

Nekton The larger fishes and squids, whose swimming speed considerably exceeds that of the ambient currents. (Literally: swimmers – Gk.)

Otolith The earbone of a fish. The age of a fish is commonly estimated by counting the rings laid down annually in the otolith, much as the age of a tree is determined by counting the rings across its trunk. (Literally: ear stone – Gk.)

Pelagic Referring to the water column.

Paleozoic The geological era that extends from the dawn of the modern phyla in the Cambrian (about 570 million years ago) to the end of the Permian (about 245 million years ago). (Literally: ancient life – Gk.)

Photophores Light-producing organs commonly found in mesopelagic organisms.

Plankton The weakly swimming organisms that live in the water column and drift with the currents. Plankton range in size from viruses and micron-sized bacteria to single-celled algae and protists (generally less than 200 micrometers) to small crustaceans (generally up to a few millimeters), the krill, extending to a few centimeters, and jellyfishes, which may be meters long. (Literally: wanderers – Gk.)

Prokaryotes The group of organisms (bacteria and archaea) without a true nucleus. (Literally: pre-nuclear – Gk.)

Protists Single-celled eukaryotic organisms (as distinguished from the bacteria), now generally considered to comprise their own kingdom, alongside the plants and the animals. Members of this group, traditionally known as the Protozoa, are key grazers on the smaller phytoplankton.

Pycnoclines Gradients in density through the water column. Differences in density between water types are created by differences in their temperature and salt content; cooler, saltier water is more dense. Pycnoclines often coincide with thermoclines (vertical temperature gradients), but can also arise from differences in the salt content between water types. (Literally: density slope – Gk.)

Residence time The average length of time an entity remains in a particular part of its cycle, e.g. the residence time of water in the deep ocean is the average time a water parcel remains in the deep sea. The term is often used in environmental chemistry to denote the length of time an ion or compound remains in, say, the atmosphere or surface waters.

Salps Gelatinous filter feeders, at times important phytoplankton grazers. Their transparent bodies are barrel-shaped, open at both ends; they draw water in at one end and expel it at the other as they move through the water. Salps, along with larvaceans, are protochordates.

Seamounts Steep-sided sub-surface mountains, usually conical or elliptical in shape and of volcanic origin.

Sibling species Closely related species that have recently diverged from each other through evolution.

Species diversity (of a biological community) The number of species and the relative evenness of their abundance.

Species richness The number of species in a community or set of samples.

Stock A local fish population – and ideally, the basic unit managed in fisheries. A stock is a breeding unit, defined by having its own time and place for spawning, and its own pattern of migration to and from nursery, feeding, and spawning grounds. Stocks can generally be differentiated genetically, morphologically, or based on other differences.

Subduction The downward movement of one crustal plate beneath another, typically occurring within the deep ocean trenches.

Suspension feeders Invertebrates such as corals, clams, or sponges that feed on prey or detritus suspended in the water.

Thermocline A steep gradient in temperature through the water column.

Tragedy of the commons The degradation of commonly owned resources due to the lack of incentive for individual users to conserve them. 'Commons' were originally shared grazing areas, which were generally overgrazed. (The full expression was coined by Garrett Hardin in 1968.)

Trophodynamics The study of ecosystem productivity, based on food-chain theory. (Literally: nutrition dynamics – Gk.)

Wet weight The weight of an organism or sample that still retains its natural internal water content.

Zooxanthellae The symbiotic algae that reside within many species of tropical shallow-water corals, whose photosynthesis contributes to the corals' nutrition, thereby enabling them to thrive in tropical waters where there is little prey available.

Notes

Preface

1 Uchida RN, Hayasi S & Boehlert GW (1986) Environment and resources of seamounts in the North Pacific. *NOAA Technical Report NMFS*, p. 43.
2 Holdaway RN & Jacomb C (2000) Rapid extinction of the moas (Aves: Dinorinthiformes): model, test, and implications. *Science* 287, pp. 2250–54.

Chapter 1

1 Herodotus, *The Histories*, VII, p. 89.
2 Plato, *Phaedo*, trans. by H Tredennick, in *The Collected Dialogues of Plato*, E Hamilton and H Cairns (eds.), Bollingen series 71, Random House, NY, p. 110a.
3 TH Huxley stated in 1883: 'I believe that the cod fishery, the herring fishery, the pilchard fishery, the mackerel fishery, and probably all the great sea-fisheries are inexhaustible; this is to say that nothing we do seriously affects the numbers of fish.' Ironically, Huxley's statement came while Europe was already in the throes of its transition from sail- to steam-driven trawling, which would soon drive all these fisheries into decline.
4 From Galileo's *Dialogue* as quoted in D Sobell (1999) *Galileo's Daughter*, Fourth Estate, London, p. 184.
5 Pliny, *Natural History*, cited in WA Herdman (1923) *Founders of Oceanography and their Work*, Edward Arnold, London, p. 4.
6 Herodotus, *The Histories*, II. p. 4.
7 Ascribed to Posidonius by Strabo, *Geography* I, pp. 3, 9, cited in J Murray (1895) Report of the Scientific Results of the Voyage of H.M.S. *Challenger*, A Summary of the Scientific Results, Her Majesty's Stationary Office, London, p. 22.
8 Hooke R, 'Dr. Hooke's Description of some Instruments for Sounding the great Depths of the Sea, and bringing Accounts of several Kinds from the Bottom of it Being the Substance of some of his Lectures, in December, 1691', in *Philosophical Experiments and Observations of the Late Eminent Dr. Robert Hooke* (ed. W Derham) (1726), pp. 225–48, reprinted by RT Gunther in *The Life and Work of Robert Hooke*, part 2, in *Early Science in Oxford*, vol. 7 (1930), cited in Deacon (1971) *Scientists and the Sea, 1650–1900*, Academic Press, London.
9 Forbes E (1851) Report on the investigation of British marine zoology by means of the dredge. Part 1. The infralittoral distribution of marine invertebrata on the southern, western, and northern coasts of Great Britain. *Report of the British Association for the Advancement of Science*, Edinburgh, 1850, pp. 192–263.
10 Herdmann WA (1923) *Founders of Oceanography and their Work*, Edward Arnold, London, p. 9.
11 Forbes E & Austen RG (1859) *The Natural History of the European Seas*, John van Voorst, London, pp. 246–47.
12 Thomson CW (1874) *The Depths of the Sea*, Macmillan, London, p. 17.
13 Murray, Report of the Voyage of HMS *Challenger*, and Thomson, *The Depths of the Sea*.
14 Ross J (1819) *A voyage of Discovery, made under the Orders of the Admiralty, in H.M.S.* Isabella *and* Alexander, *for the purpose of exploring Baffin's Bay, and inquiring into the probability of a North-West Passage*, London, as cited in Deacon, *Scientists and the Sea*.
15 Ross JC (1847) *A Voyage of Discovery and Research in the Southern and Antarctic Regions during the Years 1839–43*, cited in Thomson, *The Depths of the Sea*, p. 20.
16 Noted by Wallich GC (1862) *The North-Atlantic Sea Bed: comprising a diary of the voyage on board H.M.S.* Bulldog, *in 1860; and observations on the presence of animal life, and the formation and nature of organic deposits, at great depths in the ocean*, London, pp. 80, 81, cited in Murray, Report of the Voyage of HMS *Challenger*, footnote p. 79.
17 Wallich, *The North-Atlantic Sea Bed*, cited in Deacon, *Scientists and the Sea*.

18 Thomson, *The Depths of the Sea*, p. 51.
19 Darwin C (1859) *On the Origin of Species*, Carlton House, NY, p. 81.
20 Darwin, *On the Origin of Species*, p. 256.
21 Thomson, *The Depths of the Sea*, pp. 66–69.
22 Carpenter WB (1868–69) Preliminary report of dredging operations in the seas to the north of the British Islands, carried on in H.M.S. *Lightning*, by Dr. Carpenter and Dr. Wyville Thomson, Professor of Natural History in Queen's College, Belfast, *Proceedings of the Royal Society of London* 17, pp. 168–200; Thomson, *The Depths of the Sea*, cited in Deacon, *Scientists and the Sea*.

Chapter 2

1 Herdman WA (1923) *Founders of Oceanography and their Work*, Edward Arnold, London, p. 17.
2 Tizard TH, Moseley HN, Buchanan JY & Murray J (1885) Narrative of the cruise of H.M.S. '*Challenger*,' with general account of the scientific results of the expedition. Narrative 1 (1, 2). Report of the Scientific Results of the Voyage of H.M.S. *Challenger*.
3 Tizard et al., Narrative of the cruise of H.M.S. '*Challenger*.'
4 Thomson CW (1874) *The Depths of the Sea*, Macmillan, London, p. 434.
5 Thomson, *The Depths of the Sea*, p. 454.
6 Thomson, *The Depths of the Sea*, pp. 470–71.
7 Thomson, *The Depths of the Sea*, p. 411.
8 Tizard et al., Narrative of the cruise of H.M.S. '*Challenger*.'
9 Moseley HN (1879) *Notes by a Naturalist: an Account of Observations made during the Voyage of H.M.S. 'Challenger.'* Live Books Resurrected, LS Jast (ed.), 1944, T. Werner Laurie, London, p. 501.
10 Murray, A Summary of the Scientific Results, p. x.
11 Broecker WS (1974) *Chemical Oceanography*, Harcourt Brace Jovanovich, NY, p. 31.
12 Schlee S (1973) *A History of Oceanography*, Robert Hale, London, footnote, p. 154.
13 Moseley, *Notes by a Naturalist*, pp. 508–509.
14 Murray, A Summary of the Scientific Results, p. 1438.
15 Jacobs DK & Lindberg DR (1998) Oxygen and evolutionary patterns in the sea: onshore/offshore trends and recent recruitment of deep-sea faunas. *Proceedings of the National Academy of Science USA* 95, pp. 9396–401.
16 Buchan A (1895) 'Report on oceanic circulation, based on the observations made on board H.M.S. *Challenger*, and other observations,' in Report on the Scientific Results of the Voyage of H.M.S. *Challenger*. Summary of the Scientific Results, Vol. 2, appendix (Physics and Chemistry, part 8), 38 pages and 16 maps.
17 Murray, A Summary of the Scientific Results.
18 Murray, A Summary of the Scientific Results, p. 1432.
19 Murray J (1913), *The Ocean: A General Account of the Science of the Sea*. Williams and Norgate, London, p. 182.
20 Compare Haedrich RL & Merrett NR (1990) Little evidence for faunal zonation or communities in deep-sea demersal fish faunas. *Progress in Oceanography* 24, pp. 234–50 with Koslow JA (1993) Community structure in North Atlantic deep-sea fishes. *Progress in Oceanography* 31, pp. 321–38.

Chapter 3

1 CW Thomson, quoted in Linklater E (1972) *The Voyage of the Challenger*, John Murray Ltd, London, p. 276.
2 Carson RL (1951) *The Sea Around Us*, Staples Press, London, p. 40.
3 The classification of pelagic environments generally follows Hedgpeth JW (1957) Classification of marine environments. *Geological Society of America*, Memoir 67, vol. 1: 17–28. Statistics of the world ocean from Oceans and Seas, *Encyclopaedia Britannica* (1983) 13: 482–484.
4 Angel MV (2003) The pelagic environment of the open ocean, in PA Tyler (ed.) *Ecosystems of the Deep Oceans*, Elsevier, Amsterdam, pp. 39–79.
5 Broad WJ (1997) *The Universe Below*, Simon and Schuster; Robison BH (1995) Light in the Ocean's midwater. *Scientific American* July, pp. 60–64.
6 BBC News Online, 2 April 2003.
7 Young RE & Mangold KM (1999) *Mesonychoteuthis* Robson, 1925, Tree of Life Web Project: http://tolweb.org/tree?group=Mesonychoteuthis&contgroup=Cranchiidae
8 Clarke MR (1983) Cephalopod biomass – estimation from predation. *Memoirs of the National Museum of Victoria* 44, pp 95–107; Cherel Y & Duhamel G (2004) Antarctic jaws: cephalopod prey of sharks in Kerguelen waters. *Deep-Sea Research* I 51, pp. 17–31.
9 Clarke, Cephalopod biomass. The research trawls had a 9 m^2 mouth opening and the mouth openings of commercial midwater trawls ranged from 230 to 700 m^2. The mean weight of squid from sperm whale stomachs from different parts of the world ranged from 0.6 to 8.0 kg, whereas those from commercial midwater trawls weighed less than 0.1 kg on average and from the research nets only 2 g (0.002 kg).
10 Agassiz A (1892) Reports on the dredging operations off the west coast of Central America to the Galapagos, to the west coast of Mexico, and in the Gulf of California. II. General sketch of the expedition of the 'Albatross,' from February to May, 1891. *Bulletin of the Museum of Comparative Zoology* 23, pp. 1–89. Recounted in Mills E (1983) Problems of deep-sea biology: an historical perspective, in GT Rowe (ed.) *Deep-sea Biology: The Sea*, Wiley-Interscience, New York, Vol. 8, pp. 1–79.
11 Woodward AS (1898) The antiquity of the deep-sea fish-fauna. *Natural Science* 12, pp. 257–260, cited in Gordon JDM, Merrett NR & Haedrich RL (1995) Environmental and biological aspects of slope-dwelling fishes of the North Atlantic, in Hopper AG (ed.) *Deep-Water Fisheries of the North Atlantic Oceanic Slope,* Kluwer, Dordrecht, p. 8.
12 Torres JJ & Somero GN (1988) Metabolism, enzymic activities and cold adaptation in Antarctic mesopelagic fishes. *Marine Biology* 98, pp. 169–180.

13 Siebenaller J & Somero GN (1978) Pressure-adaptive differences in lactate dehydrogenases of congeneric fishes living at different depths. *Science* 201, pp. 255–57.
14 Somero GN, Siebenaller JF & Hochachka PW (1983) Biochemical and physiological adaptations of deep sea animals, in Rowe (ed.) *Deep-sea Biology: The Sea*, pp. 261–330.
15 Menzies RJ & Wilson JBS (1961) Preliminary field experiments on the relative importance of pressure and temperature on the penetration of marine invertebrates into the deep sea. *Oikos* 12, pp. 302–309, reported in Angel MV (2003) The pelagic environment of the open ocean, in Tyler PA (ed.) *Ecosystems of the Deep Oceans*, Elsevier, Amsterdam, pp. 39–79.
16 Gunther AC (1887) Report on the deep-sea fishes collected by H.M.S. *Challenger* during the years 1873–1876. *Challenger* Report 22. Recounted in NB Marshall (1954) *Aspects of Deep Sea Biology*, Anchor Press, Essex, p. 15.
17 Bone Q & Marshall NB (1982) *The Biology of Fishes*, Blackie and Son, Glasgow, pp. 50–58.
18 Gulland, JA (1971) *The Fish Resources of the Ocean*, Fish News Books, Surrey, Chapter P.
19 Miller CB (2004) *Biological Oceanography*, Blackwell Science, Oxford, p. 177.
20 Zhang X & Dam HG (1998) Downward export of carbon by diel migrant mesozooplankton in the central equatorial Pacific. *Deep-Sea Research* II 44, pp. 2191–202. Angel MV (2003) The pelagic environment of the open ocean, in PV Tyler (ed.) *Ecosystems of the Deep Oceans*, Elsevier, Amsterdam, pp. 39–79.
21 Longhurst AZ & Harrison WG (1989) The biological pump: profiles of plankton production and consumption in the upper ocean. *Progress in Oceanography* 22, pp. 47–123.
22 Warner JA, Latz MI & Case JF (1979) Cryptic bioluminescence in a midwater shrimp. *Science* 203, pp. 1109–110.
23 Marshall NB (1979) *Developments in Deep-Sea Biology*, Blandford Press, Poole, Dorset, pp. 362–63.
24 Beebe W (1935) *Half Mile Down*, John Lane The Bodley Head, London, plate IV; Denton EJ & Warren FJ (1957) The photosensitive pigments in the retina of deep-sea fish. *Journal of the Marine Biological Association of the UK* 36, pp. 651–62, quoted in Marshall NB (1979) *Developments in Deep-Sea Biology*, Blandford Press, Poole, Dorset, p. 383.
25 Beebe, *Half Mile Down*, p. 103.
26 Kloser RJ, Ryan T, Sakov P, Williams A & Koslow JA (2002) Acoustic species identification in deep water using multiple frequencies. *Canadian Journal of Fisheries and Aquatic Sciences* 59, pp. 1065–77.
27 Barham EG (1966) Deep scattering layer migration and composition: observations from a diving saucer. *Science* 151, pp. 1399–403.
28 Robison, Light in the Ocean's midwater; Broad (1997) *The Universe Below*, p. 203 ff.
29 Robison, Light in the Ocean's midwater; Widder EA et al. (1989) Bioluminescence in the Monterey Submarine Canyon: image analysis of video recordings from a midwater submersible. *Marine Biology* 100, pp. 541–51.
30 Van-Tee J in Beebe, *Half Mile Down*, Appendix E, p. 263.
31 Barnes AT & Case JF (1972) Bioluminescence in the mesopelagic copepod *Gaussia princeps* (T. Scott). *Journal of Experimental Marine Biology and Ecology* 15, pp. 203–221.
32 Herring PJ (1976) Bioluminescence in decapod Crustacea. *Journal of the Marine Biological Association of the UK* 56, pp. 1029–48. For a more general discussion see NB Marshall, *Developments in Deep-Sea Biology*, Chapter 10.
33 Robison, Light in the Ocean's midwater.
34 Robison, Light in the Ocean's midwater.
35 Nealson KH & Hastings JW (1979) Bacterial bioluminescence: its control and ecological significance. Microbiological Reviews 43, pp. 496–518.
36 Widder et al., Bioluminescence in the Monterey Submarine Canyon.
37 Childress JJ & Nygaard MH (1973) The chemical composition of midwater fishes as a function of depth of occurrence off southern California. *Deep-Sea Research* 20, pp 1093–1109.
38 Childress JJ, Barnes AT, Quetin LB & Robison BH (1978) Thermally protecting cod-ends for recovery of living deep-sea animals. *Deep-Sea Research* 25, 415–22; Robison BH (1973) A system for maintaining midwater fishes in captivity. *Journal of the Fisheries Research Board of Canada* 30, pp. 126–28.
39 Torres JJ, Belman BW & Childress JJ (1979) Oxygen consumption rates of midwater fishes as a function of depth of occurrence. *Deep-Sea Research* 26A, pp. 185–97.
40 Torres JJ, Aarset AV, Donnelly J, Hopkins TL, Lancraft TM & Ainley DG (1994) Metabolism of Antarctic micronektonic Crustacea as a function of depth of occurrence and season. *Marine Ecology Progress Series* 113, pp. 207–219; Torres JJ & Somero GN (1988) Metabolism, enzymic activities and cold adaptation in Antarctic mesopelagic fishes. *Marine Biology* 98, pp. 169–80; Torres JJ & Somero GN (1988) Vertical distribution and metabolism in Antarctic mesopelagic fishes. *Comparative Biochemistry and Physiology* 90B, pp. 521–28.
41 Childress JJ (1975) The respiratory rates of midwater crustaceans as a function of depth of occurrence and relation to the oxygen minimum layer off Southern California. *Comparative Biochemistry and Physiology* 50, pp. 787–99.
42 Thuesen EV & Childress JJ (1993) Enzymatic activities and metabolic rates of pelagic chaetognaths: lack of depth-related declines. *Limnology and Oceanography* 38, pp. 935–48; Thuesen EV & Childress JJ (1993) Metabolic rates, enzyme activities and chemical compositions of some deep-sea pelagic worms, particularly *Nectonemertes mirabilis* (Nemertea: Hoplonemertinea) and *Poeobius meseres* (Annelida: Polychaeta). *Deep-Sea Research* 40, pp. 937–51; Thuesen EV & Childress JJ (1994) Oxygen consumption rates and metabolic enzyme

activities of oceanic California medusae in relation to body size and habitat depth. *Biological Bulletin* 187, pp. 84–98.

43 Childress JJ (1995) Are there physiological and biochemical adaptations of metabolism in deep-sea animals? *Trends in Ecology and Evolution* 10, pp. 30–36.

44 Childress JJ, Taylor SM, Cailliet GM & Price MH (1980) Patterns of growth, energy utilization and reproduction in some meso- and bathypelagic fishes off Southern California. *Marine Biology* 61, pp. 27–40.

45 Childress et al., Patterns of growth, energy utilization and reproduction.

46 Murray J & Hjort J (1912) *The Depths of the Ocean*, Macmillan, London.

47 Clarke WD (1961) A giant specimen of *Gnathophausia ingens* (Dohrn, 1870) (Mysidacea) and remarks on the asymmetry of the paragnaths in the suborder Lophogastrida. *Crustaceana* 2 (4), pp. 313–24, cited in Marshall, *Developments in Deep-Sea Biology*.

48 Silver MW, Coale SL, Pilskaln CH & Steinberg DR (1998) Giant aggregates: importance as microbial centers and agents of flux in the mesopelagic zone. *Limnology and Oceanography* 43, pp. 498–507.

49 Acuna JL (2001) Pelagic tunicates: why gelatinous? *American Naturalist* 158, pp. 100–107.

50 Cohen DM (1970) How many recent fishes are there? *Proceedings of the California Academy of Sciences* 38, pp. 341–45; Bone Q & Marshall NB (1986) *Biology of Fishes*, Blackie, Glasgow; Nelson JS (1976) *Fishes of the World*, John Wiley & Sons, New York.

Chapter 4

1 Wolff T (1960) The hadal community, an introduction. *Deep-Sea Research* 6, pp. 95–124.

2 Bruun AF (1956) The abyssal fauna: its ecology, distribution and origin. *Nature* 177, pp. 1105–108; Wolff, The hadal community, and Wolff T (1970) The concept of the hadal or ultra-abyssal fauna. *Deep-Sea Research* 17, pp. 983–1003.

3 The epic story of the Piccards' development and proving of the bathyscaph is told by Jacques Piccard and his American co-aquanaut, Robert Dietz (1961) in *Seven Miles Down: the story of the bathyscaph Trieste*, Longmans, Green, London. The French side of the tale is recounted in Houot GS & Willm PH (1955) *2000 fathoms down*, E.P. Dutton, New York.

4 More than 50 years on, the so-called '^{14}C method' remains a standard method for measuring primary production in the sea: Steeman-Nielsen E (1952) The use of radioactive carbon (^{14}C) for measuring organic production in the sea. *Journal du conseil international pour l'exploration de la mer* 18, pp. 117–40. The method involves spiking a seawater sample with radioactively labeled carbonate (CO_3^{2-}) and incubating it for a set period. Photosynthesis involves the uptake of carbon dioxide, which forms a chemical equilibrium with the carbonate ion. Following the incubation, the phytoplankton is retained on a filter and the amount of ^{14}C incorporated is measured with a scintillation counter. The results of the *Galathea* expedition were published in Steeman-Nielsen E & Aabye Jensen E (1957) Primary oceanic production, the autotrophic production of organic matter in the oceans. *Galathea* Reports 1, pp. 49–126, quoted in Berger WH (1989) Global maps of oceanic productivity, in Berger WH, Smetacek VS & Wefer G (eds.) *Productivity of the Ocean*, Wiley, Chichester, pp. 429–55.

5 Rowe GT (1983) Biomass and production of the deep-sea macrobenthos, in GT Rowe (ed.) *Deep-Sea Biology*, John Wiley & Sons, New York, pp. 97–121.

6 The regions of low productivity in the central oceanic gyres are also regions of high species diversity, a major difference between marine and terrestrial ecosystems. Scripps zooplankton ecologist John McGowan studied the exceptionally diverse and stable pelagic ecosystem of the central North Pacific, which in many aspects mirrors the even greater stability and diversity of the seafloor fauna. See McGowan JA & Walker PW (1985) Dominance and diversity maintenance in an oceanic ecosystem. *Ecological Monographs* 55, pp. 103–118.

7 Key papers outlining this synthesis include: Reid JL Jr (1962) On circulation, phosphate-phosphorus content, and zooplankton volumes in the upper part of the Pacific Ocean. *Limnology and Oceanography* 7, pp. 287–306; Ryther JH (1969) Photosynthesis and fish production in the sea. *Science* 166, pp. 72–77; Gulland JA (ed.) *The Fish Resources of the Ocean*, Fishing News (Books), Surrey, England.

8 Ekman S (1967) *Zoogeography of the Sea*, 2nd edition, Sidgwick and Jackson, London, pp. 266–303.

9 For a recent review of Soviet studies, see NG Vinogradova (1997) Zoogeography of the abyssal and hadal zones. *Advances in Marine Biology* 32, pp. 326–86. A comprehensive synthesis of the deep-sea benthic literature is found in Gage JD & Tyler PA (1991) *Deep-Sea Biology*. Cambridge University Press, Cambridge.

10 Gage & Tyler, *Deep-Sea Biology*, Chapter 13.

11 For a good general description of the deep seafloor megafauna, see Gage & Tyler, *Deep-Sea Biology*, Chapter 4. For a description of the 'monster' camera, see Isaacs JD (1969) The nature of oceanic life. *Scientific American* 221, pp. 146–62; Isaacs JD & Schwartzlose RA (1975) Active animals of the deep-sea floor. *Scientific American* 233, pp. 84–91.

12 The taxonomy of the Gadiformes and of the Macrouridae, in particular, is in a state of flux, with almost as many schemes to sort out the order and potential sub-orders, families, and sub-families as there are taxonomists who work with the group. I follow here Nelson JS (1994) *Fishes of the World*, 3rd edition, John Wiley & Sons, New York.

13 Bergstad OA (1990) Distribution, population structure, growth and reproduction of the roundnose grenadier *Coryphaenoides rupestris* (Pisces: Macrouridae) in the deep waters of the Skagerrak. *Marine Biology* 107, pp. 25–39; Bergstad OA (1995) Age determination of deep-

water fishes: experiences, status and challenges for the future, in AG Hopper (ed.) *Deep-water Fisheries of the North Atlantic Oceanic Slope*, pp. 267–83; Andrews AH, Caillet GM & Coale KH (1999) Age and growth of the Pacific grenadier (*Coryphaenoides acrolepis*) with age estimate validation using an improved radiometric ageing technique. *Canadian Journal of Fisheries and Aquatic Sciences* 56, pp. 1339–50; Caillet GM et al. (2001) Age determination and validation studies of marine fishes: do deep-dwellers live longer? *Experimental Gerontology* 36, pp. 739–64.
14 Kurlansky M (1998) *Cod*, Jonathan Cape, London, p. 49.
15 Murray J (1913) *The Ocean*, Williams and Norgate, London, p. 182.
16 Murray, *The Ocean*, p. 183.
17 Ekman, *Zoogeography of the Sea*, p. 274.
18 Sanders HL, Hessler RR & Hampson GR (1965) An introduction to the study of deep-sea benthic faunal assemblages along the Gay Head–Bermuda transect. *Deep-Sea Research* 12, pp. 845–67; Hessler RR & Sanders HL (1967) Faunal diversity in the deep sea. *Deep-Sea Research* 14, pp. 65–78.
19 Sanders, Hessler & Hampson, Deep-sea benthic faunal assemblages along the Gay Head–Bermuda transect; Hessler & Saunders, Faunal diversity in the deep sea.
20 Thiel H (1983) Meiobenthos and nanobenthos of the deep sea, in GT Rowe (ed.) *Deep-Sea Biology: The Sea*, Vol. 8, Wiley-Interscience, New York, pp. 167–229.
21 Grassle JF & Maciolek NJ (1992) Deep-sea species richness: regional and local diversity estimates from quantitative bottom samples. *American Naturalist* 139, pp. 313–41.
22 May RM (1992) Bottoms up for the oceans. *Nature* 357, pp. 278–79.
23 Koslow JA, Williams A & Paxton JR (1997) How many demersal fish species in the deep sea? A test of a method to extrapolate from local to global diversity. *Biodiversity and Conservation* 6, pp. 1523–32.
24 Erwin TL (1982) Tropical forests: their richness in Coleoptera and other Arthropod species. *Coleopterists' Bulletin* 36, pp. 74–75.
25 Honjo S (1980) Material fluxes and modes of sedimentation in the mesopelagic and bathypelagic zones. *Journal of Marine Research* 38, pp. 53–95; Gage & Tyler, *Deep-Sea Biology*, p. 273.
26 Smith CR (1985) Food for the deep sea: utilization, dispersal, and flux of nekton falls at the Santa Catalina Basin floor. *Deep-Sea Research* 32, pp. 417–42.
27 Gage & Tyler, *Deep-Sea Biology*, p. 269. This figure may also be derived from considering the proportion of surface production that reaches the seafloor (1–3 percent) and the annual rates of primary production in the open ocean, which range from about 150 g carbon/m^2 in the oligotrophic central gyres to about twice that in the productive region of equatorial upwelling, yields an estimated annual vertical flux to the seafloor of about 1–10 g C/m^2. Production data from Miller CB (2004) *Biological Oceanography*, Blackwell Science, Malden, USA, pp. 223, 229.
28 Bruun, The abyssal fauna: its ecology, distribution and origin.
29 *Encyclopedia Britannica* (1983), Macropedia, Vol. 2, article on Lake Baikal.
30 Myers N, Mittermeier RA, Mittermeier CG, da Fonseca GAB & Kent J (2000) Biodiversity hotspots for conservation priorities. *Nature* 403, pp. 853–58.
31 McGowan JA (1971) Oceanic biogeography of the Pacific, in Funnell B & Riedel W (eds.) *The Micropaleontology of the Oceans*, Cambridge University Press, Cambridge, pp. 3–74; McGowan & Walker, Dominance and diversity maintenance in an oceanic ecosystem; McGowan JA & Walker PW (1978) Structure in the copepod community of the north Pacific central gyre. *Ecological Monographs* 49, pp. 195–226.
32 Jannasch HW, Eimlijellen K, Wirsen CO & Farmanfarmaian A (1971) Microbial degradation of organic matter in the deep sea. *Science* 171, pp. 672–75; Jannasch HW & Wiesen CO (1973) Deep-sea microorganisms: in situ response to nutrient enrichment. *Science* 180, pp. 641–43.
33 Smith KL Jr & Hessler RR (1974) Respiration of benthopelagic fishes: *In situ* measurements at 1230 meters. *Science* 184, pp. 72–73. Smith KL Jr (1978) Metabolism of the abyssopelagic rattail *Coryphaenoides armatus* measured *in situ*. *Nature* 274, pp. 362–64.
34 Hutchinson GE (1961) The paradox of the plankton. *American Naturalist* 95, pp. 137–45.
35 Grice GD & Hart AD (1962) The abundance, seasonal occurrence, and distribution of the epizooplankton between New York and Bermuda. *Ecological Monographs* 32, pp. 287–309; McGowan & Walker, Dominance and diversity maintenance in an oceanic ecosystem.
36 Snelgrove PVR & Smith CR (2002) A riot of species in an environmental calm: the paradox of the species-rich deep-sea floor. *Oceanography and Marine Biology Annual Review* 40, pp. 311–42.
37 Sanders HL (1968) Marine benthic diversity: a comparative study. *American Naturalist* 102, pp. 243–82; Sanders HL & Hessler RR (1969) Ecology of the deep-sea benthos. *Science* 163, pp. 1419–24.
38 Frank PW (1957) Coactions in laboratory populations of two species of *Daphnia*. *Ecology* 38, 510–19; Park T (1962) Beetles, competition, and populations. *Science* 138, pp. 1369–75.
39 Diamond J (1975) in Cody ML & Diamond J (eds.), *Ecology and Evolution of Communities*, Belknap, Cambridge, Mass., p. 343.
40 Connell JH (1978) Diversity in tropical rain forests and coral reefs. *Science* 199, pp. 1302–310.
41 Dayton PK & Hessler RR (1972) Role of biological disturbance in maintaining diversity in the deep sea. *Deep-Sea Research* 19, pp. 199–208.
42 Grassle JF & Sanders HL (1973) Life histories and the role of disturbance. *Deep-Sea Research* 20, pp. 643–59.
43 Bergstad OA (1990) Distribution, population structure, growth and reproduction of the roundnose grenadier *Coryphaenoides rupestris* (Pisces: Macrouridae) in the

deep waters of the Skagerrak. *Marine Biology* 107, pp. 25–39. Caillet GM et al (2001) Age determination and validation studies of marine fishes: do deep-dwellers live longer? *Experimental Gerontology* 36, pp. 739–64.
44 Turekian KK et al. (1975) Slow growth rate of a deep-sea clam determined by ^{228}Ra chronology. *Proceedings of the National Academy of Science, USA* 72, pp. 2829–32. For a contrasting view, see Gage JD & Tyler PA (1991) *Deep-Sea Biology*, pp. 320 ff. However the evidence for extreme longevity in deep-sea organisms, both fish and invertebrates, has mounted in recent years (see Chapter 6).
45 Isaacs JD & Schwartzlose RA (1975) Active animals of the deep-sea floor. *Scientific American* 233, pp. 84–91.
46 Hollister CD & McCave IN (1984) Sedimentation under deep-sea storms. *Nature* 309, pp. 220–25.
47 Deuser WG & Ross EH (1980) Seasonal change in the flux of organic carbon to the deep Sargasso Sea. *Nature* 283, pp. 364–65; Billett DSM, Lampitt RS, Rice AL & Mantoura RFC (1983) Seasonal sedimentation of phytoplankton to the deep-sea benthos. *Nature* 302, pp. 520–22; Tyler PA (1988) Seasonality in the deep sea. *Oceanography and Marine Biology Annual Review* 26, pp. 227–58; Beaulieu, S.E., 2002. Accumulation and fate of phytodetritus on the sea floor. *Oceanography and Marine Biology Annual Review*, 40, 171–232.
48 Snelgrove PVR, Grassle JF & Petrecca RF (1992) The role of food patches in maintaining high deep-sea diversity: field experiments with hydrodynamically unbiased colonization trays. *Limnology and Oceanography* 37, pp. 1543–50; Snelgrove PVR, Grassle JF & Petrecca RF (1994) Macrofaunal response to artificial enrichments and depressions in a deep-sea habitat. *Journal of Marine Research* 52, pp. 345–69.
49 Smith CR (1994) Tempo and mode in deep-sea benthic ecology: punctuated equilibrium revisited. *Palaios* 9, pp. 3–13.
50 Smith KL & Baldwin RJ (1984) Seasonal fluctuations in deep-sea sediment community oxygen consumption: central and eastern North Pacific. *Nature* 307, pp. 624–26; Smith KL, Kaufmann RS, Baldwin RJ, Carlucci AF (2001) Pelagic-benthic coupling in the abyssal eastern North Pacific: an 8-year time-series study of food supply and demand. *Limnology and Oceanography* 46, pp. 543–56; Smith KL Jr & Kaufmann RS (1999) Long-term discrepancy between food supply and demand in the deep eastern North Pacific. *Science* 284, pp. 1174–77.
51 Billett DSM et al. (2002) Long-term change in the megabenthos of the Porcupine Abyssal Plain (NE Atlantic). *Progress in Oceanography* 50, pp. 325–48.
52 Murray, *The Ocean*, pp. 180 ff.
53 Snelgrove PVR & Grassle JF (1995) The deep sea: desert AND rainforest. *Oceanus* 38, pp. 25–29; Snelgrove & Smith (2002) A riot of species in an environmental calm, pp. 311–42.
54 Haedrich RL, Rowe GT & Polloni PT (1975) Zonation and faunal composition of epibenthic populations on the continental slope south of New England. *Journal of Marine Research* 23, pp. 191–212; Haedrich RL & Merrett NR (1990) Little evidence for faunal zonation or communities in deep sea demersal fish faunas. *Progress in Oceanography* 24, pp. 234–50; Gage & Tyler, *Deep-Sea Biology*; Koslow JA (1993) Community structure in North Atlantic deep-sea fishes. *Progress in Oceanography* 31, pp. 321–38.
55 McGowan, Oceanic biogeography of the Pacific; McGowan & Walker, Structure in the copepod community of the north Pacific central gyre.
56 See the CeDAMar (Census of Diversity of Abyssal Marine Life) website: www.cedamar.org.

Chapter 5

1 Sverdrup HU, Johnson MW & Fleming RH (1942) *The Oceans: their Physics, Chemistry, and General Biology*, Prentice-Hall, Englewood Cliffs, New Jersey, p. 14.
2 Bacon F (1620) *Novum organum,* cited in JP Kennett (1982) *Marine Geology,* Prentice-Hall, New Jersey, p. 106.
3 Stanley SM (1986) *Earth and Life through Time,* WH Freeman, New York, pp. 165–69.
4 Murray J & Hjort J (1912) *The Depths of the Ocean.* Macmillan and Co, London, p. 135.
5 Stanley, *Earth and Life through Time.*
6 Hess H (1962) History of Ocean Basins, in Engel, AEJ, James HL & Leonard BF (eds.) *Petrologic Studies: A Volume in Honor of A.F. Buddington,* Geological Society of America, Boulder, pp. 599–620. Kuhn T (1962) *The Structure of Scientific Revolutions,* University of Chicago Press, Chicago.
7 Kennett, *Marine Geology,* pp. 109, 114; Stanley, *Earth and Life through Time,* p. 185.
8 Hess, History of Ocean Basins, pp. 599–620.
9 Elder JW (1965) American Geophysical Union Monograph 8, p. 211, cited in: Corliss JB, Dymond J et al. (1979) Submarine thermal springs on the Galapagos Rift. *Science* 203, pp. 1073–83.
10 Crane K (2003) *Sea Legs,* Westview Press, Boulder, Colorado, pp. 54–55; Kennett, *Marine Geology,* p. 495.
11 Herzig PM & Petersen S (2002) Polymetallic massive sulphide deposits at the modern seafloor and their resource potential, in *Polymetallic massive sulphides and cobalt-rich ferromanganese crust: status and prospects*, International Seabed Authority, Kingston, Jamaica, ISA Technical Study 2, pp. 7–35.
12 Murray J & Renard AF (1891) Report on Deep-Sea Deposits. Challenger Expedition Reports 3, Her Majesty's Stationery Office, London.
13 Bostrom K & Peterson MNA (1966) Precipitates from hydrothermal exhalations on the East Pacific rise, *Economic Geology* 61, pp. 1258–65.
14 *Encyclopedia Britannica* (1983), article on Copper Products and Production, Macropedia Vol. 5, p. 148.
15 van Dover CL (2000) *The Ecology of Deep-Sea Hydrothermal Vents,* Princeton University Press, Princeton, pp. 61–62.
16 Corliss JB (1973) The sea as alchemist. *Oceanus* 17, pp. 38–43; Edmond JM & Von Damm K (1983) Hot springs

on the ocean floor. *Scientific American*, April, pp. 78–93; Edmond JM, Von Damm K, McDuff RE & Measures C (1982) Chemistry of hot springs on the East Pacific Rise and their effluent dispersal. *Nature* 297, pp. 187–91.

17 The pH scale is a logarithmic measure of the concentration of hydrogen ions in solution. Seawater with a pH of 6 contains 10 times more hydrogen ions and is 10 times more acidic than seawater with a pH of 7 – and seawater with a pH of 3 is 31 600 times more acidic than seawater with a pH of 7.5.

18 MacIntyre F (1970) Why the sea is salt. *Scientific American* November.

19 Edmond, Chemistry of hot springs on the East Pacific Rise.

20 Kaharl VA (1990) *Water Baby: the Story of Alvin,* Oxford University Press, Oxford, p. 45.

21 Kaharl, *Water Baby,* Chapters 2–3.

22 Crane, *Sea Legs,* Chapter 12; Corliss JB and 10 co-authors (1979) Submarine thermal springs on the Galàpagos Rift. *Science* 203, pp. 1073–82.

23 Crane, *Sea Legs,* Chapter 12.

24 Lonsdale P (1977) Clustering of suspension-feeding macrobenthos near abyssal hydrothermal vents at oceanic spreading centers. *Deep-Sea Research* 24, pp. 857–63. 'Suspension feeding' is a term for feeding on small prey, such as unicellular organisms and detritus, suspended in the water, as opposed to raptorial feeding on individual prey. Clams, mussels, sponges, and many copepods are typical suspension feeders.

25 Cone J (1991), *Fire Under the Sea: The Discovery of the Most Extraordinary Environment on Earth – Volcanic Hot Springs on the Ocean Floor,* William Morrow, New York, Chapter 4.

26 Mayr E (1982) *The Growth of Biological Thought*, Harvard University Press, Cambridge, Mass., p. 152.

27 Hessler RR & Smithey, WM (1983) The distribution and community structure of megafauna at the Galapagos Rift hydrothermal vents, in Rona PA, Bostrom K, Laubier L & Smith KL (eds.) *Hydrothermal Processes at Seafloor Spreading Centers*, NATO Conference Series IV, Plenum Press, New York, pp. 735–70, cited in Gage JD & Tyler PA (1999) *Deep-Sea Biology*, Cambridge University Press, Cambridge, p. 369.

28 Lutz RA and 6 co-authors (1994) Rapid growth at deep-sea vents. *Nature* 371, pp. 663–64.

29 Jones ML (1981) *Riftia pachyptila*, new genus, new species, the vestimentiferan from the Galapagos Rift geothermal vents (Pogonophora). *Proceedings of the National Academy of Science, USA* 93, pp. 1295–313.

30 Brusca RC & Brusca GJ (1990) *Invertebrates*, Sinauer, Sunderland, Mass., p. 453.

31 Gage JD & Tyler PA (1999) *Deep-Sea Biology*, Cambridge University Press, Cambridge, p. 124.

32 Hyman LH (1959) *The Invertebrates. Vol V: Smaller coelomate groups.* McGraw-Hill Book Co, New York, pp. 208–210.

33 Jones ML (1985) On the Vestimentifera, new phylum: six new species, and other taxa, from hydrothermal vents and elsewhere. *Bulletin of the Biological Society of Washington* 6, pp. 117–58. The controversy is reviewed in van Dover, *The Ecology of Deep-Sea Hydrothermal Vents*, pp. 316–18.

34 Uschakov PV (1933) Eine neue Form aus der Familie Sabellidae (Polychaeta). *Zoologischer Anzeiger* 104, pp. 205–208, cited in Schulze A & Halanych KM (2003) Siboglinid evolution shaped by habitat preference and sulfide tolerance. *Hydrobiologia* 496, pp. 199–205.

35 Hyman LH, *The Invertebrates. Vol V: Smaller coelomate groups*, p. 224. The terms 'protostome' (meaning primary mouth) and 'deuterostome' (secondary mouth) are derived from the embryology of these two lineages. In the protostomes, the mouth develops from the embryonic blastopore; in the deuterostomes, the anus develops from the blastopore and the mouth develops from a secondary opening.

36 Meglitsch PA (1972) *Invertebrate Zoology*, 2nd edition, Oxford University Press, London, pp. 758–60; Brusca & Brusca, *Invertebrates*, pp. 450–57.

37 Black MB et al. (1997) Molecular systematics of vestimentiferan tubeworms from hydrothermal vents and cold-water seeps. *Marine Biology* 130, pp. 141–49; Schulze A & Halanych KM (2003) Siboglinid evolution shaped by habitat preference and sulfide tolerance. *Hydrobiologia* 496, pp. 199–205; van Dover CL *The Ecology of Deep-Sea Hydrothermal Vents*, Chapter 11; van Dover CL, German CR, Speer KG, Parson LM & Vrijenhoek RC (2002) Evolution and biogeography of deep-sea vent and seep invertebrates. *Science* 295: pp. 1253–57.

38 Cone, *Fire Under the Sea*, p. 86.

39 Winogradsky first observed chemosynthesis in 1887: van Dover, *The Ecology of Deep-Sea Hydrothermal Vents*, p. 117. The fixing of organic carbon in photosynthesis is summarized as: $CO_2 + H_2O \rightarrow [CH_20] + O_2$. Chemosynthesis can be summarized as: $CO_2 + H_2O + H_2S + O_2 \rightarrow [CH_20] + H_2SO4$.

40 Brock TD & Freeze H (1969) *Thermus aquaticus* gen. n. and sp. n., a nonsporulating extreme thermophile. *Journal of Bacteriology* 98, pp. 289–97; for his memoir, see Brock TD (1995) The road to Yellowstone – and beyond. *Annual Review of Microbiology* 49, pp. 1–28.

41 Cone, *Fire Under the Sea*, p. 87; Crane, *Sea Legs*, p. 145.

42 Cone, *Fire Under the Sea*, pp. 170–73; Kunzig R (2000) *Mapping the Deep*, Sort of Books, London, pp. 141–43.

43 Cavanaugh CM et al. (1981) Prokaryotic cells in the hydrothermal vent tube worm *Riftia pachyptila* Jones: possible chemoautotrophic symbionts. *Science* 213, pp. 340–42.

44 van Dover, *The Ecology of Deep-Sea Hydrothermal Vents*, Chapter 7.2.2.

45 van Dover, *The Ecology of Deep-Sea Hydrothermal Vents,* Chapter 7.4.

46 Childress JJ, Felbeck H & Somero GN (1987) Symbiosis in the deep sea. *Scientific American* 255, pp. 114–120; van Dover CL, *The Ecology of Deep-Sea Hydrothermal Vents*, Table 7.2.

47 van Dover, *The Ecology of Deep-Sea Hydrothermal Vents*, Chapter 8.

48 van Dover, *The Ecology of Deep-Sea Hydrothermal Vents*, p. 56.

49 van Dover, *The Ecology of Deep-Sea Hydrothermal Vents*, Figure 2.22; http://www.ocean.udel.edu/deepsea/level-2/geology/vents.html

50 van Dover, *The Ecology of Deep-Sea Hydrothermal Vents*, Chapter 2.5.2.

51 van Dover, *The Ecology of Deep-Sea Hydrothermal Vents*, pp. 202–203.

52 Chevaldonné P, Desbruyères D & Childress JJ (1992) … and some even hotter. *Nature* 359, pp. 593–94.

53 van Dover, *The Ecology of Deep-Sea Hydrothermal Vents*, pp. 56–57. Some sulfide mounds on the Juan de Fuca Ridge off the US Pacific Northwest have been aged to more than 200 years old.

54 van Dover, *The Ecology of Deep-Sea Hydrothermal Vents*, Chapters 5.2.6, 10.4.

55 van Dover, *The Ecology of Deep-Sea Hydrothermal Vents*, pp. 300–301, 340.

56 Tunnicliffe V, McArthur AG & McHugh D (1998) A biogeographical perspective of the deep-sea hydrothermal vent fauna. *Advances in Marine Biology* 34, pp. 353–442.

57 van Dover, *The Ecology of Deep-Sea Hydrothermal Vents*, Chapter 2.

58 Tunnicliffe et al., A biogeographical perspective of the deep-sea hydrothermal vent fauna.

59 van Dover, *The Ecology of Deep-Sea Hydrothermal Vents*, pp. 150–52.

60 Tunnicliffe V & Fowler CMR (1996) Influence of sea-floor spreading on the global hydrothermal vent fauna. *Nature* 379, pp. 531–33.

61 Mayr, *The Growth of Biological Thought*, p. 244.

62 Whittaker RH (1969) New concepts of kingdoms of organisms. *Science* 163, pp. 150–60; Pace NR (1997) A molecular view of microbial diversity and the biosphere. *Science* 276, pp. 734–40.

63 16S refers to a particular size of RNA in the ribosomes, S standing for the Svedburg unit, a measure of the speed at which the molecule sediments out in a centrifuge. rRNA stands for the ribosomal RNA (ribonucleic acid). Further references: Howland JL (2000) The *Surprising Archaea*, Oxford University Press, New York, Chapter 2; Woese CR & Fox GE (1977) Phylogenetic structure of the prokaryotic domain: the primary kingdoms. *Proceedings of the National Academy of Science, USA* 74, pp. 5088–90; Woese CR (1981) Archaebacteria. *Scientific American* June, pp. 98–122.

64 Of course there are other characteristics, besides the structure of their rRNA, that link the archaea and separate them from other bacteria: fundamental differences in the biochemical structure of their cell wall and of their transfer RNA, among others. For a more recent account, see Pace NR (1997) A molecular view of microbial diversity and the biosphere. *Science* 276, pp. 734–40.

65 Miller SL (1953) A production of amino acids under possible primitive Earth conditions. *Science* 117, pp. 258–59.

66 van Dover, *The Ecology of Deep-Sea Hydrothermal Vents*, Chapter 13; Kasting JF (1993) Earth's earliest atmosphere. *Science* 259, pp. 920–26.

67 van Dover, *The Ecology of Deep-Sea Hydrothermal Vents*, Chapter 13; Maher KA & Stevenson DJ (1988) Impact frustration of the origin of life. *Nature* 331, pp. 612–14.

68 The development of this hypothesis is described in Cone, *Fire Under the Sea*, Chapter 9. The hypothesis is developed in Corliss JB, Baross JA & Hoffman SE (1981) An hypothesis concerning the relationship between submarine hot springs and the origin of life on Earth. *Oceanologica Acta* 4, pp. 59–69; Baross JA & Hoffman SE (1985) Submarine hydrothermal vents and associated gradient environments as sites for the origin and evolution of life. *Origins of Life* 15, pp. 3327–45. See also van Dover CL, *The Ecology of Deep-Sea Hydrothermal Vents*, Chapter 13.

69 General explications: Waldrop MM (1990) Goodbye to the warm little pond? *Science* 250, pp. 1078–80; van Dover CL, *The Ecology of Deep-Sea Hydrothermal Vents*, pp. 404–407. Also, Wächtershäuser G (1988) Before enzymes and templates: theory of surface metabolism. *Microbiology Review* 52, pp. 452–84; Wächtershäuser G (1988) Pyrite formation, the first energy source for life: a hypothesis. *Systematics and Applied Microbiology* 10, pp. 207–210.

70 Deming JW & Baross JA (1993) Deep-sea smokers; windows to a subsurface biosphere? *Geochimica et Cosmochimica Acta* 57, pp. 3219–30.

71 Gold T (1992) The deep, hot biosphere. *Proceedings of the National Academy of Science, USA* 89, pp. 6045–49.

72 Sibuet M & Olu K (1998) Biogeography, biodiversity and fluid dependence of deep-sea cold-seep communities at active and passive margins. *Deep-Sea Research* II 45, pp. 517–67.

73 Bergquist DC, Williams FM & Fisher CR (1999) Longevity record for deep-sea invertebrate. *Nature* 403, pp. 499–500.

74 Sibuet & Olu, Biogeography, biodiversity and fluid dependence. Thirteen chemosynthetic mussel species, 20 vesicomyids and 14 pogonophorans are known from seeps compared with eight, three, and ten, respectively, from vents.

75 van Dover, *The Ecology of Deep-Sea Hydrothermal Vents*, Chapter 12.

76 Narrative of the dive from Kaharl, *Water Baby*, pp. 315–17.

77 Smith CR, Kukert H, Wheatcroft RA, Jumars PA & Deming JW (1989) Vent fauna on whale remains. *Nature* 341, pp. 27–28; Smith CR & Baco AR (2003) Ecology of whale falls at the deep-sea floor. *Oceanography and Marine Biology, Annual Review* 41, pp. 311–54; Smith CR (2006) Bigger is better: The role of whales as detritus in marine ecosystems. In JA Estes, DP DeMaster, RL Brownell Jr., DF Doak & TM Williams (eds) *Whales, Whaling and Ocean Ecosystems*, University of California Press, Berkeley, USA, in press.

78 Butman CA, Carlton JT & Palumbi SR (1995) Whaling effects on deep-sea biodiversity. *Conservation Biology* 9, pp. 462–64. See also Jelmert A, Oppen-Berntsen DO (1996) Whaling and deep-sea biodiversity. *Conservation Biology* 10, pp. 653–54; Butman CA, Carlton JT & Palumbi

SR (1995) Whales don't fall like snow. *Conservation Biology* 10, pp. 655–57; Smith, Bigger is better: The role of whales as detritus in marine ecosystems.
79 Squires RL, Goedert JL & Barnes LG (1991) Whale carcasses. *Nature* 349, p. 574; Goedert JL, Squires RL & Barnes LG (1995) Paleoecology of whale-fall habitats from deep-water Oligocene rocks, Olympic Peninsula, Washington state. *Palaeogeography, Palaeoclimatology, Palaeoecology* 118, pp. 151–58.
80 Gage & Tyler, *Deep-Sea Biology*, pp. 124–25.
81 Distel DL, Baco AR, Chuang E, Morrill W, Cavanaugh C & Smith CR (2000) Do mussels take wooden steps to deep-sea vents? *Nature* 403, pp. 725–26.
82 McArthur AG & Tunnicliffe V (1998) Relics and antiquity revisited in the modern vent fauna, in Harrison K & Mills R (eds.) *Modern Ocean Floor Processes and the Geological Record*, pp. 271–91; McArthur AG & Koop BF (1999) Partial 28S rDNA sequences and the antiquity of hydrothermal vent endemic gastropods. *Molecular Phylogenetics and Evolution* 13, pp. 255–74.
83 Grassle JF (1986) The ecology of deep sea hydrothermal vent communities. *Advances in Marine Biology* 23, pp. 301–362; Tunnicliffe V (1991) The biology of hydrothermal vents: ecology and evolution. *Oceanography and Marine Biology Annual Review* 29, pp. 319–407.
84 Little CTS, Herrington RJ, Maslennikov VV, Morris NJ & Zaykov VV (1997) Silurian hydrothermal-vent community from the southern Urals, Russia. *Nature* 385, pp. 146–48.
85 Schulze & Halanych, Siboglinid evolution shaped by habitat preference and sulfide tolerance.
86 Black et al. (1997) Molecular systematics of vestimentiferan tubeworms; van Dover et al., Evolution and biogeography of deep-sea vent and seep invertebrates.
87 Newman WA (1985) The abyssal hydrothermal vent invertebrate fauna: a glimpse of antiquity? *Bulletin of the Biological Society of Washington* 6, pp. 231–42.
88 Isozaki Y (1997) Permo-Triassic boundary superanoxia and stratified superocean: records from lost deep sea. *Science* 276, pp. 235–38; Wignall PB & Twitchett RJ (1996) Oceanic anoxia and the end Permian mass extinction. *Science* 272, pp. 1155–58.
89 Craddock CWR et al. (1995) Evolutionary relationships among deep-sea mytilids (Bivalvia: Mytilidae) from hydrothermal vents and cold-water methane/sulfide seeps. *Marine Biology* 121, pp. 477–85; Peek AS, Gustafson RG, Lutz RA & Vrijenhoek RC (1997) Evolutionary relationships of deep-sea hydrothermal vent and cold-water seep clams (Bivalvia: Vesicomyidae): results from the mitochondrial cytochrome oxidase subunit I. *Marine Biology* 130, pp. 151–61; van Dover et al., Evolution and biogeography of deep-sea vent and seep invertebrates.
90 Shank TM, Black MB, Halanych KM, Lutz RA & Vrijenhoek RC (1999) Miocene radiation of deep-sea hydrothermal vent shrimp (Caridea: Bresiliidae): evidence from mitochondrial cytochrome oxidase subunit I. *Molecular Phylogenetics and Evolution* 13, pp. 244–54.

Chapter 6

1 Cairns SD (2005) Deep-water corals: a primer. The 3rd International Symposium on Deep-sea Corals, Miami.
2 Cairns SD (2001) A brief history of taxonomic research on azooxanthellate *Scleractinia* (Cnidaria: Anthozoa). *Bulletin of the Biological Society of Washington* 10, pp. 191–203; Cairns SD (1999) Species richness of recent Scleractinia. *Atoll Research Bulletin* 459, pp. 1–46.
3 Wilson JB (2001) *Lophelia* 1700 to 2000 and beyond, in Willison JHM et al. (eds.) *Proceedings of the First International Symposium on Deep-Sea Corals*, Ecology Action Centre, Nova Scotia Museum, Halifax, Canada, pp. 1–5.
4 Fossa JH, Mortensen PB & Furevik DM (2002) The deep-water coral *Lophelia pertusa* in Norwegian waters: distribution and fishery impacts. *Hydrobiologia* 471, pp. 1–12.
5 Rogers A (1999) The biology of *Lophelia pertusa* (Linnaeus 1758) and other deep-water reef-forming corals and impacts from human activities. *Internationale Revue der Gesamten Hydrobiologie* 84, pp. 315–406; Fossa JH, Mortensen PB & Furevik DM (2002) The deep-water coral *Lophelia pertusa* in Norwegian waters: distribution and fishery impacts. *Hydrobiologia* 471, pp. 1–12; Hovland M, Vasshus S, Indreeide A, Austdal L & Nilsen Ø (2002) Mapping and imaging deep-sea coral reefs off Norway, 1982–2000. *Hydrobiologia* 471, pp. 13–17.
6 Gage JD & Tyler PA (1991) *Deep-Sea Biology*, Cambridge University Press, Cambridge; Rogers AD (1994) The biology of seamounts. *Advances in Marine Biology* 30, pp. 305–350.
7 Moseley HN (1879) *Notes by a naturalist: An account of observation made during the voyage of H.M.S. 'Challenger'*, Macmillan, London, p. 511 in republished edition Jast LS (ed.) (1944), Live Books resurrected.
8 Murray HW (1941) Submarine mountains in the Gulf of Alaska. *Bulletin of the Geological Society of America* 52, pp. 333–62.
9 Hess HH (1946) Drowned ancient islands of the Pacific basin. *American Journal of Science* 244, pp. 772–91.
10 Menard HW (1959) Geology of the Pacific sea floor. *Experimentia* 15, pp. 205–213; Menard HW & Ladd HS (1963) Oceanic islands, seamounts, guyots and atolls. In Hill MN (ed.) *The Sea*, vol. 3, Interscience, pp. 365–85.
11 Smith DK & Jordan TH (1988) Seamount statistics in the Pacific ocean. *Journal of Geophysical Research* 93, pp. 2899–919; Epp D & Smoot NC (1989) Distribution of seamounts in the North Atlantic. *Nature* 337, pp. 254–57; Smith DK (1991) Seamount abundances and size distributions, and their geographic variations. *Reviews in Aquatic Sciences* 5, pp. 197–210; Wessel P (1997) Sizes and ages of seamounts using remote sensing: implications for intraplate volcanism. *Science* 277, pp. 802–805.
12 Roden GI (1986) Aspects of oceanic flow and thermohaline structure in the vicinity of seamounts, in Uchida RN, Hayasi S & Boehlert GW (eds) *Environment*

and resources of Seamounts in the North Pacific. NOAA Technical Report NMFS 43, pp. 3–12; Haney JC, Haury LR, Mullineaux LS & Fey CL (1995) Sea-bird aggregation at a deep North Pacific seamount. *Marine Biology* 123, pp. 1–9; Gubbay S (2003) *Seamounts of the North-East Atlantic*. Report to OASIS & WWF, Germany, Frankfurt am Main.

13 For reviews of the influence of seamounts on ocean circulation, see Roden GI (1987) Effect of seamounts and seamount chains on ocean circulation and thermohaline structure, in Keating B, Fryer P, Batiza R & Boehlert G (eds.) *Seamounts, Islands, and Atolls*. American Geophysical Union, *Geophysical Monographs* 43, pp. 335–54. There is a growing empirical and theoretical literature on the subject. The following are a few examples: Genin A, Noble M & Lonsdale PF (1989) Tidal currents and anticyclonic motions on two North Pacific seamounts. *Deep-Sea Research* 36, pp. 1803–815; Brink, KH (1990) On the generation of seamount-trapped waves. *Deep-Sea Research* 37, pp. 1569–82; Eriksen CC (1991) Observations of amplified flows atop a large seamount. *Journal of Geophysical Research* 96, pp. 15 227–36; Haidvogel DB, Beckmann A, Chapman DC & Lin R-Q (1993) Numerical simulation of flow around a tall isolated seamount: Part II: Resonant generation of trapped waves. *Journal of Physical Oceanography* 23, pp. 2373–91.

14 Hubbs CL (1959) Initial discoveries of fish faunas on seamounts and offshore banks in the Eastern Pacific, *Pacific Science* 13, pp. 311–16.

15 Isaacs JD & Schwartzlose RA (1965) Migrant sound scatterers: interaction with the sea floor. *Science* 150, pp. 1810–13.

16 Wilson RR & Kaufmann RS (1987) Seamount biota and biogeography, in Keating BH, Fryer P, Batiza R & Boehlert GW, *Seamounts, Islands and Atolls*. American Geophysical Union, *Geophysical Monographs* 43, pp. 355–77.

17 Iles TD & Sinclair M (1982) Atlantic herring: stock discreteness and abundance. *Science* 215, pp. 627–33; Sinclair M (1988) *Marine Populations*, University of Washington Press, Seattle.

18 Ekman S (1967) *Zoogeography of the Sea*. Sidgwick & Jackson, London, Chapter 13; Gage & Tyler, *Deep-Sea Biology*, Chapter 10.

19 Reid JL (1981) On the mid-depth circulation of the world ocean. Warren BA & Wunsch C (eds.) *Evolution of Physical Oceanography*, MIT Press, Cambridge, Mass., pp. 70–111; Koslow JA, Bulman CM & Lyle JM (1994) The mid-slope community off southeastern Australia, *Deep Sea Research* 41, pp. 113–41.

20 Richer de Forges B, Koslow JA & Poore GCB (2000) Diversity and endemism of the benthic seamount fauna in the southwest Pacific. *Nature* 405, pp. 944–47. The coefficient of community, a statistic measuring the proportion of species in common between two seamounts, was on average 0.21 for any two seamounts along either the Norfolk Ridge or Lord Howe Rise, but only 0.04 for seamounts compared between these ridges. The ridges run parallel to each other in a north–south direction in the north Tasman Sea, separated by less than 10° of longitude, or about 1000 km.

21 Parker T & Tunnicliffe V (1994) Dispersal strategies of the biota on an oceanic seamount: implications for ecology and biogeography. *Biological Bulletin* 187, pp. 336–45.

22 Parin NV, Mironov AN & Nesis KN (1997) Biology of the Nazca and Sala y Gómez submarine ridges, an outpost of the Indo-West Pacific fauna in the Eastern Pacific Ocean: composition and distribution of the fauna, its communities and history. *Advances in Marine Biology* 32, pp. 145–242.

23 Samadi S, Bottan L, Macpherson E, Richer de Forges B & Boisselier M-C (2006) Seamount endemism questioned by the geographic distribution and population genetic structure of marine invertebrates. *Marine Biology* 149, pp. 1463–75.

24 Freiwald et al., *Cold-water coral reefs: out of sight – no longer out of mind*.

25 Freiwald et al., *Cold-water coral reefs: out of sight – no longer out of mind*.

26 Jensen A & Frederiksen R (1992) The fauna associated with the bank-forming deepwater coral *Lopehlia pertusa* (Scleractinaria) on the Faroe shelf. *Sarsia* 77, pp. 53–69.

27 Reed, Comparison of deep-water coral reefs and lithotherms.

28 Guinotte J, Orr J, Cairns S, Freiwald A, Morgan L, George R (2006) Will human-induced changes in seawater chemistry alter the distribution of deep-sea scleractinian corals? Frontiers in Ecology and the Environment 4, pp. 141–46.

29 Broecker WS (1974) *Chemical Oceanography*. Harcourt Brace Jovanovich, New York, Chapter 2.

30 Jensen & Frederiksen, The fauna associated with the bank-forming deepwater coral *Lophelia pertusa*; Freiwald et al., *Cold-water coral reefs: out of sight – no longer out of mind*, p. 35.

31 Bett BJ (2001) UK Atlantic Margin Environmental Survey: introduction and overview of bathyal benthic ecology. *Continental Shelf Research* 21, pp. 917–56; Duncan C & Roberts JM (2001) Darwin mounds: deep sea biodiversity 'hotspots'. *Marine Conservation* 5, pp. 12–13.

32 Champion A (2003) A life line for the Darwin mounds. *Marine Conservation* 6 (3), p. 15

33 Schroeder WW (2002) Observations of *Lophelia pertusa* and the surficial geology at a deep-water site in the northeastern Gulf of Mexico. *Hydrobiologia* 471, pp. 29–33; Freiwald et al., *Cold-water coral reefs: out of sight – no longer out of mind*.

34 Freese L, Auster PJ, Heifetz J & Wing BL (1999) Effects of trawling on seafloor habitat and associated invertebrate taxa in the Gulf of Alaska. *Marine Ecology Progress Series* 182, pp. 119–26; Krieger KJ & Wing BL (2002) Megafauna associations with deepwater corals (*Primnoa* spp.) in the Gulf of Alaska. *Hydrobiologia* 471, pp. 83–90; Fossa et al., The deep-water coral *Lophelia pertusa*.

35 Isaacs & Schwartzlose, Migrant sound scatterers; Genin A, Greene C, Haury L, Wiebe P, Gal G, Kaartvedt S, Meir E, Fey C & Dawson J (1994) Zooplankton patch

dynamics: daily gap formation over abrupt topography. *Deep-Sea Research* 41, pp. 941–51; Genin A, Haury L & Greenblatt P (1988) Interactions of migrating zooplankton with shallow topography: predation by rockfishes and intensification of patchiness. *Deep-Sea Research* 35, pp. 151–75; Tseitlin VB (1985) The energetics of fish populations inhabiting seamounts. *Oceanology* 25, pp. 237–39; Koslow JA (1997) Seamounts and the ecology of deep-sea fisheries. *American Scientist* 85, pp. 168–76.

36 Childress JJ & Nygaard M (1973) The chemical composition of midwater fishes as a function of depth of occurrence off Southern California. *Deep-Sea Research* 20, pp. 1093–1109; Smith KL Jr & Hessler RR (1974) Respiration of benthopelagic fishes: *In situ* measurements at 1230 meters. *Science* 184, pp. 72–73; Childress JJ (1975) The respiratory rates of midwater crustaceans as a function of depth of occurrence and relation to the oxygen minimum layer off Southern California. *Comparative Biochemistry and Physiology* 50, pp. 787–99; Smith KL Jr (1978) Metabolism of the abyssopelagic rattail *Coryphaenoides armatus* measured *in situ*. *Nature* 274, pp. 362–64; Torres JJ, Belman BW & Childress JJ (1979) Oxygen consumption rates of midwater fishes as a function of depth of occurrence. *Deep-Sea Research* 26A, pp. 185–97; Childress JJ, Taylor SM, Cailliet GM & Price MH (1980) Patterns of growth, energy utilization and reproduction in some meso- and bathypelagic fishes off Southern California. *Marine Biology* 61, pp. 27–40; Smith KL & Brown NO (1983) Oxygen consumption of pelagic juveniles and demersal adults of the deep sea fish *Sebastolobus altivelis*, measured at depth. *Marine Biology* 76, pp. 325–32.

37 Koslow JA (1996) Energetic and life-history patterns of deep-sea benthic, benthopelagic and seamount-associated fish. *Journal of Fish Biology* 49 Supplement A, pp. 54–74.

38 van den Broek WLF & Tracey DM (1981) Concentration and distribution of mercury in flesh of orange roughy (*Hoplostethus atlanticus*). *New Zealand Journal of Marine and Freshwater Research* 15, pp. 255–60; Vlieg P (1983) Transmission oil from fish? *Catch* 10, pp. 21–22; Elliott N, Bakes M & Nichols P (1994) Oil content of Australian and North Atlantic oreos and orange roughy. *Australian Fisheries* 53, pp. 30–32.

39 Rosecchi E, Tracey DM & Webber WR (1988) Diet of orange roughy, *Hoplostethus atlanticus* (Pisces: Trachichthyidae) on the Challenger Plateau, New Zealand. *Marine Biology* 99, pp. 293–306; Bulman CM & Koslow JA (1992) Diet and food consumption of a deep-sea fish, orange roughy *Hoplostethus atlanticus* (Pisces: Trachichthyidae), off southeastern Australia. *Marine Ecology Progress Series* 82, pp. 115–29; Merrett NR & Haedrich RL (1997) *Deep-sea Demersal Fish and Fisheries*, Chapman and Hall, London.

40 Koslow, Energetic and life-history patterns of deep-sea … fish.

41 Koslow JA, Bell J, Virtue P & Smith DC (1995) Fecundity and its variability in orange roughy: effects of population density, condition, egg size, and senescence. *Journal of Fish Biology* 47, pp. 1063–80.

42 Humphreys RL Jr, Winans GA & Tagami DT (1989) Synonymy and life history of the North Pacific pelagic armorhead, *Pseudopentaceros wheeleri* Hardy (Pisces: Pentacerotidae). *Copeia* 1989, pp. 142–53; Humphreys RL Jr (2000) Otolith-based assessment of recruitment variation in a North Pacific seamount population of armorhead *Pseudopentaceros wheeleri*. *Marine Ecology Progress Series* 204, pp. 213–23.

43 Childress et al., Patterns of growth, energy utilization and reproduction.

44 Bergstad OA (1990) Distribution, population structure, growth and reproduction of the roundnose grenadier *Coryphaenoides rupestris* (Pisces: Macrouridae) in the deep waters of the Skagerrak. *Marine Biology* 107, pp. 25–39; Bergstad OA (1995) Age determination of deep-water fishes: Experiences, status and challenges for the future, in Hopper AG (ed.) *Deep-water fisheries of the North Atlantic oceanic slope*, pp. 267–83; Andrews AH, Caillet GM & Coale KH (1999) Age and growth of the Pacific grenadier (*Coryphaeoides acrolepis*) with age estimate validation using an improved radiometric ageing technique. *Canadian Journal of Fisheries and Aquatic Sciences* 56, pp. 1339–50; Caillet GM et al. (2001) Age determination and validation studies of marine fishes: do deep-dwellers live longer? *Experimental Gerontology* 36, pp. 739–64.

45 Fenton GE, Short SA & Ritz DA (1991) Age determination of orange roughy, *Hoplostethus atlanticus* (Pisces: Trachichthyidae) using $^{210}Pb/^{226}Ra$ disequilibria. *Marine Biology* 109, pp. 197–202; Smith DC & Stewart BD (1994) Development of methods to age commercially important dories and oreos. Final Report, Fisheries Research & Development Corporation; Smith DC, Fenton GE, Robertson SG & Short SA (1995) Age determination and growth of orange roughy (*Hoplostethus atlanticus*): a comparison of annulus counts with radiometric ageing. *Canadian Journal of Fisheries and Aquatic Sciences* 52, pp. 391–401.

46 Barham EG (1966) Deep scattering layer migration and composition: observations from a diving saucer. *Science* 151, pp. 1399–403.

47 Koslow JA, Kloser R & Stanley CA (1995) Avoidance of a camera system by a deepwater fish, the orange roughy (*Hoplostethus atlanticus*). *Deep-Sea Research* 42, pp. 233–44.

48 Bulman CM & Koslow JA (1995) Development and depth distribution of the eggs of orange roughy, *Hoplostethus atlanticus* (Pisces: Trachichthyidae). *Marine and Freshwater Research* 46, pp. 697–705.

49 The orange roughy (family Trachichthyidae) and alfonsino (family Berycidae) belong to the order Beryciformes. The oreos (family Oreosomatidae) are in the order Zeiformes, and the pelagic armorhead (family Pentacerotidae) are in the most diverse order of modern teleost fishes, the Perciformes.

Chapter 7

1 Norse EA (ed.) (1993) *Global Marine Biological Diversity*, Island Press, Washington D.C., pp. 127–29 and references cited therein: Fowler CA (1987) *A review of seal and sea lion entanglement in marine fishing debris.* Paper presented at the North Pacific Rim Fishermen's Conference on Marine Debris, October, Kailua-Kona, Hawaii; Shomura RS & Yoshida HO (eds.)(1985) *Proceedings of the Workshop on the Fate and Impact of Marine Debris, 27–29 November 1984, Honolulu, Hawaii.* US Dept of Commerce. NOAA Technical Memorandum, NMFS, NOAA-TM-NMFS-SWFC-54; Shomura RS & Godfrey ML (eds.) (1990) *Proceedings of the Second International Conference on Marine Debris, 2–7 April 1989, Honolulu, Hawaii.* US Dept of Commerce. NOAA Technical Memorandum, NMFS, NOAA-TM-NMFS-SWFC-154.

2 Thiel H (2003) Anthropogenic impacts on the deep sea, in Tyler PA (ed.) *Ecosystems of the Deep Oceans.* Elsevier, Amsterdam, pp. 427–71.

3 Thiel H, Angel MV, Foell EJ, Rice AL & Shriever G (1998) *Marine science and technology – environmental risks from large-scale ecological research in the deep sea,* Luxembourg, European Commission.

4 Thiel et al., *Environmental risks from large-scale ecological research in the deep sea.*

5 Thiel et al., *Environmental risks from large-scale ecological research in the deep sea.*

6 A curie is the basic unit of radioactivity. Named after Marie Curie, it is equal to the radioactivity emitted by a gram of radium, or 37 billion disintegrations per second. The Three Mile Island nuclear accident released about 50 curies.

7 Broad WJ (1997) *The Universe Below*, Simon and Schuster, New York, Chapter 7.

8 Glover AG & Smith CR (2003) The deep-sea floor ecosystem: current status and prospects of anthropogenic change by the year 2025. *Environmental Conservation* 30, pp. 219–41.

9 International Atomic Energy Agency (1991) *Inventory of radioactive material entering the marine environment: sea disposal of radioactive waste*, IAEA, Vienna, as cited in Broad, *The Universe Below*; Reports of the Nuclear Energy Agency (NEA) cited and compiled in Thiel et al., *Environmental risks from large-scale ecological research in the deep sea*, especially Table 3.2.7; also Glover and Smith, The deep-sea floor ecosystem.

10 Davis WJ et al. (1982) *Evaluation of oceanic radioactive dumping programs,* Environmental Studies Institute, University of California, Santa Cruz, as cited in Broad, *The Universe Below*; Thiel, Anthropogenic impacts on the deep sea; Smith CR, Present TMC & Jumars PA (1988) *Development of benthic biological monitoring criteria for disposal of low-level radioactive waste in the abyssal deep sea.* Final report for EPA Contract 68-02-4303, Washington, DC, as cited in Glover and Smith, The deep-sea floor ecosystem.

11 Yablokov AV (2001) Radioactive Waste Disposal in Seas Adjacent to the Territory of the Russian Federation. *Marine Pollution Bulletin* 43, pp. 8–18.

12 Broad, *The Universe Below*, Chapter 7.

13 Hollister CD & Nadis S (1998) Burial of radioactive waste under the seabed. *Scientific American*, January, pp. 60–65.

14 Hollister & Nadis, Burial of radioactive waste under the seabed.

15 Champ MA, Gomez LS, Makeyev VM, Brooks JM, Palmer HD & Betz F (2001) Ocean storage of nuclear wastes? Experiences from the Russian Arctic. *Marine Pollution Bulletin* 43, pp. 1–7.

16 Thiel, *Anthropogenic impacts on the deep sea.*

17 International Maritime Organization (1991), *The London Dumping Convention. The First Decade and Beyond.* IMO, London, as cited in Thiel, Anthropogenic impacts on the deep sea.

18 van Dover CL, Grassle JF, Fry B, Garritt RH & Starczak VR (1992) Stable isotope evidence for entry of sewage-derived organic material into a deep-sea food web. *Nature* 360, pp. 153–56.

19 National Research Council (1975) Marine litter, in *Assessing potential ocean pollutants. A report of the study panel on assessing potential ocean pollutants to the Ocean Affairs Board.* Commission on Natural Resource, NRC, NAS, Washington, DC, pp. 405–433, as cited in Norse, EA (ed.) (1993) *Global Marine Biological Diversity*, Island Press, Washington, DC, p. 127.

20 Galil BS, Golik A & Türkay M (1995) Litter at the bottom of the sea: a sea bed survey in the eastern Mediterranean. *Marine Pollution Bulletin* 30, pp. 22–24.

21 Goldberg, ED (1986) TBT: an environmental dilemma. *Environment* 28, pp. 17–44, cited in Stewart C & Thompson JAJ (1994) Extensive butyltin contamination in southwestern coastal British Columbia, Canada. *Marine Pollution Bulletin* 28, pp. 601–606.

22 Compare Ten Hallers-Tjabbes CC, Kemp JF & Boon JP (1994) Imposex in whelks (*Buccinum undatum*) from the open North Sea: relation to shipping traffic intensities. *Marine Pollution Bulletin* 28, pp. 311–13 and Ide I, Witten EP, Fischer J, Kalbfus W, Zellner A, Stroben E & Watermann B (1997) Accumulation of organotin compounds in the common whelk *Buccinum undatum* and the red whelk *Neptunea antiqua* in association with imposex. *Marine Ecology Progress Series* 152, pp. 197–203.

23 Ten Hallers-Tjabbes et al., Imposex in whelks.

24 Stewart & Thompson, Extensive butyltin contamination.

25 Ten Hallers-Tjabbes et al., Imposex in whelks.

26 Takahashi S, Tanabe S & Kubodera T (1997) Butyltin residues in deep-sea organisms collected from Sugara Bay, Japan. *Environmental Science and Technology* 31, pp. 3103–109.

27 Stewart & Thompson, Extensive butyltin contamination.

28 Woodwell GM, Craig PP & Johnson HA (1971) DDT in the biosphere: where does it go? *Science* 174, pp. 1101–107.

29 MacGregor JS (1974) Changes in the amount and proportions of DDT and its metabolites DDE and DDD, in the marine environment off southern California, 1949–

72. *Fishery Bulletin* 72, pp. 275–93.
30 MacGregor, Changes in the amount and proportions of DDT and its metabolites.
31 Clark RB, Frid C & Attrill M (1997) *Marine Pollution*, Clarendon Press, Cambridge, UK, p. 91.
32 Clark et al., *Marine Pollution*, p. 89.
33 Price NM, Harrison PJ, Landry MR, Azam F & Hall KJF (1986) Toxic effects of latex and Tygon tubing on marine phytoplankton, zooplankton and bacteria. *Marine Ecology Progress Series* 34, pp. 41–49.
34 Harvey GR, Miklas HP, Bowen VT & Steinhauer WG (1974) Observations on the distribution of chlorinated hydrocarbons in Atlantic Ocean organisms. *Journal of Marine Research* 32, pp. 103–118.
35 Studies cited in Woodwell GM (1967) Toxic substances and ecological cycles. *Scientific American* 216 (3), pp. 24–31.
36 Osterberg C, Carey AG & Curl H (1963) Acceleration of sinking rates of radionuclides in the ocean. *Nature* 200, p. 1276.
37 Osterberg C (1962) Fallout radionuclides in euphausiids. *Science* 138, pp. 529–30.
38 Fowler SW, Buat-Menard P, Yokoyama Y, Ballastra S, Holm E & Huu Van Nguyen (1987) Rapid removal of Chernobyl fallout from Mediterranean surface waters by biological activity. *Nature* 329, pp. 56–58.
39 Higgo JJW, Cherry M, Heyraud RD & Fowler SW (1977) Rapid removal of plutonium from the oceanic surface layer by zooplankton fecal pellets. *Nature* 266, pp. 623–24.
40 Cherry RD & Heyraud M (1982) Evidence of high natural radiation doses in certain mid-water oceanic organisms. *Science* 218, pp. 54–56.
41 Cherry & Heyraud, Evidence of high natural radiation doses; also, Cherry RD & Heyraud M (1981) Polonium-210 content of marine shrimp: variation with biological and environmental factors. *Marine Biology* 65, pp. 165–75; Heyraud M & Cherry RD (1979) Polonium-210 and lead-210 in marine food chains. *Marine Biology* 52, pp. 227–36.
42 Heyraud M, Domanski P, Cherry RD & Fasham MJR (1988) Natural tracers in dietary studies: data for ^{210}Po and ^{210}Pb in decapod shrimp and other pelagic organisms in the Northeast Atlantic Ocean. *Marine Biology* 97, pp. 507–519.
43 Elder DL & Fowler SW (1977) Polychlorinated biphenyls: penetration into the deep ocean by zooplankton fecal pellet transport. *Science* 197, pp. 459–61.
44 Woodwell et al., DDT in the biosphere.
45 Tanabe S & Tatsukawa R (1983) Vertical transport and residence time of chlorinated hydrocarbons in the open ocean water column. *Journal of the Oceanographical Society of Japan* 39, pp. 53–62.
46 Harvey GR & Steinhauer WG (1974) Atmospheric transport of polychlorobiphenyls to the North Atlantic. *Atmospheric Environment,* cited in Harvey et al., Observations on the distribution of chlorinated hydrocarbons.
47 Harvey et al., Observations on the distribution of chlorinated hydrocarbons.
48 Gage JD & Tyler PA (1999) *Deep-Sea Biology*, Cambridge University Press, Cambridge, UK, p. 342.
49 Harvey GR, Steinhaver WG & Miklas HP (1974) Decline of PCB concentrations in North Atlantic surface water. *Nature* 252, pp. 387–88.
50 Tanabe & Tatsukawa, Vertical transport and residence time of chlorinated hydrocarbons.
51 Duke TW & Wilson AJ (1971) Chlorinated hydrocarbons in livers of fishes from the Northeastern Pacific Ocean. *Pesticide Monitoring Journal* 5, pp. 228–32, cited in MacGregor, Changes in the amount and proportions of DDT and its metabolites.
52 Baird RC, Thompson NP, Hopkins TL & Weiss WR (1975) Chlorinated hydrocarbons in mesopelagic fishes of the eastern Gulf of Mexico. *Bulletin of Marine Science* 25, pp. 473–81.
53 Harvey et al., Decline of PCB concentrations in North Atlantic surface water.
54 Haedrich RL & Henderson NR (1974) Pelagic food of *Coryphaenoides armatus*, a deep benthic rattail. *Deep-Sea Research* 21, pp. 739–44; Haedrich RL (1974) Pelagic capture of the epibenthic rattail *Coryphaenoides rupestris*. *Deep-Sea Research* 21, pp. 977–79.
55 Barber RT & Warlen SM (1979) Organochlorine insecticide residues in deep sea fish from 2500 m in the Atlantic Ocean. *Environmental Science and Technology* 13, pp. 1146–48; Arima S, Marchaud M & Martin JLM (1980) Pollutants in deep sea organisms and sediments. *Ambio Special Report* 6, pp. 97–100.
56 Mason RP, Fitzgerald WF & Morel FMM (1994) The biogeochemical cycling of elementary mercury: anthropogenic influences. *Geochima et Cosmochimica Acta* 58, pp. 3191–98; Cossa D, Martin J-M, Takayanagi K, Sanjuan J (1997) The distribution and cycling of mercury species in the western Mediterranean. *Deep-Sea Research* II 44, pp. 721–40.
57 Slemr F & Langer E (1992) Increase in global atmospheric concentrations of mercury inferred from measurements over the Atlantic Ocean. *Nature* 355, pp. 434–36.
58 Clark et al., *Marine Pollution*, pp. 64, 68.
59 Slemr & Langer, Increase in global atmospheric concentrations of mercury.
60 Compare Weiss HV, Koide M & Godberg ED (1971) Mercury in a Greenland ice sheet: evidence of recent input by man. *Science* 174, p. 692, with Mason RP, Fitzgerald WF & Morel FMM (1994) The biogeochemical cycling of elementary mercury: anthropogenic influences. *Geochima et Cosmochimica Acta* 58, pp. 3191–98. The time series of measurements over the ocean are found in: Slemr & Langer, Increase in global atmospheric concentrations of mercury.
61 Clark et al., *Marine Pollution*.
62 Slemr & Langer, Increase in global atmospheric concentrations of mercury.
63 Slemr F, Junkermann W, Schmidt RWH & Sladkovic R (1995) Indication of change in the global and regional trends of atmospheric mercury concentrations. *Geophysical Research Letters* 22, pp. 2143–46; Temme C,

Slemr F, Ebinghaus R & Einax JW (2003) Distribution of mercury over the Atlantic Ocean in 1996 and 1999–2001. *Atmospheric Environment* 37, pp. 1889–97.
64 Mason RP, Reinfelder JR & Morel FMM (1995) Bioaccumulation of mercury and methylmercury. *Water, Air, and Soil Pollution* 80, pp. 915–21.
65 Cossa, The distribution and cycling of mercury.
66 Mason RP, Rolfhus KR & Fitzgerald WF (1998) Mercury in the North Atlantic. *Marine Chemistry* 61, pp. 37–53.
67 Mason et al., Mercury in the North Atlantic.
68 Leatherland TM, Burton JDCF, McCartney MJ & Morris RJ (1973) Concentrations of some trace metals in pelagic organisms and of mercury in Northeast Atlantic Ocean water. *Deep-Sea Research* 20, pp. 679–85; Kress N, Hornung H & Herut B (1998) Concentrations of Hg, Cd, Cu, Zn, Fe and Mn in deep sea benthic fauna from the southeastern Mediterranean Sea: a comparison study between fauna collected at a pristine area and at two waste disposal sites. *Marine Pollution Bulletin* 36, pp. 911–21; Kobayashi R, Hirata E, Shiomi K, Yamanaka H & Kikuchi T (1979) Heavy metal contents in deep-sea fishes. *Bulletin of the Japanese Society of Scientific Fisheries* 45, pp. 493–97; Aston SR & Fowler SW (1985) Mercury in the open Mediterranean: evidence of contamination. *The Science of the Total Environment* 43, pp. 13–26.
69 Kobayashi et al., Heavy metal contents in deep-sea fishes; Arima S, Marchaud M & Martin JLM (1980) Pollutants in deep sea organisms and sediments. *Ambio Special Report* 6, pp. 97–100; Monteiro LR, Costa V, Furness RW & Santos RS (1996) Mercury concentrations in prey fish indicate enhanced bioaccumulation in mesopelagic environments. *Marine Ecology Progress Series* 141, pp. 21–25; Cronin M, Davies IM, Newton A, Pirie JM, Topping G & Swan S (1998) Trace metal concentrations in deep sea fish from the North Atlantic. *Marine Environmental Research* 45, pp. 225–38; Mormede S & Davies IM (2001) Heavy metal concentrations in commercial deep-sea fish from the Rockall Trough. *Continental Shelf Research* 21, pp. 899–916.
70 Cutshall NH, Naidu JR & Pearcy WG (1978) Mercury concentrations in Pacific hake, *Merluccius productus* (Ayres), as a function of length and latitude. *Science* 200, pp. 1489–91; van den Broek WLF & Tracey DM (1981) Concentration and distribution of mercury in flesh of orange roughy (*Hoplostethus atlanticus*). *New Zealand Journal of Marine and Freshwater Research* 15, pp. 255–60; Aston SR & Fowler SW (1985) Mercury in the open Mediterranean: evidence of contamination. *The Science of the Total Environment* 43, pp. 13–26.
71 Hornung H, Krom MD, Cohen Y & Bernhard M (1993) Trace metal content in deep-water sharks from the eastern Mediterranean Sea. *Marine Biology* 115, pp. 331–38.
72 Clark, *Marine Pollution*, p. 65.
73 Kania HJ & O'Hara J (1974) Behavioral alterations in a simple predator-prey system due to sub-lethal exposure to mercury. *Transactions of the American Fisheries Society* 103, pp. 134–36.
74 Cossa et al., The distribution and cycling of mercury.
75 Schober SE et al. (2003) Blood mercury levels in US children and women of childbearing age, 1999–2000. *Journal of the American Medical Association* 289, pp. 1667–74.
76 Martin JH & Broenkow WW (1975) Cadmium in plankton: elevated concentrations off Baja California. *Science* 190, pp. 884–85.
77 Leatherland et al., Concentrations of some trace metals in pelagic organisms; Ridout PS, Willcocks AD, Morris RJ, White SL & Rainbow PS (1985) Concentrations of Mn, Fe, Cu, Zn and Cd in the mesopelagic decapod *Systellaspis debilis* from the East Atlantic Ocean. *Marine Biology* 87, pp. 285–88; White SL & Rainbow PS (1987) Heavy metal concentrations and size effects in the mesopelagic decapod crustacean *Systellaspis debilis. Marine Ecology Progress Series* 37, pp. 147–51; Ridout PS, Rainbow PS, Roe HSJ & Jones HR (1989) Concentrations of V, Cr, Mn, Fe, Ni, Co, Cu, Zn, As and Cd in mesopelagic crustaceans from the North East Atlantic Ocean. *Marine Biology* 100, pp. 465–71.
78 Mormede & Davies, Heavy metal concentrations in commercial deep-sea fish.
79 Ballschmiter KH, Froescheis O, Jarman WM & Caillet G (1997) Contamination of the deep sea. *Marine Pollution Bulletin* 34, pp. 288–89.
80 Wurster Jr CF (1968) DDT reduces photosynthesis by marine phytoplankton. *Science* 159, pp. 1474–75; Menzel DW, Anderson J & Randtke A (1970) Marine phytoplankton vary in their response to chlorinated hydrocarbons. *Science* 167, pp. 1724–26; Mosser JL, Fisher NS & Wurster CF (1972) Polychlorinated biphenyls and DDT alter species composition in mixed cultures of algae. *Science* 176, pp. 533–35; O'Connors Jr HB, Wurster CF, Powers CD, Biggs DC & Rowland RG (1978) Polychlorinated biphenyls may alter marine trophic pathways by reducing phytoplankton size and production. *Science* 201, pp. 737–39.
81 Slemr & Langer, Increase in global atmospheric concentrations of mercury; Mason et al., The biogeochemical cycling of elementary mercury.
82 Brewer PG (1997) Ocean chemistry of the fossil fuel CO_2 signal: the haline signal of 'business as usual.' *Geophysical Research Letters* 24, pp. 1367–69; Intergovernmental Panel on Climate Change (2001) Climate change 2001: the scientific basis, Figure 3.1.
83 Sarmiento JL & Orr JC (1991) Three-dimensional simulations of the impact of Southern Ocean nutrient depletion on atmospheric CO_2 and ocean chemistry. *Limnology and Oceanography* 36, pp. 1928–50.
84 Parson EA & Keith DW (1998) Fossil fuels without CO_2 emissions. *Science* 282, pp. 1053–54.
85 Marchetti C (1977) On geoengineering and the CO_2 problem. *Climatic Change* 1, pp. 59–68.
86 Martin JH et al. (1994) Testing the iron hypothesis in ecosystems of the equatorial Pacific Ocean. *Nature* 371, pp. 123–29; Coale KH et al. (1996) A massive phytoplankton bloom induced by an ecosystem-scale

iron fertilization experiment in the equatorial Pacific Ocean. *Nature* 383, pp. 495–501; Boyd PW et al. (2000) A mesoscale phytoplankton bloom in the polar Southern Ocean stimulated by iron fertilization. *Nature* 407, pp. 695–701; Boyd PW et al. (2004) The decline and fate of an iron-induced subarctic phytoplankton bloom. *Nature* 428, pp. 549–53.
87 Martin JH & Fitzwater WE (1988) Iron deficiency limits phytoplankton growth in the north-east Pacific subarctic. *Nature* 331, pp. 341–43; Martin JH (1990) Glacial-interglacial CO_2 change: the iron hypothesis. *Paleoceanography* 5, pp. 1–13.
88 Schiemeier Q (2003) The oresmen. *Nature* 421, pp. 109–110; Buesseler KO & Boyd PW (2003) Will ocean fertilization work? *Science* 300, pp. 67–68.
89 Buesseler & Boyd, Will ocean fertilization work?; Buesseler KO, Andrews JE, Pike SM & Charette MA (2004) The effects of iron fertilization on carbon sequestration in the Southern Ocean. *Science* 304, pp. 414–20.
90 Sarmiento & Orr, Three-dimensional simulations of the impact of Southern Ocean nutrient depletion on atmospheric CO_2 and ocean chemistry; Chisholm SW, Falkowski PG & Cullen JJ (2001) Dis-crediting ocean fertilization. *Science* 294, pp. 309–310.
91 Peng T-H & Broecker WS (1991) Dynamic limitations on the Antarctic iron fertilization strategy. *Nature* 349, pp. 227–29; Peng T-H & Broecker WS (1991) Factors limiting the reduction of atmospheric CO_2 by iron fertilization. *Limnology and Oceanography* 36, pp. 1919–27; Sarmiento & Orr, Three-dimensional simulations of the impact of Southern Ocean nutrient depletion on atmospheric CO_2 and ocean chemistry.
92 On the other hand, see the proposal, based on the increased density of CO_2-laden seawater, by Haugan PM & Drange H (1992) *Nature* 357, pp. 318–20.
93 Brewer PG, Friederich G, Peltzer ET & Orr Jr. FM (1999) Direct experiments on the ocean disposal of fossil fuel CO_2. *Science* 284, pp. 943–45.
94 Seibel BA & Walsh PJ (2001) Potential impacts of CO_2 injection on deep-sea biota. *Science* 294, pp. 319–20.
95 Tamburri MN, Peltzer ET, Friederich GE, Aya I, Yamane K & Brewer PG (2000) A field study of the effects of CO_2 ocean disposal on mobile deep-sea animals. *Marine Chemistry* 72, pp. 95–101.
96 Vetter EW & Smith CR (2005) Insights into the ecological effects of deep ocean CO_2 enrichment: the impacts of natural CO_2 venting at Loihi seamount on deep sea scavengers. *Journal of Geophysical Research C* 110 (C9), pp. 1–10.
97 Drange H, Alendal G & Johannessen OM (2002) Ocean sequestration of CO_2. *Ocean Challenge* 12, pp. 33–39.
98 Parson & Keith, Fossil fuels without CO_2 emissions.

Chapter 8

1 Murray J & Renard A (1891) Deep-sea deposits. Report on the Scientific Results of the Voyage of H.M.S. *Challenger*. Geology 1, pp. 1–583; Thiel H (2003) Anthropogenic impacts on the deep sea, in Tyler PA (ed.) *Ecosystems of the Deep Oceans*, Elsevier, Amsterdam, pp. 427–71.
2 Broecker WS (1974) *Chemical Oceanography*, Harcourt, Brace, Jovanovich, New York, pp. 89–113.
3 Murray J & Renard A (1891) Manganese nodules, Report on the Scientific Results of the Voyage of H.M.S. *Challenger*, 5, pp. 341–78.
4 Mullineaux LS (1987) Organisms living on manganese nodules and crusts: distribution and abundance at three North Pacific sites. *Deep-Sea Research* 34, pp. 165–84.
5 Thiel H, Schriever G, Bussau C & Borowski C (1993) Manganese nodule crevice fauna. *Deep-Sea Research* I, 40, pp. 419–23.
6 Mullineaux, Organisms living on manganese nodules and crusts, reviews these hypotheses and associated studies.
7 Mero JL (1965) *The Mineral Resources of the Sea*, Elsevier, Amsterdam.
8 Mero JL (1977) in Glasby GP (ed.) *Manganese nodule deposits*, Elsevier, Amsterdam, pp. 327–55, cited in Glasby GP (2000) Lessons learned from deep-sea mining. *Science* 289, pp. 551–53.
9 Morgan CL (2000) Resource estimates of the Clarion-Clipperton manganese nodule deposits, in Cronan DS (ed.) *Handbook of Marine Mineral Deposits.* CRC Press, Boca Raton, pp. 145–70.
10 Pardo A (1967) UN Doc A/C.1/PV.1515, Nov 1, 1967 at 6, as cited in Anand RP (1993) Changing concepts of freedom of the seas: a historical perspective, in van Dyke JM, Zaelke D, Hewison G (eds.) *Freedom for the seas in the 21st century*, Island Press, Washington DC, pp. 72–86.
11 Rona PA (2003) Resources of the sea floor. *Science* 299, pp. 673–74.
12 Borowski C & Thiel H (1998) Deep-sea macrofaunal impacts of a large-scale physical disturbance experiment in the Southeast Pacific. *Deep-Sea Research* II 45, pp. 55–82.
13 Thiel H (2003) Anthropogenic impacts on the deep sea, in Tyler PA (ed.) *Ecosystems of the Deep Oceans*, Elsevier, Amsterdam, pp. 427–71.
14 Glover AG & Smith CR (2003) The deep-sea floor ecosystem: current status and prospects of anthropogenic change by the year 2025. *Environmental Conservation* 30, pp. 219–41; Smith CR, Levin LA, Koslow JA, Tyler PA & Glover AG (2003) The near future of the deep-sea floor ecosystems. Paper presented at the 5th International Conference on Environmental Future, Zurich, Switzerland.
15 Hein J (2002) Cobalt-rich ferromanganese crusts: global distribution, composition, origin and research activities, in *Polymetallic massive sulphides and cobalt-rich ferromanganese crusts: status and prospects.* ISA Technical Study 2, pp. 36–89.
16 Rona, Resources of the sea floor.
17 Rona, Resources of the sea floor; Hein, Cobalt-rich ferromanganese crusts; Glasby, Lessons learned from deep-sea mining.
18 Hein, Cobalt-rich ferromanganese crusts.
19 Herzig PM & Petersen S (2002) Polymetallic massive sulphide deposits at the modern seafloor and their

resource potential, in *Polymetallic massive sulphides and cobalt-rich ferromanganese crust: status and prospects.* ISA Technical Study 2, pp. 7–35.
20 Herzig & Petersen, Polymetallic massive sulphide deposits.
21 Herzig & Petersen, Polymetallic massive sulphide deposits; Rona, Resources of the sea floor; Thiel, Anthropogenic impacts on the deep sea; Glasby, Lessons learned from deep-sea mining.
22 *Encyclopedia Brittanica* (1983). Macropedia, Vol. 14, Petroleum, p. 173.
23 Smith et al., The near future of the deep-sea floor ecosystems.
24 Olsgard F & Gray JS (1995) A comprehensive analysis of the effects of offshore oil and gas exploration and production on the benthic communities of the Norwegian continental shelf. *Marine Ecology Progress Series* 122, pp. 277–306.
25 GESAMP (1993) Impact of oil and related chemicals and wastes on the marine environment. *GESAMP Reports and Studies* 50, pp. 1–180.
26 Olsgard & Gray, A comprehensive analysis of the effects of offshore oil and gas exploration and production.
27 Bett BJ (2001) UK Atlantic Margin Environmental Survey: introduction and overview of bathyal benthic ecology. *Continental Shelf Research* 21, pp. 917–56; Smith et al., The near future of the deep-sea floor ecosystems.
28 Kvenvolden K (1993) Gas hydrates – geological perspective and global change. *Reviews of Geophysics* 31, pp. 173–88.
29 US Geological Survey (1992): http://marine.usgs.gov/fact-sheets/gas-hydrates/title.html; DeLong EF (2000) Microbiology: resolving a methane mystery. *Nature* 407, pp. 577–79; Adam D (2002) Methane hydrates: fire from ice. *Nature* 418, pp. 913–14.
30 Kvenvolden, Gas hydrates.
31 Dickens GR, O'Neil RR, Rea DK & Owen RM (1995) *Paleoceanography* 10, pp. 965–71, cited in Adam, Methane hydrates: fire from ice; see also, Chapter 9.

Chapter 9

1 Revelle R & Seuss H (1957) Carbon dioxide exchange between atmosphere and ocean and the question of an increase of atmospheric CO_2 during the past decades, *Tellus* 8, pp. 140–54.
2 The following background history of the CO_2 issue is from Revelle R (1985) Introduction: the scientific history of carbon dioxide, in Sundquist ET & Broecker WS (eds.) *The carbon cycle and atmospheric CO_2: natural variations Archean to Present.* American Geophysical Union, Washington DC, pp. 1–4. See also Revelle & Suess, Carbon dioxide exchange between atmosphere and ocean; Revelle R (1982) Carbon dioxide and world climate. *Scientific American* 247, pp. 335–43.
3 Arrhenius S (1896) On the influence of carbonic acid in the air upon the temperature on the ground. *The London, Edinburgh, and Dublin Philosophical Magazine and Journal of Science* 41, pp. 237–79.
4 Karl TR & Trenberth KE (2003) Modern global climate change. *Science* 302, pp. 1719–23.
5 Chamberlin TC (1898) An attempt to frame a working hypothesis of the cause of glacial periods on an atmospheric basis. *Journal of Geology* 7, pp. 545–84.
6 Walker JCG, Hays PB & Kasting JF (1981) A negative feedback mechanism for the long-term stabilization of Earth's surface temperature. *Journal of Geophysical Research* 86, pp. 9776–82; Broecker WS & Sanyal A (1998) Does atmospheric CO_2 police the rate of chemical weathering? *Global Biogeochemical Cycles* 12, pp. 403–408.
7 Kasting JF, Toon OB & Pollack JB (1988) How climate evolved on the terrestrial planets. *Scientific American* 258, pp. 90–97.
8 Owen T, Cess RD & Ramanathan V (1979) Enhanced CO_2 greenhouse to compensate for reduced solar luminosity on early earth. *Nature* 277, pp. 640–41.
9 Revelle, Carbon dioxide and world climate.
10 Intergovernmental Panel on Climate Change (IPCC) (2001) *Climate Change 2001: The Scientific Basis*, Cambridge University Press, Cambridge, Fig. 3.1, pp. 197 ff. Available from http://www.ipcc.ch. Total CO_2 emissions include emissions from burning fossil fuels (5.4 Gt) and from changes in land use, especially deforestation (1.9 Gt).
11 Revelle & Seuss, Carbon dioxide exchange between atmosphere and ocean.
12 Intergovernmental Panel on Climate Change (IPCC), *Climate Change 2001.*
13 Bubbles trapped in glacial ice provide samples of the atmosphere from the time when the ice was formed. The longest ice-core records from the Antarctic now provide a record of climate and atmospheric composition extending back 420 000 years: Petit JR et al. (1999) Climate and atmospheric history of the past 420 000 years from the Vostok ice core, Antarctica. *Nature* 399, pp. 429–36.
14 IPCC, *Climate Change 2001*; Manabe S & Stouffer RJ (1993) Century-scale effects of increased atmospheric CO_2 on the ocean-atmosphere system. *Nature* 364, pp. 215–18; Stocker TF & Schmittner A (1997) Influence of CO_2 emission rates on the stability of the thermohaline circulation. *Nature* 388, pp. 862–65.
15 Frakes LA (1986) Mesozoic–Cenozoic climatic history and causes of the glaciation, in Hsu KJ (ed.) *Mesozoic and Cenozoic Oceans*, American Geophysical Union, Washington DC, pp. 33–48.
16 Hays JD, Imbrie J & Shackleton NJ (1976) Variations in the earth's orbit: pacemaker of the Ice Ages. *Science* 194, pp. 1121–32.
17 Zachos JC, Flower BP & Paul H (1997) Orbitally paced climate oscillations across the Oligocene/Miocene boundary. *Nature* 388, pp. 567–70.
18 Stable isotopes refer to non-radioactive isotopes. Oxygen occurs naturally in two forms. The more common, ^{16}O, has 8 neutrons and 8 protons (and hence an atomic weight of 16); the less common, ^{18}O, has two additional

neutrons. The heavier isotope, ^{18}O, is preferentially precipitated into the shells, and the lighter one respired, but this effect decreases with increasing temperature. Thus the deviations from a particular ratio (commonly written as $\delta^{18}O$), provide a proxy record of temperature. This was first applied to paleoceanography in Emiliani C (1955) Pleistocene temperatures, *Journal of Geology* 63, pp. 538–78. Oxygen isotopes are also fractionated during the hydrological cycle, water with the lighter oxygen isotope being preferentially evaporated, and therefore accumulating in glaciers. The ocean therefore becomes increasingly enriched in heavier water as the ice sheets build, so during glacial periods, the oxygen isotope ratio provides a mixed signal of global ice volume and ocean temperature. For a more general review of paleoceanographic proxies, see Berger WH (1981) Paleoceanography: the deep-sea record, in Emiliani C (ed.) *The Sea*, vol. 7, John Wiley & Sons, New York, pp. 1437–520.

19 Dansgaard W et al. (1993) Evidence for general instability of past climate from a 250-kyr ice-core record. *Nature* 364, pp. 218–20; Petit et al. Climate and atmospheric history of the past 420 000 years from the Vostok ice core.

20 Dansgaard et al., Evidence for general instability of past climate; Alley RB et al. (1993) Abrupt increase in Greenland snow accumulation at the end of the Younger Dryas event. *Nature* 362, p. 527; Broecker WS (1997) Thermohaline circulation, the Achilles heel of our climate system: will man-made CO_2 upset the current balance? *Science* 278, pp. 1582–88.

21 Dansgaard et al. Evidence for general instability of past climate; for a review, see Clark PU, Pisias NG, Stocker TF & Weaver AJ (2002) The role of the thermohaline circulation in abrupt climate change. *Nature* 415, pp. 863–69.

22 For a review, see Clark et al., The role of the thermohaline circulation.

23 Zachos JC, Pagani M, Sloan IC, Chomas E & Billups K (2001) Trends, rhythms, and aberrations in global climate 65 Ma to present. *Science* 292, pp. 686–93.

24 Manabe S & Stouffer RJ (1993) Century-scale effects of increased atmospheric CO_2 on the ocean-atmosphere system. *Nature* 364, pp. 215–18; Stocker & Schmittner, Influence of CO_2 emission rates; Broecker WS, Sutherland S & Peng T-H (1999) A possible 20th-century slowdown of Southern Ocean deep water formation. *Science* 286, pp. 1132–35.

25 Levitus S, Antonov JI, Boyer TP & Stephens C (2000) Warming of the world ocean. *Science* 287, pp. 2225–29; Levitus S, Antonov JI, Wang J, Delworth TL, Dixon KW & Broccoli AJ (2001) Anthropogenic warming of the Earth's climate system. *Science* 292, pp. 267–70.

26 Dickson B, Yashayaev I, Meicke J, Turrell B, Dye S & Holfort J (2002) Rapid freshening of the deep North Atlantic Ocean over the past four decades. *Nature* 416, pp. 832–37; Curry R, Dickson B & Yashayaev I (2003) A change in the freshwater balance of the Atlantic Ocean over the past four decades. *Nature* 426, pp. 826–29; Wong APS, Bindoff NL & Church JA (1999) Large-scale freshening of intermediate waters in the Pacific and Indian oceans. *Nature* 400, pp. 440–43.

27 Broecker et al., A possible 20th-century slowdown of Southern Ocean deep water formation.

28 Sarmiento JL, Hughes TMC, Stouffer RJ & Manabe S (1998) Simulated response of the ocean carbon cycle to anthropogenic climate warming. *Nature* 393, pp. 245–49; Boyd PW & Doney SC (2002) Modelling regional responses by marine pelagic ecosystems to global climate change. *Geophysical Research Letters* 29: 10.1029/2001GL014130.

29 Smith KL Jr & Druffel ERM (1998) Long time-series monitoring of an abyssal site in the NE Pacific: an introduction. *Deep-Sea Research II* 45, pp. 573–86.

30 Smith KL Jr & Kaufmann RS (1999) Long-term discrepancy between food supply and demand in the deep eastern North Pacific. *Science* 284, pp. 1174–77.

31 Roemmich D & McGowan J (1995) Climatic warming and the decline of zooplankton in the California Current. *Science* 267, pp. 1324–26.

32 Ruhl HA & Smith KL Jr (2004) Shifts in deep-sea community structure linked to climate and food supply. *Science* 305, pp. 513–15.

33 Billett DSM, Bett BJ, Rice AL, Thurston MH, Galeron J, Sibuet M & Wolff GA (2002) Long-term change in the megabenthos of the Porcupine Abyssal Plain (NE Atlantic). *Progress in Oceanography* 50, pp. 325–48.

34 Behl RJ & Kennett JP (1996) Brief interstadial events in the Santa Barbara Basin, NE Pacific, during the past 60 kyr. *Nature* 379, pp. 243–46; Cannariato KG, Kennett JP & Behl RJ (1999) Biotic response to late Quaternary rapid climate switches in Santa Barbara Basin: ecological and evolutionary implications. *Geology* 27, pp. 63–66.

35 There is a rapidly growing body of evidence that examines changes in North Atlantic deep-sea benthic communities on various glacial time scales, all linked to changes in deepwater ocean circulation. Some of the key studies are: Streeter SS (1973) Bottom water and benthonic foraminifera in the North Atlantic: glacial-interglacial contrasts. *Quaternary Research* 3, pp. 131–41; Streeter SS & Shackleton NJ (1979) Paleocirculation of the deep North Atlantic: 150 000 year record of benthic foraminfera and oxygen-18. *Science* 203, pp. 168–170; Caralp M-H (1987) Deep-sea circulation in the northeastern Atlantic over the past 30 000 years: the benthic foraminiferal record. *Oceanologica Acta* 10, pp. 27–40; Cronin TM, DeMartino DM, Dwyer GS & Rodriguez-Lazaro J (1999) Deep-sea ostracode species diversity: response to late Quaternary climate change. *Marine Micropaleontology* 37, pp. 231–49; Cronin TM & Raymo ME (1997) Orbital forcing of deep-sea benthic species diversity. *Nature* 385, pp. 624–27.

36 Cronin & Raymo, Orbital forcing of deep-sea benthic species diversity; Cannariato et al., Biotic response to late Quaternary rapid climate switches.

37 Roy K, Valentine JW, Jablonski D & Kidwell SM (1996) Scales of climatic variability and time averaging in Pleistocene biotas: implications for ecology and evolution. *Trends in Ecology and Evolution* 11, pp. 458–63.

38 Reviewed in Flower BP & Kennett JP (1994) The middle Miocene climatic transition: east Antarctic ice sheet development, deep ocean circulation and global carbon cycling. *Palaeogeography, palaeoclimatology, palaeoecology* 108, pp. 537–55.
39 Wright JD, Miller KG & Fairbanks RG (1992) Early and middle Miocene stable isotopes: Implications for deepwater circulation and climate. *Paleoceanography* 7, pp. 357–89.
40 Woodruff F (1985) Changes in Miocene deep-sea benthic foraminiferal distribution in the Pacific Ocean: relationship to paleoceanography. In: Kennett JP (ed.) *The Miocene Ocean: Paleoceanography and Biogeography.* Geological Society of America, Boulder, pp. 131–76.
41 Stanley SM (1989) *Earth and Life through Time*, 2nd edition, WH Freeman and Company, New York, pp. 604–607.
42 Vincent E & Berger WH (1985) Carbon dioxide and polar cooling in the Miocene: the Monterey hypothesis, in Sundquist ET & Broecker WS (eds.) *The carbon cycle and atmospheric* CO_2*: natural variations Archaean to present.* American Geophysical Union, Washington DC, pp. 455–68.
43 Woodruff, Changes in Miocene deep-sea benthic foraminiferal distribution.
44 Kennett JP & Stott LD (1991) Abrupt deep-sea warming, palaeoceanographic changes and benthic extinctions at the end of the Palaeocene. *Nature* 353, pp. 225–29.
45 Dickens GR, O'Neil JR, Rea DK & Owen RM (1995) Dissociation of oceanic methane hydrate as a cause of the carbon isotope excursion at the end of the Paleocene. *Paleoceanography* 10, pp. 965–72.
46 Bice KL & Marotzke J (2002) Could changing ocean circulation have destabilized methane hydrate at the Paleocene/Eocene boundary? *Paleoceanography* 17, p. 1018.
47 Dickens, Dissociation of oceanic methane hydrate; Dickens GR, Castillo MM, Walker JCG (1997) A blast of gas in the latest Paleocene: Simulating first-order effects of massive dissociation of oceanic methane hydrate. *Geology* 25, pp. 259–62; Norris RD & Röhl U (1999) Carbon cycling and chronology of climate warming during the Palaeocene/Eocene transition. *Nature* 401, pp. 775–78. Bains S, Corfield RM & Norris RD (1999) Mechanisms of climate warming at the end of the Paleocene. *Science* 285, pp. 724–27.
48 Katz ME, Pak DK, Dickens GR & Miller KG (1999) The source and fate of massive carbon input during the latest Paleocene thermal maximum. *Science* 286, pp. 1531–33.
49 Kvenvolden KA (1999) Potential effects of gas hydrate on human welfare. *Proceedings of the National Academy of Science, USA* 96, pp. 3420–26.
50 Dickens et al., Dissociation of oceanic methane hydrate.
51 Kennett JP, Cannariato KG, Hendy IL & Behl RJ (2000) Carbon isotopic evidence for methane hydrate instability during Quaternary interstadials. *Science* 288, pp. 128–33.
52 MacDonald IR, Guinasso NL, Sassen R, Brooks JM, Lee L & Scott KT (1994) Gas hydrate that breaches the sea floor on the continental slope of the Gulf of Mexico. *Geology* 22, pp. 699–702.
53 MacDonald G (1990) Role of methane clathrates in past and future climates. *Climatic Change* 16, pp. 247–81; Nisbet, EG (1990) The end of the ice age. *Canadian Journal of Earth Science* 27, pp. 148–57; Paull CK, Ussler W & Dillon WP (1991) Is the extent of glaciation limited by marine gas-hydrates? *Geophysical Research Letters* 18, pp. 432–34.
54 Kvenvolden KA (1993) Gas hydrates – geological perspective and global change. *Reviews of Geophysics* 31, pp. 173–88; MacDonald, Role of methane clathrates.
55 Judd AG, Hovland M, Dimitrov LI, García Gil S & Jukes V (2002) The geological methane budget at continental margins and its influence on climate change. *Geofluids* 2, p. 109.
56 Dickens GR (2001) The potential volume of oceanic methane hydrates with variable external conditions. *Organic geochemistry* 32, pp. 1179–93.
57 Hesselbo SP, Grocke DR, Jenkyns HC, Bjerrum CJ, Farrimond P, Morgans Bell HS & Green OR (2000) Massive dissociation of gas hydrate during a Jurassic oceanic anoxic event. *Nature* 406, pp. 392–95; Padden M, Weissert H & deRafelis M (2001) Evidence for late Jurassic release of methane from gas hydrate. *Geology* 29, pp. 223–26; Jahren AH, Arens NC, Sarmiento G, Guerrero J & Amundson R (2001) Terrestrial record of methane hydrate dissociation in the early Cretaceous. *Geology* 29, pp. 159–62.
58 Dickens GR (1999) The blast in the past. *Nature* 401, pp. 752–55.
59 Norris & Röhl, Carbon cycling and chronology of climate warming.
60 Pearson PN & Palmer MR (2000) Atmospheric carbon dioxide concentrations over the past 60 million years. *Nature* 406, pp. 695–99.
61 Manabe S & Stouffer RJ (1993) Century-scale effects of increased atmospheric CO_2 on the ocean-atmosphere system. *Nature* 364, pp. 215–18; Ganopolski A & Rahmstorf S (2001) Rapid changes of glacial climate simulated in a coupled climate model. *Nature* 409, pp. 153–58; Clark PU, Pisias NG, Stocker TF & Weaver AJ (2002) The role of the thermohaline circulation in abrupt climate change. *Nature* 415, pp. 863–69; Alley RB et al. (2003) Abrupt climate change. *Science* 299, pp. 2005–2010.
62 Manabe & Stouffer, Century-scale effects of increased atmospheric CO_2; Stocker & Schmittner, Influence of CO_2 emission rates.
63 For a review, see Broecker, Thermohaline circulation.
64 Sarmiento et al., Simulated response of the ocean carbon cycle; Boyd & Doney, Modelling regional responses by marine pelagic ecosystems.
65 Manabe & Stouffer RJ, Century-scale effects of increased atmospheric CO_2.
66 Caldeira K & Wickett ME (2003) Anthropogenic carbon and ocean pH. *Nature* 425, pp. 365–67; Orr JC et al. (2005) Anthropogenic ocean acidification over the twenty-first century and its impact on calcifying organisms. *Nature* 437, pp. 681–86.
67 Feely et al. (2004) Impact of anthropogenic CO_2 on the $CaCO_3$ system in the oceans, *Science* 305, pp. 362–66.

Chapter 10

1 Vinogradov ME (1968) *Vertical distribution of the oceanic zooplankton*, Academy of Sciences of the USSR, Institute of Oceanography; Rowe GT (1983) Biomass and production of the deep-sea macrobenthos. In Rowe GT (ed.) *The Sea*, vol 8, *Deep-Sea Biology*, Wiley-Interscience, New York.

2 Haedrich RL, Rowe GT & Polloni PT (1980) The megabenthic fauna in the deep sea south of New England, USA. *Marine Biology* 57, pp. 165–79; Rowe (1983) Biomass and production of the deep-sea macrobenthos.

3 Examples of leading, otherwise excellent, comprehensive treatments of deep-sea fishes and deep-sea life that altogether disregarded the seamount fauna include: Marshall NB (1979) *Developments in Deep-Sea Biology*, Blandford Press, Poole; and Rowe (1983) Biomass and production of the deep-sea macrobenthos. Similarly there is no mention of seamount fisheries in the comprehensive review of world fisheries by the leading fishery scientist and head of the UN Food and Agriculture Organization (FAO), Gulland JA (ed.) (1971) *The Fish Resources of the Ocean*, Fishing News (Books), Surrey, England.

4 Gudger E (1927) Wooden hooks used for catching sharks and *Ruvettus* in the South Seas: a study of their variation and distribution. *Anthropological Papers of the American Museum of Natural History* 28, pp. 199–355, as described in Merrett NR & Haedrich RL (1997) *Deep-Sea Demersal Fish and Fisheries*, Chapman and Hall, London, pp. 204–207.

5 The fishery is described, along with mention of the Portuguese poem by Manuel Tomas, in Merrett & Haedrich (1997) *Deep-Sea Demersal Fish and Fisheries*, pp. 174 ff.

6 Bowering WR & Brodie WB (1995) Greenland halibut (*Reinhardtius hippoglossoides*), a review of the dynamics of its distribution and fisheries off eastern Canada and Greenland. In Hopper AG (ed.) *Deep-Water Fisheries of the North Atlantic Oceanic Slope*. Kluwer Academic Publishers, Dordrecht, Netherlands, pp. 113–60.

7 Bowering W & Nedreaas KH (2000) A comparison of Greenland halibut (*Reinhardtius hippoglossoides* (Walbaum)) fisheries and distribution in the Northwest and Northeast Atlantic. *Sarsia* 85, pp. 61–76.

8 Gordon JDM, Berstad OA, Figueiredo I & Menezes G (2003) Deep-water Fisheries of the Northeast Atlantic: I. Description and Current Trends. *Journal of Northwest Atlantic Fishery Science* 31, pp. 137–50; Large PA, Hammer C, Bergstad OA, Gordon JDM & Lorance P (2003) Deep-water fisheries of the Northeast Atlantic: II. Assessment and management approaches. *Journal of Northwest Atlantic Fishery Science* 31, pp. 151–63.

9 FAO Fishery Yearbooks and website: http://www.fao.org.

10 Hardin G (1968) The tragedy of the commons. *Science* 162, pp. 1243–48.

11 Deimling EA & Liss WJ (1994) Fishery development in the eastern North Pacific: a natural-cultural system perspective, 1888–1976, *Fisheries Oceanography* 3, pp. 60–77.

12 FAO Fishery Yearbooks and website: http://www.fao.org; Gulland (ed.) (1971) *The Fish Resources of the Ocean*.

13 Humphreys RL Jr, Winans GA & Tagami DT (1989) Synonymy and life history of the North Pacific pelagic armorhead, *Pseudopentaceros wheeleri* Hardy (Pisces: Pentacerotidae), *Copeia* 1989, pp. 142–53.

14 McEvoy AF (1979) Economy, law, and ecology in the California fisheries to 1925. PhD dissertation, University of California, San Diego.

15 Sasaki T (1986) Development and present status of Japanese trawl fisheries in the vicinity of seamounts. In Uchida RN, Hayasi S & Boehlert GW (eds) *Environment and Resources of Seamounts in the North Pacific*. NOAA Technical Report NMFS 43, pp. 21–30.

16 Borets LA (1975) Some results of studies on the biology of the boarfish (*Pentaceros richardsoni* Smith), *Investigations of the Biology of Fishes and Fishery Oceanography*, TINRO, Vladivostok, pp. 82–90.

17 Uchida et al. (1986) *Environment and Resources of Seamounts in the North Pacific*.

18 Wilson RR & Kaufmann RS (1987) Seamount biota and biogeography. In Keating BH, Fryer P, Batiza R & Boehlert GW (eds.) *Seamounts, Islands and Atolls*, American Geophysical Union, Geophysical Monographs 43, pp. 355–77.

19 Grigg RW (1993) Precious coral fisheries of Hawaii and the U.S. Pacific Islands. *Marine Fisheries Review* 55(2), pp. 50–60.

20 Genin A, Dayton PK, Lonsdale PF & Spiess FN (1986) Corals on seamounts provide evidence of current acceleration over deep sea topography. *Nature* 322, pp. 59–61.

21 Grigg (1993) Precious coral fisheries of Hawaii and the U.S.; Koslow JA, Boehlert GW, Gordon JDM, Haedrich RL, Lorance P & Parin N (2000) The impact of fishing on continental slope and deep-sea ecosystems. *ICES Journal of Marine Science* 57, pp. 548–57; Grigg RW (2002) Precious corals in Hawaii: discovery of a new bed and revised management measures for existing beds. *Marine Fisheries Review* 64, pp. 13–20.

22 Grigg (1993) Precious coral fisheries of Hawaii and the U.S.

23 Druffel ERM, King LL, Belastock RA & Buesseler KO (1990) Growth rate of a deep-sea coral using ^{210}Pb and other isotopes. *Geochima et Cosmochimica Acta* 54, pp. 1493–500; Druffel ERM, Griffin S, Witter A, Nelson E, Southon J, Kashgarian M & Vogel J (1995) *Gerardia*: bristlecone pine of the deep-sea? *Geochima et Cosmochimica Acta* 59, pp. 5031–36; Tracey D, Neil H, Gordon D & O'Shea S (2003) Chronicles of the deep: ageing deep-sea corals in New Zealand waters. *Water and Atmosphere* 11, pp. 22–24; Thresher R, Rintoul SR, Koslow JA, Weidman C, Adkins J & Proctor C (2004) Oceanic evidence of climate change in southern Australia over the last three centuries. *Geophysical Research Letters* 31:L07212, doi:10. 1029/2003GL018869.

24 In acoustic surveys, the biomass of the species in the survey area is estimated, based on the relative proportion of species in the area (usually determined from net sampling) and the species' acoustic reflectance (otherwise known as their *target strength*). The effectiveness of the method lies in the precision of the acoustic data. Its weaknesses are the imprecision of estimates of the species mix, which is biased by species avoiding the net or slipping through its meshes. Egg surveys use plankton sampling to estimate the total number of eggs spawned by the population. The number of female fish is estimated by dividing the total number of eggs by the average *fecundity* (the number of eggs produced per female). The strength of this method lies in the ease of sampling plankton and estimating fecundity (i.e. counting the eggs in female fish before spawning). Its major drawback is its imprecision, due to the patchy distribution of the eggs.

25 Kotylar AN (1980) Age and growth speed of the big heads *Hoplostethus atlanticus* Collett and *H. mediterraneus* Cuvier (Trachichthyidae, Beryciformes). In Shirston PV (ed.) *Fishes of the open ocean*, Moscow, pp. 68–88.

26 Beamish RJ (1979) New information on the longevity of Pacific Ocean perch (*Sebastes alutus*). *Journal of the Fisheries Research Board of Canada* 36, pp. 1395–400; Beamish RJ & Chilton DE (1982) Preliminary evaluation of a method to determine the age of sablefish (*Anoplopoma fimbria*). *Canadian Journal of Fisheries and Aquatic Sciences* 39, pp. 277–87; Bennett JT, Boehlert GW & Turekian KK (1982) Confirmation of longevity in *Sebastes diploproa* (Pisces: Scorpaenidae) using ^{210}Pb/^{226}Ra measurements in otoliths. *Marine Biology* 71, pp. 209–15.

27 Fenton GE, Short SA & Ritz DA (1991) Age determination of orange roughy, *Hoplostethus atlanticus* (Pisces: Trachichthyidae) using ^{210}Pb/^{226}Ra disequilibria. *Marine Biology* 109, pp. 197–202; Smith DC, Fenton GE, Robertson SG & Short SA (1995) Age determination and growth of orange roughy (*Hoplostethus atlanticus*): a comparison of annulus counts with radiometric ageing. *Canadian Journal of Fisheries and Aquatic Sciences* 52, pp. 391–401.

28 Smith DC & Stewart BD (1994) Development of methods to age commercially important dories and oreos, Final Report, Fisheries Research & Development Corporation, Canberra; Stewart BD, Fenton GE, Smith DC & Short SA (1995) Validation of otolith increment age estimates for a deepwater fish species, warty oreo, *Allocyttus verrucosus*, by radiometric analysis, *Marine Biology* 123, pp. 29–38. Tracey DM, George K & Gilbert DJ (2000) Estimation of age, growth, and mortality parameters of black cardinalfish (*Epigonus telescopus*) in QMA2 (east coast North Island), *New Zealand Fisheries Assessment Report* 2000/27: 1–21. For a review, see Caillet GM, Andrews AH, Burton EJ, Watters DL, Kline E & Ferry-Graham LA (2001) Age determination and validation studies of marine fishes: do deep-dwellers live longer? *Experimental Gerontology* 36, pp. 739–64.

29 Koslow JA, Bax NJ, Bulman CM, Kloser RJ, Smith ADM & Williams A (1997) Managing the fishdown of the Australian orange roughy resource. In Hancock DA, Smith DC, Grant A & Beumer JP (eds.) *Developing and Sustaining World Fisheries Resources: the state of science and management: 2nd World Fisheries Congress proceedings*. CSIRO Publishing, Collingwood, Australia, pp. 558–62.

30 Wayte S & Bax N (2001) Draft assessment of Eastern Zone orange roughy. Report submitted to the Southeast Trawl Management Advisory Group (SETMAC), CSIRO. Available as 'ORAG 2001 Stock assessment report.pdf' from http://www.marine.csiro.au/orag/publications/index.htm. For a good review of orange roughy fisheries in Australia, New Zealand and the North Atlantic, see Lack M, Short K & Willock A (2003) Managing risk and uncertainty in deep-sea fisheries: lessons from orange roughy. TRAFFIC Oceania and WWF Australia. Available from http://assets.panda.org/downloads/oranger0.pdf.

31 Clark M (1999) Fisheries for orange roughy (*Hoplostethus atlanticus*) on seamounts in New Zealand. *Oceanologica Acta* 22, pp. 593–602; Clark M (2001) Are deepwater fisheries sustainable? – the example of orange roughy in New Zealand. *Fisheries Research* 51, pp. 123–35.

32 Andrae D (Oct 2000) NZ makes savage cuts in roughy. *Fishing News International* 39(10), p. 7.

33 Tempo heading, Joplin S (1907) The Nonpareil. In Lawrence VB (ed.)(1971) *Scott Joplin: Collected Piano Works*, New York Public Library, p. 164.

34 Bax N (2000) Final draft stock assessment report 2000, orange roughy (*Hoplostethus atlanticus*). Report submitted to the Australian Fisheries Management Authority. South East Fishery Stock Assessment Group.

35 Bax N (2000) Final draft stock assessment report 2000, orange roughy.

36 Bax N (2000) Final draft stock assessment report 2000, orange roughy; Koslow et al., Managing the fishdown of the Australian orange roughy resource.

37 Ludwig D, Hilborn R & Walters C (1993) Uncertainty, resource exploitation, and conservation: lessons from history. *Science* 260, pp. 17, 36.

38 Francis RICC (1992) Use of risk analysis to assess fishery management strategies: a case study using orange roughy (*Hoplostethus atlanticus*) on the Chatham Rise, New Zealand. *Canadian Journal of Fisheries and Aquatic Sciences* 49, pp. 922–30.

39 Koslow JA & Tuck G (2001) The boom and bust of deepwater fisheries: why haven't we done better? NAFO SCR Document 141, Series 4535. Presented at the NAFO Symposium on Deepwater Fisheries, Varadero, Cuba.

40 Koslow JA, Bulman CM & Lyle JM (1994) The mid-slope demersal fish community off southeastern Australia. *Deep-Sea Research* 41, pp. 113–41.

41 Lack et al. (2003) Managing risk and uncertainty in deep-sea fisheries: lessons from orange roughy.

42 Branch TA (2001) A review of orange roughy *Hoplostethus atlanticus* fisheries, estimation methods, biology and stock structure. *South African Journal of Marine Science* 23 (A Decade of Namibian Fisheries Science), pp. 181–203.

43 Horn PL (2001) Age and growth of Patagonian toothfish (*Dissostichus eleginoides*) and Antarctic toothfish (*D. mawsoni*) in waters from the New Zealand subantarctic to the Ross Sea, Antarctica. *Fisheries Research* 57, pp. 275–87.

44 Kock KH (2000) A brief description of the main species exploited in the Southern Ocean, Annex I to *Understanding CCAMLR's Approach to Management*, CCAMLR website: http://www.ccamlr.org/pu/e/e_pubs/am/p7.htm.

45 FAO website and fishery statistics: http://www.fao.org. Lack M & Sant G (2001) Patagonian toothfish: are conservation and trade measures working? *TRAFFIC Bulletin* 19, pp. 1–19.

46 Lord P (21 April 2003) HK firm 'will not buy fish caught illegally.' Pillage and piracy in Antarctic seas. *The Standard*. See http://www.asoc.org/media/04.21.03TheStandard.htm; http://www.abc.net.au/4corners/stories/s689740.htm.

47 Lack, & Sant, Patagonian toothfish: are conservation and trade measures working?

48 Kuo CTL & Tanaka S (1984) Feeding of hoki *Macruronus novaezelandiae* (Hector) in waters around New Zealand. *Bulletin of the Japanese Society of Scientific Fisheries* 50, pp. 783–86; Bulman CM & Blaber SJM (1986) Feeding ecology of *Macruronus novaezelandiae* (Hector) (Teleostii: Merlucciidae) in south east Australia. *Australian Journal of Marine and Freshwater Research* 37, pp. 621–68.

49 Horn PL & Sullivan KJ (1996) Validated aging methodology using otoliths, and growth parameters for hoki (*Macruronus novaezelandiae*) in New Zealand waters. *New Zealand Journal of Marine and Freshwater Research* 30, pp. 161–74.

50 Bulman CM, Koslow JA & Haskard KA (1999) Estimation of the spawning stock biomass of blue grenadier (*Macruronus novaezelandiae*) in south-eastern Australia based upon the annual egg production method. *Marine and Freshwater Research* 50, pp. 197–207; Coombs RF & Cordue PL (1995) Evolution of a stock assessment tool: acoustic surveys of spawning hoki (*Macruronus novaezelandiae*) off the west coast of South Island, New Zealand, 1985–91. *New Zealand Journal of Marine and Freshwater Research* 29, pp. 175–94; Zeldis JR, Murdoch RC, Cordue PL & Page MJ (1998) Distribution of hoki (*Macruronus novaezelandiae*) eggs, larvae, and adults off Westland, New Zealand, and the design of an egg production survey to estimate hoki biomass. *Canadian Journal of Fisheries and Aquatic Sciences* 55, pp. 1682–94.

51 Sverdrup HU, Johnson MW & Fleming RH (1942) *The Oceans: Their Physics, Chemistry, and General Biology*. Prentice-Hall, Englewood Cliffs, NJ, Table 5.

52 Gordon JDM (2001a) Deep-water fisheries at the Atlantic frontier. *Continental Shelf Research* 21, pp. 987–1003.

53 For reviews of modern developments in European deepwater fisheries, see: Gordon JDM (2001b) Deep-water fish and fisheries: introduction. *Fisheries Research* 51, pp. 105–11; Gordon JDM (2001a) Deep-water fisheries at the Atlantic frontier; Gordon JDM, Bergstad OA, Figueiredo I & Menezes G (2003) Deep-water fisheries of the Northeast Atlantic: I. Description and current trends. *Journal of Northwest Atlantic Fishery Science* 31, pp. 137–50; Large PA, Hammer C, Bergstad OA, Gordon JDM & Lorance P (2003) Deep-water fisheries of the Northeast Atlantic: II. Assessment and management approaches. *Journal of Northwest Atlantic Fishery Science* 31, pp. 151–63; Gordon JDM (2003) The Rockall Trough, North East Atlantic: the cradle of deep-sea biological oceanography that is now being subjected to unsustainable fishing activity. *Journal of Northwest Atlantic Fishery Science* 31, pp. 57–83.

54 Data from FAO data for marine capture fisheries (fish, crustaceans and mollusks) available from http://www.fao.org, and analyzed with the FAO software FISHSTAT.

55 Atkinson DB (1995) The biology and fishery of roundnose grenadier (*Coryphaenoides rupestris* Gunnerus, 1765) in the North West Atlantic. In Hopper (ed.) *Deep-Water Fisheries of the North Atlantic Oceanic Slope*, pp. 51–111; Bowering WR & Brodie WB (1995) Greenland halibut; Troyanovsky FM & Lisovsky SF (1995) Russian (USSR) fisheries research in deep waters (below 500 meters) in the North Atlantic. In Hopper (ed.) *Deep-Water Fisheries of the North Atlantic Oceanic Slope*, pp. 357–65.

56 Bergstad OA (1995) Age determination of deep-water fishes: experiences, status and challenges for the future. In Hopper (ed.) *Deep-Water Fisheries of the North Atlantic Oceanic Slope*, pp. 267–83.

57 Russian scientists maintain that there is only one population: Troyanovsky & Lisovsky, Russian (USSR) fisheries research in deep waters (below 500 meters) in the North Atlantic, while others consider that the groups fished on the mid-Atlantic Ridge and in the northeast and northwest Atlantic are separate stocks: Atkinson, The biology and fishery of roundnose grenadier. This is critical for management and conservation, since each stock must be managed separately.

58 Devine JA, Baker KD & Haedrich RL (2006) Deep-sea fishes qualify as endangered. *Nature* 439, p. 29.

59 Bowering & Nedreaas (2000) A comparison of Greenland halibut fisheries and distribution in the Northwest and Northeast Atlantic.

60 Bowering WR & Nedreaas KH (2001) Age validation and growth of Greenland halibut (*Reinhardtius hippoglossoides* (Walbaum)): a comparison of populations in the Northwest and Northeast Atlantic. *Sarsia* 86, pp. 53–68; Bowering W & Nedreaas KH (2000) A comparison of Greenland halibut fisheries and distribution in the Northwest and Northeast Atlantic. Also see http://www.fishbase.org for a synthesis of the species' biological characteristics.

61 Merrett & Haedrich, *Deep-Sea Demersal Fish and Fisheries*, pp. 207–10; Atkinson, The biology and fishery of roundnose grenadier. Also Bowering & Nedreaas, A comparison of Greenland halibut fisheries and distribution in the Northwest and Northeast Atlantic; Bowering & Nedreaas, Age validation and growth of Greenland halibut.

62 Scott WB & Scott MG (1988) Atlantic Fishes of Canada, *Canadian Bulletin of Fisheries and Aquatic Sciences* 219, pp. 482–87; Magnusson J & Magnusson JV (1995) Oceanic redfish (*Sebastes mentella*) in the Irminger Sea and adjacent waters. *Scientia Marina* 59, pp. 241–54.
63 Koslow et al. (2000) The impact of fishing on continental slope and deep-sea ecosystems.
64 Johansen T, Naevdal G, Daniélsdóttir AK & Hareide NR (2000) Genetic characterisation of giant *Sebastes* in the deep water slopes in the Irminger Sea. *Fisheries Research* 45, pp. 207–16.
65 Hareide N-R, Garnes G & Langedal G (2001) The boom and bust of the Norwegian longline fishery for redfish (*Sebastes marinus* 'Giant') on the Reykjanes Ridge. NAFO SCR Doc 01/126.
66 Campana SE, Zwanenburg KCT & Smith JN (1990) ^{210}Pb/^{226}Ra determination of longevity in redfish. *Canadian Journal of Fisheries and Aquatic Sciences* 47, pp. 163–65; Avila de Melo A, Alpoim R & Saborido-Rey F (2002) The present status of beaked redfish (*S. mentella* and *S. fasciatus*) in NAFO Division 3M and medium term projections under a low commercial catch/high shrimp fishery by-catch regime. NAFO SCR Document 02-54; see also http://www.fishbase.org.
67 Fishery data from http://www.fao.org.
68 Koslow et al., The impact of fishing on continental slope and deep-sea ecosystems.
69 ICES (1997) Report of the ICES Advisory Committee on Fishery Management 1996. *ICES Cooperative Research Report* 221, as cited in Koslow et al., The impact of fishing on continental slope and deep-sea ecosystems.
70 Bjornsson H & Sigurdsson T (2003) Assessment of golden redfish (*Sebastes marinus* L) in Icelandic waters. *Scientia Marina* 67, pp. 301–14.
71 Gordon JDM (2001) Deep-water fisheries at the Atlantic frontier. *Continental Shelf Research* 21, pp. 987–1003.
72 Gunnarsson G (1995) Deep-water trawling techniques used by Icelandic fishermen. In Hopper (ed.) *Deep-Water Fisheries of the North Atlantic Oceanic Slope*, pp. 385–95. Fishery statistics from http://www.fao.org. Fishery summary also from http://www.fisheries.is/stocks/redfish.htm.
73 Hareide et al. (2001) The boom and bust of the Norwegian longline fishery for redfish.
74 Magnusson JV & Magnusson J (1995) The distribution, relative abundance, and biology of the deep-sea fishes of the Icelandic slope and Reykjanes Ridge. In Hopper (ed.) *Deep-Water Fisheries of the North Atlantic Oceanic Slope*, pp. 161–99; Large et al. (2003) Deep-water fisheries of the Northeast Atlantic: II. Assessment and management approaches.
75 Gordon (2001) Deep-water fisheries at the Atlantic frontier.
76 Large et al. (2003) Deep-water fisheries of the Northeast Atlantic: II. Assessment and management approaches.
77 Report quoted in Gordon et al. (2003) Deep-water fisheries of the Northeast Atlantic: I. Description and current trends.
78 Press release, Outcome of the EU Agriculture and Fisheries Council: meeting 21–22 December 2004. Reference: MEMO/04/306 Date: 22/12/2004.
79 Hutchings JA (2000) Collapse and recovery of marine fishes. *Nature* 406, pp. 882–85.
80 Gianni M (2004) *High seas bottom trawl fisheries and their impacts on the biodiversity of vulnerable deep-sea ecosystems: options for international action.* IUCN, Gland, Switzerland.
81 Safina C (1995) The world's imperiled fish. *Scientific American* 273(5), pp. 46–53.
82 Heinberg R (2003) *The Party's Over: Oil, War and the Fate of Industrial Societies.* New Society, Gabriola Island, Canada.
83 March EJ (1978) *Sailing Trawlers: The Story of Deep-Sea Fishing with Longline and Trawl.* David & Charles, Devon, UK, p. 33.
84 Merrett & Haedrich (1997) *Deep-Sea Demersal Fish and Fisheries*, p. 168.
85 Anderson OF & Clark MR (2003) Analysis of bycatch in the fishery for orange roughy, *Hoplostethus atlanticus*, on the South Tasman Rise. *Marine Freshwater Research* 54, pp. 643–52.
86 Koslow JA, Gowlett-Holmes K, Lowry J, O'Hara T, Poore G & Williams A (2001) The seamount benthic macrofauna off southern Tasmania: community structure and impacts of trawling. *Marine Ecology Progress Series* 213, pp. 111–25; Clark MR & O'Driscoll R (2003) Deepwater fisheries and aspects of their impact on seamount habitat in New Zealand. *Journal of Northwest Atlantic Fishery Science* 31, pp. 441–58.
87 Fosså JH, Mortensen PB & Furevik DM (2002) The deep-water coral *Lophelia pertusa* in Norwegian waters: distribution and fishery impacts. *Hydrobiologia* 471, pp. 1–12.
88 Freiwald A, Fosså JH, Grehan A, Koslow T & Roberts JM (2004) *Cold-water coral reefs: out of sight – no longer out of mind.* Cambridge, UK, UNEP-WCMC.
89 Champion A (2003) A life line for the Darwin Mounds. *Marine Conservation* 6(3), p. 15.
90 Freiwald et al. (2004) *Cold-water coral reefs: out of sight – no longer out of mind.*
91 Freiwald et al. (2004) *Cold-water coral reefs: out of sight – no longer out of mind*; Mortensen PB, Buhl.-Mortensen L, Gordon DC Jr, Fader GBJ, McKeown DL & Fenton DG (2005) Effects of fisheries on deep-water gorgonian corals in the Northeast Channel, Nova Scotia (Canada). In Barnes P & Thomas J (eds) *Benthic Habitats and the Effects of Fishing,* American Fisheries Society, Bethesda, Maryland; Mortensen PB & Buhl-Mortensen L (2005) Deep-water corals and their habitats in The Gully, a submarine canyon off Atlantic Canada. In Freiwald A & Roberts JM (eds) *Cold-water Corals and Ecosystems.* Springer Publishing, Heidelberg, Germany, pp. 247–77.
92 Shester G & Ayers J (2005) A cost effective approach to protecting deep-sea coral and sponge ecosystems with an application to Alaska's Aleutian Islands region. In Freiwald & Roberts (eds) *Cold-water Corals and Ecosystems,* pp. 1151–69.
93 Welch C (11 Feb 2005) Coral concerns spur vast trawling ban. *Seattle Times.*

94 Final Environmental Impact Statement, Preferred Alternative Groundfish Essential Fish Habitat, Pacific Fishery Management Council: http://www.pcouncil.org/groundfish/gfefheis/pfmc_efheis_pa.pdf.
95 Haedrich RL, Heppell SS, Koslow JA, Myers RA, Pikitch EK, Roberts JM, Smith MD & Sumaila UR (submitted) Liquidating deep-sea capital? *Science.*

Chapter 11

1 See Chapter 4; Grassle JF & Maciolek NJ (1992) Deep-sea species richness: regional and local diversity estimates from quantitative bottom samples. *American Naturalist* 139, pp. 313–341.
2 See Chapter 6; Hovland M, Vasshus S, Indreeide A, Austdal L & Nilsenó (2002) Mapping and imaging deep-sea coral reefs off Norway, 1982–2000. *Hydrobiologia* 471, pp. 13–17.
3 See Chapter 6; Richer de Forges B, Koslow JA & Poore GCB (2000) Diversity and endemism of the benthic seamount fauna in the southwest Pacific. *Nature* 405, pp. 944–47.
4 Grotius H (1609) *Mare Liberum (The Freedom of the Seas)*, trans. Ralph van Daman Magoffin (1916), Oxford University Press, London, available online from Batoche Books, Kitchener, Canada (2000), http://socserv2.mcmaster.ca/~econ/ugcm/3ll3/grotius/Seas.pdf.
5 Grotius, *Mare Liberum*, pp. 23–24.
6 van Dyke JM (1993) International governance and stewardship of the high seas and its resources, in van Dyke JM, Zaelke D & Hewison G (eds.) *Freedom for the Seas in the 21st Century*, Island Press, Washington DC, pp. 13–22.
7 Hardin G (1968) The tragedy of the commons. *Science* 162, pp. 1243–48. See Chapter 10.
8 The text of the FAO Code of Conduct, the non-technical booklet, *What is the Code of Conduct for Responsible Fisheries*, and other relevant literature, are available at: http://www.fao.org/fi/agreem/codecond/codecon.asp.
9 Ludwig D, Hilborn R & Walters C (1993) Uncertainty, resource exploitation, and conservation: lessons from history. *Science* 260, pp. 17, 36.
10 Fisheries and Agriculture Organization (UN)(2004) *The state of world fisheries and aquaculture, 2004*, FAO, Rome.
11 Kurlansky M (1998) *Cod*, Jonathan Cape, London.
12 See CCAMLR website: http://www.ccamlr.org.
13 From CCAMLR Rules of Procedure: http://www.ccamlr.org/pu/e/e_pubs/bd/pt3.pdf.
14 Gianni M (2004) *High seas bottom fisheries and their impact on the biodiversity of vulnerable deep-sea ecosystems: summary findings,* IUCN, Gland, Switzerland, p. 67.
15 Press releases of the NEAFC of 15 November 2004, available at http://www.neafc.org, and of the WWF at http://www.wwf.org, 17 November 2004: NE Atlantic Fisheries Commission protects cold-water corals from trawling.
16 See press release, 2004: http://www.nafo.ca.
17 Gianni, *High seas bottom fisheries*, p. 61.
18 Gianni, *High seas bottom fisheries*, p. 55.
19 The statement is available at: http://www.mcbi.org/DSC_statement/sign.htm. (So that our biases are known, we note that Tony Koslow is the lead signatory.)
20 Gianni, *High seas bottom fisheries*, p. 75.
21 UN General Assembly resolution A/RES/58/240, para 52, available at http://www.un.org/Depts/los/general_assembly/general_assembly_resolutions.htm.
22 Gianni, *High seas bottom fisheries*, pp. 48–51.
23 A record of these initiatives can be found at: http://www.savethehighseas.org.
24 See previous chapter. Gianni, *High seas bottom fisheries.*
25 This concluding section, with small changes, is from an unpublished report: Koslow JA & Smith CR (2005) Report to the Secretary-General of the United Nations on Marine Biodiversity in Areas beyond National Jurisdiction: Threats to Ecosystems and Biodiversity caused by Human Activities.
26 This is not a new idea. In the 1830s, a Latin American jurist, A. Bello, argued that things which cannot be held by one nation without detriment to the others ought to be considered by the international community as 'common patrimony.' In the early 1900s A.G. de Lapradelle, a French jurist, advanced the idea that the oceans should be 'le patrimoine de l'humanité.' This idea was also taken up (without success) by Prince Waithayaken of Thailand, the President of the first Law of the Sea Convention in 1958.

Figures

a Sources: Hedgpeth J (1957) Classification of marine environments. *Geological Society of America* Memoir 57, Vol. 1, fig.1, p. 18, as modified by Angel MV (2003) The pelagic environment of the open ocean, in PA Tyler (ed.) *Ecosystems of the Deep Oceans*, Elsevier, Amsterdam, pp. 39–79; Isaacs JD (1969) The Nature of Oceanic Life. *Scientific American* 221, pp. 146–62.
b Drawing of *Diaphus effulgens* courtesy of BG Nafpaktitis et al. (1977) Family Myctophidae, in *Fishes of the Western North Atlantic*, Memoir Sears Foundation for Marine Research. No. 1, Part 7: pp. 13–265.
c Brauer A (1906) Die Tiefsee-Fische. I. Systematischer Teil, in C Chun, Wissenschaftl. Ergebnisse der deutschen Tiefsee-Expedition 'Valdivia,' 1898–99. *Jena* 15, pp. 1–432, Pls. 1–18.
d Vinogradov ME (1968) *Vertical distribution of the oceanic zooplankton*, Academy of Sciences of the USSR, Institute of Oceanography.
e Belyaev GM, Vinogradova NG, Levenshyteyn NGPFA, Sokolova MN & Filatova ZA (1973) Distribution patterns of deep-water bottom fauna related to the idea of the biological structure of the ocean. *Oceanology* 13, pp. 114–20.
f Heezen BC & Hollister CD (1971) *The Face of the Deep*, Oxford University Press, New York.

g Marshall NB (1979) *Developments in Deep-Sea Biology*, Blandford Press, Poole, Dorset, figure 94.

h Cohen DM, Inada T, Iwamoto T & Scialabba N (1990) *FAO Species Catalogue*, vol. 10, FAO Fisheries Synopses 10 (125).

i Wolff T (1956) Crustacea Tanaidacea from depths exceeding 6000 meters *Galathea Report* 2, pp. 187–241.

j Hartman O (1965) Deep-water benthic polychaetous annelids of New England to Bermuda and other North Atlantic areas. *Allan Hancock Foundation Publications* 28, pp. 1–378.

k Lincoln RJ & Boxshall GA (1983) Deep-sea asellote isopods of the north-east Atlantic: the family Haploniscidae. *Journal of Natural History* 19, pp. 655–95.

l Barnard JL (1973) Deep-sea Amphipoda of the genus *Lepechinella* (Crustacea). *Smithsonian Contributions to Zoology* 133, p. 15.

m Sources: http://oceanexplorer.noaa.gov/explorations/02fire/background/vent_chem/media/chemistry_600.jpg; Massoth GJ et al. (1988) The geochemistry of submarine venting fluids at Axial Volcano, Juan de Fuca Ridge: new sampling methods and a vents program rationale. NOAA National Undersea Research Program, Research Report 88-4, pp. 29–59.

n Sibuet M & Olu K (1998) Biogeography, biodiversity and fluid dependence of deep-sea cold-seep communities at active and passive margins. *Deep-Sea Research* II 45, pp. 517–67; van Dover CL, German CR, Speer KG, Parson LM & Vrijenhoek RC (2002) Evolution and biogeography of deep-sea vent and seep invertebrates. *Science* 295, pp. 1253–57.

o Pace NR (1997) A molecular view of microbial diversity and the biosphere. *Science* 276, pp. 734–40. Figure modified from his Figure 1.

p Smith DK (1991) Seamount abundances and size distributions, and their geographic variations. *Reviews in Aquatic Sciences* 5, pp. 197–210.

q Beckmann A & Mohn C (2002) The upper ocean circulation at Great Meteor Seamount, *Ocean Dynamics* 52, pp. 194–204, Figure 2a, c, d, f.

r Richer de Forges B, Koslow JA & Poore GCB (2000) Diversity and endemism of the benthic seamount fauna in the southwest Pacific. *Nature* 405, pp. 944–47, Figure 2.

s Rona PA (2003) Resources of the sea floor. *Science* 299, pp. 673–74, Figure 1.

t Intergovernmental Panel on Climate Change (IPCC)(2001) *Climate Change 2001: The Scientific Basis*, Cambridge University Press, Cambridge. Available from http://www.ipcc.ch.

u Zachos JC, Pagani M, Sloan IC, Chomas E & Billups K (2001) Trends, rhythms, and aberrations in global climate 65 Ma to present. *Science* 292, pp. 686–93, Figure 2.

v Watson R & Pauly D (2001) Systematic distortions in world fisheries catch trends. *Nature* 414, pp. 534–36.

w Koslow et al. (2000) The impact of fishing on continental slope and deep-sea ecosystems, figs 1, 2.

Plates

i Brauer A (1906) *Die Tiefsee-Fische. I. Systematischer Teil*, in C Chun, *Wissenschaftl. Ergebnisse der deutschen Tiefsee-Expedition* 'Valdivia,' 1898–99. *Jena* 15, pp. 1–432, Pls. 1–18.

ii Brauer, *Die Tiefsee-Fische. I. Systematischer Teil.*

iii Behrenfeld MJ & Falkowski PG (1997) Photosynthetic rates derived from satellite based chlorophyll concentration. *Limnology and Oceanography* 42 (1), pp. 1–20, figure 9B.

iv Koslow JA (1997) Seamounts and the ecology of deep-sea fisheries. *American Scientist* 85, pp. 168–76.

v http://www.niwascience.co.nz/pubs/wa/13-2/coral

vi Smith W & Sandwell D (1997) Measured and Estimated Seafloor Topography, World Data Center for Marine Geology & Geophysics, Boulder Research Publication RP-1, poster; available at: http://www.ngdc.noaa.gov/mgg/image/walter/World.jpg

Index

Page references in *italics* relate to figures.

PLC 15, 16